Animal Nutrition Conference of Canada (ANCC 2025)

Practical Opportunities to Navigate the Future of Feed

Niagara Falls, Ontario
7 - 9 May 2025

ISBN: 979-8-3313-2313-4

Printed from e-media with permission by:

Curran Associates, Inc.
57 Morehouse Lane
Red Hook, NY 12571

proceedings.com

Some format issues inherent in the e-media version may also appear in this print version.

Printed by Curran Associates, Inc. (2025)

Animal Nutrition Society of Canada
150 Metcalfe Street, Suite 1301
Ottawa, Ontario K2P 1P1
Canada

Phone: (613) 241-6421
Fax: (613) 241-7970

info@anacan.org

Additional copies of this publication are available from:

Curran Associates, Inc.
57 Morehouse Lane
Red Hook, NY 12571 USA
Phone: 845-758-0400
Fax: 845-758-2633
Email: curran@proceedings.com
Web: www.proceedings.com

ANCC

2025 ANIMAL NUTRITION CONFERENCE OF CANADA

CNAC

2025 COLLOQUE DE NUTRITION ANIMALE DU CANADA

BOOK OF PROCEEDINGS CAHIER DE CONFÉRENCES

May 7-9 mai Niagara Falls, Ontario

Practical Opportunities to Navigate the Future of Feed

Opportunités concrètes pour naviguer l'avenir de l'alimentation animale

Pre-conference symposium sponsors
Commanditaires du symposium pré-colloque

dsm-firmenich HUVEPHARMA KEMIN NEOGEN

Table of Contents / *Table des matières*

Monogastric session / *Séance sur les monogastriques*

Ruminant Session / *Séance sur les ruminants*

Closing plenary / *Plénière de clôture*

Graduate Student Posters / *Les affiches des étudiants de cycle supérieur*

List of Advertisers / *Liste d'annonceurs*

Animal Nutrition Association of Canada
ANAC
Association de nutrition animale du Canada

ANAC welcomes you to ANCC 2025

The **Animal Nutrition Association of Canada (ANAC)** is delighted to welcome the animal nutrition community to the ninth annual Animal Nutrition Conference of Canada (ANCC). We are excited to share, once again, an exceptional program filled with meaningful learning and networking opportunities for colleagues from industry, academia and government.

The ANCC Organizing Committee recognizes that these are times of change and uncertainty. That's why, under the theme *"***Practical Opportunities to Navigate the Future of Feed***,"* the committee has prepared a program that brings forward the most relevant and timely research. From strategies to improve biosecurity to the use of AI and data in feed mills, this year's sessions are intended to keep you informed and equipped for the future. As we all work to adapt and thrive in this evolving world, we hope the conversations and insights shared at this conference will help you move forward with confidence.

An event of this calibre is only possible with the collective effort and support of many partners. Thank you to our sponsors who continue to show tremendous support for this conference. ANAC also recognizes the contributions of the Organizing Committee, whose concerted effort and commitment are instrumental to the success of ANCC 2025. Our appreciation also extends to our knowledgeable speakers for sharing their valuable time and expertise.

Almost a year ago, we saw the publication of the *Feeds Regulations, 2024* — the most significant change to the feed industry in over 40 years. We are pleased to invite you to a round table discussion where respected colleagues will share their firsthand experiences including key challenges faced, practical solutions developed, and lessons learned during the transition to the new regulatory requirements. This is a valuable opportunity to learn from your peers and reflect on how our industry is moving forward together.

We hope you enjoy your time with us here in Niagara Falls and we are already looking forward to meeting with you again at next year's conference in **Edmonton, Alberta**!

Sarah Hopkins
Government Affairs Manager
Animal Nutrition Association of Canada

Animal Nutrition Association of Canada

ANAC

Association de nutrition animale du Canada

L'ANAC vous souhaite la bienvenue au CNAC 2025

L'**Association de nutrition animale du Canada (ANAC)** est ravie d'accueillir la communauté de la nutrition animale à la neuvième édition du Colloque annuel de nutrition animale du Canada (CNAC). Nous sommes heureux de partager, encore une fois, un programme exceptionnel rempli d'opportunités d'apprentissage et de réseautage significatives avec des collègues de l'industrie, du milieu universitaire et du gouvernement.

Le comité organisateur du CNAC est conscient que nous traversons une période de changement et d'incertitude. C'est pourquoi, sous le thème « **Opportunités concrètes pour naviguer l'avenir de l'alimentation animale** », le comité a préparé un programme qui met en évidence les recherches les plus pertinentes et les plus récentes. Des stratégies d'amélioration de la biosécurité à l'utilisation de l'IA et des données dans les meuneries, les sessions de cette année visent à vous tenir informés et équipés pour l'avenir. Alors que nous travaillons tous à nous adapter et à prospérer dans ce monde en évolution, nous espérons que les conversations et les idées partagées lors de ce colloque vous aideront à aller de l'avant avec confiance.

Un événement de cette envergure n'est possible que grâce à l'effort collectif et au soutien de nombreux partenaires. Merci à nos commanditaires qui continuent à apporter un soutien considérable à cet événement. L'ANAC reconnaît également les contributions du comité organisateur, dont les efforts concertés et l'engagement sont essentiels au succès du CNAC 2025. Nous exprimons également notre gratitude à nos conférenciers chevronnés d'avoir partagé leur temps et leur expertise.

Il y a près d'un an, nous avons assisté à la publication du *Règlement de 2024 sur les aliments du bétail*, le changement le plus important apporté à l'industrie de l'alimentation animale depuis plus de 40 ans. Nous avons le plaisir de vous inviter à une table ronde où des collègues respectés partageront leurs expériences de première main, notamment les principaux défis rencontrés, les solutions pratiques développées et les leçons apprises lors de la transition vers les nouvelles exigences réglementaires. C'est une occasion précieuse d'apprendre de vos pairs et de réfléchir à la manière dont notre industrie progresse ensemble.

Nous espérons que vous apprécierez votre séjour parmi nous ici à Niagara Falls et nous avons hâte de vous retrouver au colloque de l'année prochaine à **Edmonton, en Alberta**!

Sarah Hopkins
Gestionnaire des affaires gouvernementales
Association de nutrition animale du Canada

ANCC Organizing Committee / *Comité organisateur du CNAC*

Welcome everyone!

On behalf of the Organizing Committee, it is my honour to welcome you all to the 9th annual Animal Nutrition Conference of Canada (ANCC) in Niagara Falls, Ontario. We are looking forward to a dynamic and informative week ahead, as we learn together from an incredible group of industry and academic experts speaking on the theme of "Practical Opportunities to Navigate the Future of Feed."

It has been my pleasure to serve as your ANCC Program Chair in previous years, and I can confidently state that no past event presented as much of a challenge when building a relevant and timely theme as the 2025 ANCC. The Organizing Committee puts significant thought and effort into curating a theme and speaker list that encompasses the wants of the membership while remaining relevant to the current trends of animal agriculture in Canada; however, it is no secret that the times are changing faster than the "current trends" can keep up.

Tariffs, trade wars, snap elections, and shifting trade alliances and agreements are all causing their fair share of chaos in an industry that is already regularly affected by factors outside the control of those who work within it such as weather or disease. This year's theme was designed to provide valuable insights, perspectives, and knowledge to help the animal agriculture sector become more efficient, profitable, and sustainable in the face of so much global uncertainty.

Speakers will cover a broad range of practical topics from the use of machine learning and artificial intelligence that support the feed industry to species-specific topics covering biosecurity and nutrient digestion and homeostasis. This year, in an ANCC first, the Organizing Committee has also included speakers on the topics of fish and companion animal nutrition and how they can influence our understanding of farm animal nutrition.

In such uncertain times, ANCC serves an important role in bringing our industry together to learn from and support one another. As always, we are incredibly grateful for the generous support of our industry partners and sponsors, without whom this event would not be possible. We hope that the program we have built with their support provides everyone with a take home message that they can directly apply within their roles.

Thank you again for attending the 2025 Animal Nutrition Conference of Canada. We wish you all a fun informative week!

Bienvenue à tous!

Au nom du comité organisateur, j'ai l'honneur de vous souhaiter la bienvenue au 9e Colloque de nutrition animale du Canada (CNAC) à Niagara Falls, en Ontario. Nous attendons avec impatience une semaine dynamique et instructive à venir, au cours de laquelle nous apprendrons ensemble d'un groupe incroyable d'experts de l'industrie et du monde universitaire qui s'exprimeront sur le thème « Opportunités concrètes pour naviguer l'avenir de l'alimentation animale ».

J'ai eu le plaisir de servir en tant que votre présidente du programme du CNAC au cours des années précédentes, et je peux affirmer en toute confiance qu'aucun événement passé n'a présenté autant de défis lors de l'élaboration d'un thème pertinent et opportun que le CNAC 2025. Le comité organisateur a consacré beaucoup de réflexion et d'efforts à l'élaboration d'un thème et d'une liste de conférenciers qui répondent aux attentes des membres tout en restant pertinents par rapport aux tendances actuelles de l'agriculture animale au Canada ; cependant, ce n'est un secret pour personne que les temps changent plus vite que les « tendances actuelles » ne peuvent suivre.

Droits de douane, guerres commerciales, élections anticipées et changements d'alliances et d'accords commerciaux sont autant de facteurs qui provoquent leur part de chaos dans un secteur déjà régulièrement affecté par des facteurs indépendants de la volonté de ceux qui y travaillent, tels que les conditions météorologiques ou les maladies. Le thème de cette année a été conçu pour fournir des informations, des perspectives et des connaissances précieuses afin d'aider le secteur de l'agriculture animale à devenir plus efficace, plus rentable et plus durable face à tant d'incertitudes mondiales.

Les conférenciers aborderont un large éventail de sujets pratiques, allant de l'utilisation de l'apprentissage automatique et de l'intelligence artificielle qui soutiennent l'industrie de l'alimentation animale à des sujets spécifiques aux espèces, tels que la biosécurité, la digestion des nutriments et l'homéostasie. Cette année, pour la première fois, le comité organisateur du CNAC a également inclus des intervenants sur les thèmes de la nutrition des poissons et des animaux de compagnie et de la manière dont ils peuvent influencer notre compréhension de la nutrition des animaux d'élevage.

En ces temps incertains, le CNAC joue un rôle important en rassemblant notre industrie pour apprendre les uns des autres et se soutenir mutuellement. Comme toujours, nous sommes extrêmement reconnaissants du généreux soutien de nos partenaires et commanditaires de l'industrie, sans lesquels cet événement ne serait pas possible. Nous espérons que le programme que nous avons élaboré avec leur soutien fournira à chacun un message à retenir qu'il pourra appliquer directement dans le cadre de ses fonctions.

Merci encore d'avoir participé au Colloque de nutrition animale au Canada de 2025. Nous vous souhaitons à tous une semaine amusante instructive!

Holly McGill, Diamond V®
Program Chair / *Présidente du programme*

Elyas Tesfaye, Agribution
Sponsorship Chair / *Président des Commandites*

Katie Wood, University of Guelph
Academic Chair / *Présidente de l'académie*

Amanda Kerr, Wallenstein Feed and Supply Ltd.
Student Events Chair / *Présidente des événements pour étudiants*

Trevor Devries, University of Guelph

Marcio Duarte, University of Guelph

Marie France Duckett, Lallemand

Bertrand Medina, Probiotech

Keshia Paddick, More Than Just Feed

Greg Page PhD, Huvepharma

Stephanie Terry, Agriculture and Agri-Food Canada

Philip Zhang, BASF

ANCC CNAC

MARK YOUR CALENDARS!

The Animal Nutrition Conference of Canada 2026 will take place **May 5 to 7** at Westin Hotels & Resorts, **Edmonton, Alberta**

MARQUEZ VOS CALENDRIERS!

Le Colloque de nutrition animale du Canada 2026 aura lieu du **5 au 7 mai** au Westin Hotels & Resorts, **Edmonton (Alberta)**

Practical Opportunities to Navigate the Future of Feed

WEDNESDAY, MAY 7

Student Networking Event – ***Strategy Room 2*** **– 3:00 to 4:30 pm**
This is an opportunity for students to learn about careers in the feed industry and network with potential employers.

Welcome Cocktail – ***Fallsview Studio ABC*** **– 4:30 to 6:30 pm**
Join us for an evening of networking and animated discussions with industry colleagues. Registration packages will also be available for collection.

THURSDAY, MAY 8

NEOGEN · HUVEPHARMA · dsm-firmenich · KEMIN

Pre-Conference Symposium
EMERGING TRENDS IN ANIMAL NUTRITION
Great Room BC
Sponsored by: Neogen, Huvepharma, dsm-firmenich and Kemin

7:15	Hot breakfast	
8:00	Introduction	
8:05	Monitoring and verification technologies to support preventive control programs – *Feed Regulations, 2024* - ***sponsored by Neogen***	**Brandon Wong and Dr. Daniel Aubert** Neogen
8:55	Poultry Ca & P nutrition as influenced by vitamin D3 and phytase application - ***sponsored by Huvepharma***	**Dr. Doug Korver** Alpine Poultry Nutrition Inc.
9:45	Health break	
10:20	Recent evidence to support modernized, optimum vitamin D nutrition for efficient swine production - ***sponsored by dsm-firmenich***	**Dr. Jon Bergstrom** dsm-firmenich
11:10	Utilization of diet and enteric neurochemistry as a new mechanistic-based feed strategy to accelerate the application of the gut microbiome in poultry and livestock production - ***sponsored by Kemin***	**Dr. Joshua Lyte** United States Department of Agriculture (USDA)
12:00	Lunch	

Opening Plenary – ***Great Room BC***

1:00	Opening remarks from ANAC	
1:05	Organizing Committee welcome	
1:10	Regulation of digestion kinetics in swine	**Dr. Walter Gerrits** Wageningen University and Research
1:55	From bowl to barn: A shared future for companion and farm animal nutritional management	**Dr. Anna-Kate Shoveller** University of Guelph
2:40	**ANAC Scholarship Lecture:** Maximizing the utilization of wheat straw with canola or flax screenings supplementation in finishing beef cattle diets: growth performance, carcass characteristics and economic analysis	**Beatriz Montenegro** University of Saskatchewan
3:00	Health break	
3:30	Use of machine learning in the modern commercial feed mill	**Dr. Jennifer Ellis** University of Guelph
4:15	More than a moment: Sustained impacts of sire nutrition on offspring via paternal programming	**Dr. Carl Dahlen** North Dakota State University
5:00 - 7:00	**ANCC 2025 Reception in the *Fallsview Studio ABC*:** Enjoy an evening of food, drinks and networking. Visit the **Industry Partner Showcase** and **Graduate Student Poster Exhibition**.	

FRIDAY, MAY 9

CONCURRENT SESSIONS

Please note that the monogastric and ruminant sessions will run in parallel.

Monogastric Session – *Great Room B*

7:15	Hot breakfast	
8:10	Opening remarks	
8:15	Practical on-farm and regional biosecurity: Where are we heading?	**Dr. Jean-Pierre Vaillancourt** Université de Montréal
8:50	Keeping track of fibres: How fibres shape digesta transit and digestion in chickens	**Dr. Sonja de Vries** Wageningen University and Research
9:25	**Graduate student presentation:** Effects of canola meal processing on the choline bioavailability in canola meal-based broiler diets	**Saluja Karki** University of Saskatchewan
9:45	Health break	
10:15	Impacts of feeding probiotics and insects on the nutrition and gut microbiome of farmed fish with new insights for other monogastric animals	**Dr. David Huyben** University of Guelph
10:50	Optimizing gestation feeding strategies to maximize mammary development and milk yield in modern prolific sows	**Dr. Chantal Farmer** Agriculture and Agri-Food Canada
11:25	Up-and-coming research mini presentations	
12:00	Lunch	

Ruminant Session – *Great Room C*

7:15	Hot breakfast	
8:10	Opening remarks	
8:15	Effects of low feed intake on gastrointestinal function	**Dr. Greg Penner** University of Saskatchewan
8:50	Greenhouse gas emissions from beef production systems	**Dr. Kim Stackhouse - Lawson** Colorado State University
9:25	**Graduate student presentation:** Effects of chromium propionate supplementation for beef cows at late gestation on cow performance and skeletal muscle metabolism of the offspring	**Emily Knapper** University of Guelph
9:45	Health break	
10:15	Forage inclusion in finishing feedlot diets: Does source, quantity, and quality matter?	**Dr. Gabriel Ribeiro** University of Saskatchewan
10:50	Strategies related to calcium homeostasis across the transition period	**Dr. Laura Hernandez** University of Wisconsin-Madison
11:25	Up-and-coming research mini presentations	
12:00	Lunch	

Closing Plenary – *Great Room BC*

1:15	Can AI truly transform livestock and feed industries, or are we overpromising?	**Dr. Joao Dorea** University of Wisconsin-Madison
1:50	Round table discussion on the implementation of the *Feeds Regulations, 2024* with Danille Dunn (Floradale Feed Mill Limited), Dr. Nancy Fischer (Country Junction Feeds), Marc-Antoine Guesthier (Cargill), Dr. Lisa Hodgins (New Life Mills) and Dr. Usama Tayyab (Otter Co-Op)	Moderated by: **Erin Ross** Animal Nutrition Association of Canada
2:45	Closing remarks	

Opportunités concrètes pour naviguer l'avenir de l'alimentation animale

MERCREDI 7 MAI

Événement de réseautage pour les étudiants – *Strategy Room 2* – 15h00 à 16h30
Il s'agit d'une occasion pour les étudiants de se renseigner sur les carrières dans l'industrie de l'alimentation animale et de réseauter avec des employeurs potentiels.

Cocktail de bienvenue – *Fallsview Studio ABC* – 16h30 à 18h30
Joignez-vous à nous pour une soirée de réseautage et discussions animées avec des collègues de l'industrie. Les trousses d'inscription seront également disponibles pour être récupérées.

JEUDI 8 MAI

Symposium pré-colloque
NOUVELLES TENDANCES EN MATIÈRE DE NUTRITION ANIMALE
Great Room BC
Commandité par :
Neogen, Huvepharma, dsm-firmenich and Kemin

7h15	Petit-déjeuner chaud	
8h00	Introduction	
8h05	Technologies de surveillance et de vérification dans la mise en œuvre des plans de contrôle préventif – *Règlement de 2024 sur les aliments du bétail* – **parrainé par Neogen**	**Brandon Wong et Dr. Daniel Aubert** Neogen
8h55	Influence de la vitamine D_3 et de la phytase sur l'apport en Ca et P dans l'alimentation des volailles – **parrainé par Huvepharma**	**Dr. Doug Korver** Alpine Poultry Nutrition Inc
9h45	Pause-santé	
10h20	Données récentes en faveur de l'actualisation et de l'optimisation de la supplémentation en vitamine D pour favoriser l'efficacité de la production porcine – **parrainé par dsm-firmenich**	**Dr. Jon Bergstrom** dsm-firmenich
11h10	Utilisation de l'alimentation et de la neurochimie entérique en tant que nouvelle stratégie alimentaire basée sur la mécanistique pour accélérer l'application du microbiome intestinal à la production de volaille et de bétail – **parrainé par Kemin**	**Dr. Joshua Lyte** United States Department of Agriculture (USDA)
12h00	Dîner	

Plénière d'ouverture - *Great Room BC*

13h00	Propos d'ouverture de l'ANAC	
13h05	Bienvenue du comité organisateur	
13h10	Régulation de la cinétique de la digestion chez le porc	**Dr. Walter Gerrits** Wageningen University and Research
12h55	Du bol à l'étable : Un avenir commun pour la gestion nutritionnelle des animaux de compagnie et de ferme	**Dre. Anna-Kate Shoveller** University of Guelph
14h40	**Conférence de la bourse d'études de l'ANAC :** Maximiser l'utilisation de la paille de blé avec une supplémentation en criblures de canola ou de lin dans les rations des bovins de finition : performances de croissance, caractéristiques de carcasse et analyse économique	**Beatriz Montenegro** University of Saskatchewan
15h00	Pause-santé	
15h30	Utilisation de l'apprentissage automatique dans les meuneries commerciales modernes	**Dre. Jennifer Ellis** University of Guelph
16h15	Plus qu'un moment : Impacts durables de la nutrition du géniteur sur la progéniture par le biais de la programmation paternelle	**Dr. Carl Dahlen** North Dakota State University
17h00-19h00	**Réception du CNAC 2025 dans le *Fallsview Studio ABC* :** Venez profiter d'une soirée où gastronomie et réseautage seront à l'honneur. Visitez le **salon des partenaires de l'industrie** et **l'exposition d'affiches pour étudiants de cycle supérieur**.	

VENDREDI 9 MAI

SÉANCES SIMULTANÉES

Veuillez noter que les séances sur les monogastriques et les ruminants seront offertes simultanément.

Séance sur les monogastriques - *Great Room B*

7h15	Petit-déjeuner chaud	
8h10	Propos d'ouverture	
8h15	Biosécurité pratique à la ferme et régionale: dans quelle direction allons-nous?	**Dr. Jean-Pierre Vaillancourt** Université de Montréal
8h50	Garder les fibres à l'œil : comment les fibres influencent le transit gastro-intestinal et la digestion chez les poulets	**Dre. Sonja de Vries** Wageningen University and Research
9h25	**Présentation de l'étudiante diplômée :** Effets de la transformation du tourteau de canola sur la biodisponibilité de la choline dans les rations pour poulets à griller à base de tourteau de canola	**Saluja Karki** University of Saskatchewan
9h45	Pause-santé	
10h15	Impacts de l'alimentation avec des probiotiques et des insectes sur la nutrition et le microbiome intestinal des poissons d'élevage et nouvelles perspectives pour d'autres animaux monogastriques	**Dr. David Huyben** University of Guelph
10h50	Optimiser les stratégies alimentaires en gestation pour maximiser le développement mammaire et la production laitière des truies modernes hyperprolifiques	**Dre. Chantal Farmer** Agriculture et Agroalimentaire Canada
11h25	Courtes présentations sur les recherches en cours et à venir	
12h00	Dîner	

Séance sur les ruminants - *Great Room C*

7h15	Petit-déjeuner chaud	
8h10	Propos d'ouverture	
8h15	Effets d'une prise alimentaire réduite sur la fonction gastro-intestinale	**Dr. Greg Penner** University of Saskatchewan
8h50	Émissions de gaz à effet de serre dans les systèmes de production de viande bovine	**Dre. Kim Stackhouse - Lawson** Colorado State University
9h25	**Présentation de l'étudiante diplômée :** Effets de l'ajout de propionate de chrome dans l'alimentation des vaches de boucherie en fin de gestation sur les performances des vaches et le métabolisme des muscles squelettiques de la progéniture	**Emily Knapper** University of Guelph
9h45	Pause-santé	
10h15	Intégration des fourrages aux rations servies en parc de finition : la source, la quantité et la qualité importent-elles?	**Dr. Gabriel Ribeiro** University of Saskatchewan
10h50	Stratégies liées à l'homéostasie du calcium pendant la période de transition	**Dre. Laura Hernandez** University of Wisconsin-Madison
11h25	Courtes présentations sur les recherches en cours et à venir	
12h00	Dîner	

Plénière de clôture - *Great Room BC*

13h15	L'IA est-elle en mesure de révolutionner les secteurs de l'élevage et de l'alimentation animale, ou nos attentes sont-elles excessives?	**Dr. Joao Dorea** University of Wisconsin-Madison
13h50	Table ronde sur la mise en œuvre du *Règlement de 2024 sur les aliments du bétail* avec Danille Dunn (Floradale Feed Mill Limited), Dre. Nancy Fischer (Country Junction Feeds), Marc-Antoine Guesthier (Cargill), Dre. Lisa Hodgins (New Life Mills) et Dr. Usama Tayyab (Otter Co-Op)	Animée par : **Erin Ross** Association de nutrition animale du Canada
14h45	Propos de clôture	

ANCC

2025 ANIMAL NUTRITION CONFERENCE OF CANADA

CNAC

2025 COLLOQUE DE NUTRITION ANIMALE DU CANADA

Pre-conference Symposium
Symposium pré-colloque

Sponsored by/Parrainé par:

dsm-firmenich

HUVEPHARMA

KEMIN
Compelled by Curiosity™

NEOGEN

KEMIN
Compelled by Curiosity™
STRRRRRRRRRRRRRETCH
YOUR FEED FURTHER.
LYSOFORTE®, a unique biosurfactant, is designed to increase the absorption and digestion of energy-rich feed ingredients. Ultimately, helping you raise healthier animals while lowering your feed costs.
LYSOFORTE®
Scan to extend your feed and your budget.
kemin.com/lysoforte-ca
© Kemin Industries, Inc. and its group of companies 2025. All rights reserved. ®™ Trademarks of Kemin Industries, Inc., U.S.A.

Monitoring and verification technologies to support preventive control programs - *Feed Regulations, 2024*

Technologies de surveillance et de vérification dans la mise en œuvre des plans de contrôle préventif – Règlement de 2024 sur les aliments du bétail

D.F Aubert[1], B. Wong[1*], M. Noack[2]*
[1] Neogen Corp., Food Safety, London, ON, Canada
[2] Neogen Corp., Lansing, MI, USA
BWong@neogen.com ; DAubert@neogen.com
*co-first authors

PRE-CONFERENCE SPONSORED BY / PRÉ-COLLOQUE COMMANDITÉ PAR

NEOGEN

Abstract

The modernization of Canadian feed regulations, *Feed Regulations, 2024*, introduces a robust, outcome-based framework for livestock feed safety, emphasizing hazard analysis, preventive controls, traceability, and enhanced record-keeping. This article explores the role of monitoring and verification technologies in supporting preventive control programs (PCPs) and compliance under the updated framework, with a focus on environmental monitoring, Salmonella control and mycotoxin evaluation. By integrating these technologies into their PCPs, feed producers can enhance animal and human health, protect the environment, and meet regulatory requirements.

Résumé

La modernisation du règlement canadien sur les aliments du bétail, le *Règlement de 2024 sur les aliments du bétail*, établit un cadre robuste et axé sur les résultats pour l'innocuité des aliments pour les animaux de ferme, en mettant l'accent sur l'analyse des dangers, les contrôles préventifs, la traçabilité et les exigences accrues en matière de tenue de registres. Cet article explore le rôle des technologies de surveillance et de vérification dans la mise en œuvre des plans de contrôle préventif (PCP) et de la conformité dans le nouveau cadre, avec une attention particulière portée à la surveillance environnementale, la lutte contre les salmonelles et l'évaluation des mycotoxines.

En intégrant ces technologies dans leurs PCP, les fabricants d'aliments pour le bétail peuvent améliorer la santé animale et humaine, protéger l'environnement et répondre aux exigences réglementaires.

Introduction

The livestock feed industry plays a critical role in the global food chain, directly influencing both animal and human health (Rukambile et al., 2019). Recognizing the importance of feed in future outcomes, the Canadian Food Inspection Agency (CFIA) introduced the Feed Regulations, 2024, designed to enhance feed safety through a result-driven system (CFIA, 2024). This approach allows producers to achieve measurable goals by adopting best practices and technologies suited to their needs and demonstrate the effectiveness of their strategies while ensuring compliance. Effective June 17, 2025, these regulations apply to livestock feed manufactured, sold, imported, or exported in Canada, and extend across various types of feed, encompassing animal-derived and plant-based feeds, single ingredients and mixed formulations (CFIA, 2024). The primary goals are to ensure safety, nutritional value, sustainability and align with international standards.

Contamination can enter the feed production facility through various raw ingredients. In facilities producing animal-derived ingredients such as rendered meat and bone meal, feather meal, and fish meal, harmful pathogenic bacteria can be introduced by infected animals or improper handling during slaughter. In the field, environmental factors such as weather and pests are beyond control and can create conditions that promote crop contamination and crop loss even before harvest. Under favorable temperature and moisture conditions, microorganisms (including mycotoxin producing molds and pathogenic bacteria) can proliferate in contaminated raw materials. Poor storage conditions, continuous production cycles with inadequate processing conditions, insufficient sanitation, and poor airflow management can further increase the risk of contamination (Podolak et al., 2010). If not properly handled, treated or monitored, contaminated raw materials can directly compromise the finished product while contaminated feed contact surfaces can act as pathogen reservoirs, resulting in persistent cross-contamination within feed production facilities.

Implementing preventive measures is essential to ensuring safe feed production (Jones, 2011; Obe et al., 2023). A key requirement for feed establishments under the updated regulations is the implementation of a written Preventive Control Program (PCP) (CFIA, 2024). This systemic and structured approach is based on Hazard Analysis Critical Control Points (HACCP) principles that are designed to manage risks. Successful implementation of a PCP is essential in livestock feed manufacturing to identify, evaluate, and control hazards. A PCP includes: 1) Hazard identification and risk assessment notably to manage contaminants associated with livestock feeds such as chemical (e.g. mycotoxins), biological (e.g. *Salmonella*), and physical hazards (e.g. stones). 2) Preventive measures incorporating Good Manufacturing Practices (GMPs), sanitation, and biosecurity. 3) Monitoring procedures for Critical Control Points (CCPs) to detect deviations and maintain safety. 4) Comprehensive record-keeping to demonstrate regulatory compliance.

Many feed businesses already implemented a feed safety management program, which may already align with the PCP requirements in the Feeds Regulations, 2024. However, businesses must identify and address any gaps between these programs and regulatory requirements (CFIA, 2024). A PCP plan is unique to each facility. While a general framework is provided (CFIA, 2024), each plan must be tailored to the specific processes, equipment, raw materials, and potential hazards of a given feed manufacturing facility. Factors such as ingredient sources, production methods, facility layout have an influence in the design of the PCP. Each facility must identify its own CCPs, establish relevant critical limits, and develop monitoring and corrective action protocols that align with its unique operations. This proactive approach not only enhances feed safety but also drives continuous improvement within the facility.

This article provides insights into monitoring and verification strategies that livestock feed manufacturers can implement to detect and manage risks associated with *Salmonella* and mycotoxin contaminations, specifically. By integrating monitoring tools, manufacturers can verify the integrity of raw ingredients and maintain continuous surveillance at key control points. This approach ensures that both the production environment and processes remain under control, significantly reducing the risk of contamination.

Ensuring Feed Safety Through Mycotoxin Control

Mycotoxins are toxic secondary metabolites produced by fungi, particularly *Aspergillus*, *Fusarium*, and *Penicillium* species, that contaminate grains, nuts, fruits, and animal feed, posing significant health risks to both humans and animals upon ingestion. Exposure can lead to immunosuppression, liver damage, and, in severe cases, death, with risks influenced by toxin type, concentration, and duration of exposure. Six key mycotoxins—aflatoxins, deoxynivalenol (DON), fumonisins, zearalenone, ochratoxins, and T-2/HT-2 toxins—are subject to regulatory limits due to their impact on food and feed safety. These toxins often co-occur, either within individual grains or when multiple contaminated grain sources are combined, potentially leading to synergistic effects (Streit et al., 2013). Given their resilience to food processing and their widespread presence, effective monitoring and control are essential to mitigate risks and ensure compliance with safety standards.

Economic Impact of Mycotoxins in the feed industry

Mycotoxin contamination in feed and raw materials presents a significant economic burden on the global agricultural industry. While studies indicate that mycotoxins contaminate 25-50% of staple crop commodities (Bhat and Miller, 1991), a substantial portion of feed raw material ingredients—approximately 72%—contain multiple mycotoxins (Streit et al., 2013). In regions such as the US, Canada, and the Europe, mycotoxin-related losses in animal production amount to tens of billions of dollars annually, impacting livestock health, productivity, and overall industry profitability (Desjardins et al., 2003). These widespread contaminations highlight the urgent need for stringent monitoring and control measures to mitigate financial and agricultural losses.

Mycotoxin Formation and Control Strategies

Mycotoxins often form in the field during the growing season but can also develop or increase during harvest, drying, storage, and processing. A completely mycotoxin-free food supply is not guaranteed, as eliminating all contamination is challenging. Key factors influencing mycotoxin production include water availability and temperature in the field, while storage risks involve water activity, aeration, temperature, microbial interactions, mechanical damage, and insect infestation (Pestka and Casale, 1989). Therefore, preventive controls before and after harvest are crucial for managing contamination in the grain industry.

Mycotoxin management is systemic and includes all stages of the feed supply chain, starting from the production of raw materials to the feeding in the farm: crop phase (pre-harvest and harvest), transportation, storage, feed mill operations, and livestock production. An integrated system is based on the synchronised use of prevention and control implements at all stages of production from the field to the final consumer. Effective management involves preharvest strategies like Good Agricultural Practices (GAP), proper harvesting, storage, and processing techniques. Moisture and temperature control are key, along with cleaning, insect control, antifungal agents, and seed integrity maintenance to minimize contamination.

Mycotoxin controls in the field: Mycotoxin biosynthesis begins in the field, making effective crop management essential for limiting contamination. Good agronomic practices, including weather monitoring, help farmers make informed decisions that impact crop yield, quality, and profitability. Monitoring temperature and humidity aids in predicting and managing mycotoxin risks, while weather-based plant disease forecasts can be linked to climate change models to predict shifts in mycotoxin prevalence (Milani, 2013).

Beyond weather monitoring, selecting resistant crop varieties is another important strategy for reducing mycotoxin risk. Identifying compounds that inhibit fungal growth or interfere with mycotoxin development could enhance crop resistance and potentially save the agricultural industry billions. Research into mycotoxin biosynthetic pathways has provided insights into disrupting toxin synthesis, further aiding in contamination prevention (Kumar, 2010).

Crop rotation and tillage also help control fungal infections. Rotating host crops with non-host crops lowers infection pressure by interrupting the survival of fungal inoculum, reducing the risk of contamination (Eeckhout et al., 2013). Mold can persist in crop residue, especially when left on the soil surface, posing a higher risk when the same crop is continuously grown in the same field. Tillage has been shown to reduce mycotoxin levels, including DON, aflatoxin, and zearalenone, in crops like wheat and maize (Hooker and Schaafsma, 2005; Trenholm et al., 1989).

Applying approved fungicides during crop growth can reduce fungal infections and mycotoxin production. When used correctly, fungicides have been shown to lower DON levels by an average of 70% and other mycotoxins by more than 50% (Scarpino et al., 2015).

Harvest timing plays a crucial role in mycotoxin contamination. Factors like grain maturity, environmental conditions, and mold accumulation influence the risk. Delaying harvest can increase fungal growth and mycotoxin levels, particularly in wet and cool conditions. Studies have shown that higher aflatoxin, DON, and fumonisin levels are often associated with delayed harvest (Jones et al., 1981; Lauren et al., 2007). Additionally, ensuring proper harvesting equipment calibration is essential, as excessive grain damage weakens seed coats, increasing susceptibility to mold infection and contamination.

Cleaning and drying of grain: If incoming grains from the field are damaged, it is critically important that the grain is cleaned and properly segregated. It has been shown that even microscopic levels of damage to the grains increase the susceptibility to fungal infection (Lynch and Wilson, 1991). Sorting and cleaning of these grains have proven to reduce the risk of aflatoxin, DON, fumonisin, and zearalenone contamination (Kabak et al., 2006; Trenholm et al., 1991). It is important to note that undamaged grain can also be host to high concentrations of mycotoxins. So, removal of mold, damaged grains, and impurities will not completely eliminate mycotoxins.

In most cases drying freshly harvested grains to 14-15% is the key to adequate storage; this can differ with oily commodities (Magan and Aldred, 2007). The final safe moisture content depends on the crop and the climatic conditions where the commodity is stored. In general, the cooler and drier the grain is, the longer it can be stored safely (Magan and Aldred, 2007). Safer moisture content can resist the further growth of mold growth and mycotoxin biosynthesis, however, if contamination is already present in the grain it will not be eliminated.

Mycotoxin controls for feed processors: An effective mycotoxin management program in feed mills begins with a comprehensive quality assurance approach. This requires proactive strategies to prevent contamination throughout the whole manufacturing process. Producing safe feed is a collaborative effort that depends on the commitment of upper management, feed mill supervisors, and employees, all working together to maintain a safe, clean, and well-kept facility. Quality assurance programs in feed processing are critically important for mycotoxin risk management as these programs aim to maintain a high product quality, comply with regulations, minimize potential litigation, and maintain brand reputation. Pre-requisite programs are the keystone practices for mitigating risk associated with mycotoxin biosynthesis. However fungicidal application, chemical treatments, and other methods of inactivation have shown as supplemental practices with some coming at economical or nutritional value costs (Fumagalli et al., 2021).

Raw material contamination and quality control: A key component of the ingredient screening process is assessing the risk associated with each raw material to ensure the final feed product meets established safety limits for different animal species. An analysis conducted by Herrman et al. (2007) indicates that bulk receiving is the most prevalent critical control point for mycotoxins in existing HACCP plans. All companies that identified mycotoxins as a chemical hazard carried out quantitative testing to ensure toxin levels stayed below critical limits and to verify that suppliers adhered to established specifications.

Storage: Mycotoxin contamination can continue to develop during storage if conditions favor fungal growth and mycotoxin biosynthesis, making storage controls essential for maintaining feed safety and regulatory compliance. Taking regular, accurate measurements, alongside regular aeration of stored grain helps to control moisture and temperature, preventing "hot spots" that can promote fungal growth (Magan and Aldred, 2007). Monitoring of these environmental conditions can also help mitigate against rodent, bird, and insect infestation. Pest infestation in stored grain can create susceptibilities for mycotoxin production as grains become damaged, and temperatures and water activity increase. Prerequisite programs such as Good Storage Practices (GSP) are the foundation for reducing mycotoxin hazards during storage by eliminating sources for moisture and pests. CO_2 sensors in stored grain provide early detection of insect infestation and mold, offering a more effective strategy than traditional methods like visual inspection, smell, and temperature cables. (Ileleji et al., 2006).

Processing: Furthermore, mycotoxin biosynthesis can occur at various stages of the feed processing chain. Coradi et al. (2013) found that fungi colonies and pest infestations throughout receiving, storage, and processing was correlated with mycotoxin contamination. Infestation can be associated with inadequate use of the GMPs and/or cleaning and sanitation programs. Additionally, without proper control of moisture and temperature during final product formation—such as pellet production—mycotoxins can still develop (Coradi et al., 2013). It is increasingly important to maintain GMPs to limit mycotoxin production and implement critical limits for mycotoxins with routine monitoring throughout the whole feed processing chain.

In summary, to effectively manage mycotoxin risk in a feed mill operation, it is essential to establish an internal risk profile. This requires ongoing monitoring through regular testing of both raw materials and finished products. Corrective actions for non-compliant feed may include, isolating the affected batch and determining whether to discard, downgrade, or reprocess it. Appropriate detoxification methods can be considered based on the specific mycotoxins and the animal species involved (Janik et al., 2021). Implementing the right measures ensures both animal safety and the integrity of the overall food supply chain.

Mycotoxin Detection

Mycotoxin testing can occur at multiple stages, such as analyzing raw materials upon arrival at a feed mill using portable test kits in low-tech environments or testing final feed products after production. The type of livestock the feed is intended for, and the specific mycotoxins of concern should guide the testing process.

On-site testing offers rapid results, eliminating the need to send samples to reference laboratories, which helps maintain efficiency in feed production. Given that feed is produced both on farms and in commercial feed mills, balancing cost-effectiveness with mycotoxin control remains a challenge. The demand for simple, fast, and affordable mycotoxin detection methods continues to grow.

Rapid detection methods are more cost-effective, user-friendly, and adaptable for on-site testing. Allowing for real-time decisions to be made regarding mycotoxin results. Additionally, they help reduce labour and costs, minimize the need for complex instrumentation, and allow for operation by personnel with less technical training (Zheng, 2006). Research has demonstrated that commercially available immunoassays such as lateral flow immunochromatography and enzyme-linked immunosorbent assays (ELISA) are effective rapid detection methods for use in production settings and often relied on (Nolan et al., 2019).

Lateral flow assays are widely used for monitoring mycotoxins in practical situations. Figure 1. shows that these tests work by wicking a sample extract through a reagent zone containing antibodies specific to the target mycotoxin, which are attached to colloidal gold particles. If mycotoxins are present, they bind to the antibody-gold complex. The mixture then moves onto a membrane with a designated test zone of target mycotoxin, where any unbound antibodies are captured, forming a visible line. As mycotoxin levels increase, more antibodies bind to free mycotoxins, reducing the intensity of the test line. A control line is essential for these tests to confirm the test is functioning correctly (Neogen, 2023b; Leslie and A Logrieco, 2014). Lateral flow immunochromatography assays offer several advantages, including ease of use, fast results, and long-term stability across various climates. Their reliability and simplicity make them especially well-suited for on-site mycotoxin testing (Nolan et al., 2019).

Enzyme-linked immunosorbent assays (ELISAs) are popular monitoring tools used for the detection of mycotoxins within a laboratory setting. While they require more technical skill than some other rapid tests, ELISAs are valued for their high-throughput capability, minimal sample volume requirements, and reduced need for extensive sample preparation compared to traditional methods like Thin-Layer Chromatography (TLC) and High-Performance Liquid Chromatography (HPLC). Most commercially available ELISA kits use a competitive, heterogeneous format, displayed in Figure 2. Where mycotoxins in the sample compete with a labeled toxin (conjugate) for a limited number of antibody binding sites (Nolan et al., 2019). Higher toxin concentrations result in reduced binding of the labeled toxin, leading to a lower assay signal (Neogen, 2023b). Quantitative results are determined by measuring color changes in a microwell reader and comparing them to a standard curve.

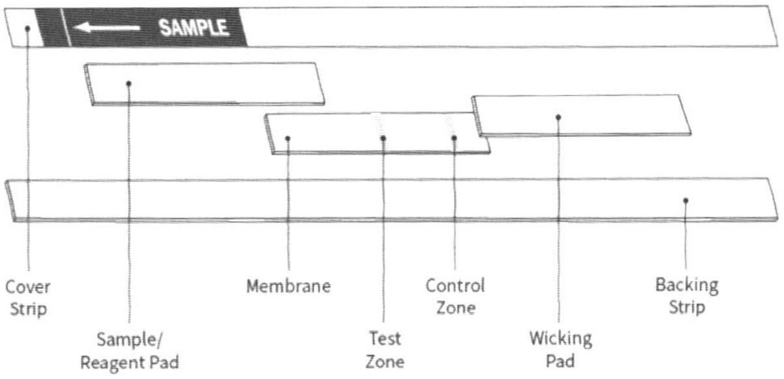

Figure 1. Diagram of components for a lateral flow test. Adapted from Neogen, 2023b.

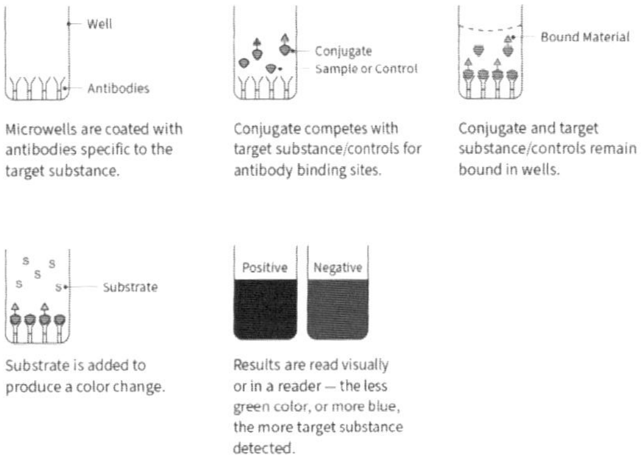

Figure 2. Methodology for competitive direct ELISA. Neogen, 2023b.

Reference methods such as HPLC, TLC, Gas chromatography (GC), and liquid chromatography-mass spectrometer (LCMS) are popular confirmation methods (Desjardins et al., 2003). However, these methods are typically limited to large commercial facilities, academic institutions, and reference laboratories due to their reliance on skilled technicians and costly equipment. Additionally, they are often time-consuming and labor-intensive.

In summary, accurate mycotoxin testing requires an understanding of proper sampling methods, sample preparation, and analysis, as well as the various sources of variability that can impact results. As mycotoxins are unevenly distributed in commodities, obtaining a representative sample and preparing it correctly is essential for reliable results. Sampling itself is the largest source of error, accounting for up to 82% of variability due to the heterogeneous and low levels of contamination (Erber and Yun, 2004). To improve accuracy and cost-effectiveness, inspection programs should be designed to minimize variability and detect contamination as precisely as possible. Testing methods also introduce variability, making it critical to study their accuracy and precision across a wide array of commodities (Zheng et al., 2005). Feed mills conducting in-house testing should also participate in proficiency or check sample programs to validate their results. Comparing internal data to certified values ensures that testing procedures, technicians, and equipment have the ability to produce reliable results (Desjardins et al., 2003).

Beyond accuracy and reproducibility, practical considerations for analytical methods include testing speed, the level of technical skill required, and whether results are qualitative or quantitative. The most effective on-site testing methods at feed mills combine all three—providing rapid, easy-to-use, and quantitative results. Ultimately, selecting the right monitoring tool depends on the specific needs and priorities of the user.

Ensuring Feed Safety Through *Salmonella* Control

Prevalence and Impact of *Salmonella* Contamination in Feed Production

Salmonella is a highly adaptable and ubiquitous pathogen capable of thriving in wet environments and persisting for weeks under dry conditions, posing a persistent threat in feed production (Humphrey, 2004; Podolak et al., 2010). It can contaminate raw ingredients and survive in dry animal feed, making feed a potential transmission vector for livestock (Juven et al., 1984; Parker et al., 2022, Sargeant et al. 2021). A meta-analysis of 85 global studies (1955–2020) estimated the prevalence of *Salmonella* at 18% in raw feed components and 9% in finished feed, highlighting the need for stringent monitoring and control in feed production facilities (Parker et al., 2022). Over time, the overall prevalence of *Salmonella* in raw feed components decreased from 25% in 1955 to 11% in 2019, likely due to advancements in technology and the adoption of more rigorous control systems (Parker et al., 2022). While *Salmonella* can be present in all types of feed, certain feeds and ingredients, are more frequently contaminated (CFIA, 2024). Reports indicate that animal-derived feed components are nearly four to six times more likely to be contaminated than plant-based components (CFIA, 2018; Ge et al., 2013; Parker et al., 2022), and mash feeds have higher contamination rates than pellet feeds (Jones et al., 2004). Consequently, high-risk feed types such as rendered products, oilseed meals, and mash feeds undergo more frequent sampling by the CFIA.

Salmonella feed contamination can negatively impact profitability, damage brand reputation, and pose regulatory challenges. It can compromise animal health and performance leading to increased veterinary costs and higher mortality rates (Lin, 2024). Contaminated feed also poses serious food safety risks due to zoonotic transmission through contaminated animal products such as meat, milk and eggs. (Crump et al., 2002; Sanchez et al., 2002). *Salmonella* is one of the leading human pathogens responsible for foodborne illnesses worldwide with nearly 88,000 cases reported annually in Canada, according to most recent estimates (Government of Canada, 2016).

Salmonella - Sources of Contamination and Preventives Measures

Raw material contamination and quality control: *Salmonella* can enter the feed production chain through contaminated raw materials such as grains, protein meals, and other ingredients. Grains are particularly susceptible to contamination from pests and wildlife, both in the field and during transportation (Jones et al., 2011). Poor pest management at feed mills further increases the risk, as rodents, birds, and insects can introduce and spread *Salmonella* in storage and production areas (Morita et al., 2006).

Raw material quality control is essential and the first step in safeguarding feed production. Supplier reviews, visual inspections and baseline microbiological testing should be conducted to detect any signs of contamination by pests and ensure the integrity of raw materials before they enter production. Early detection plays a critical role in maintaining feed safety. Studies have shown that feed samples contaminated with *Salmonella* had significantly higher Enterobacteriaceae counts compared to *Salmonella*-negative samples (Jones et al., 2004; Stewart et al., 2020). Testing for Enterobacteriaceae can therefore serve as an indicator of feed hygiene with high concentrations highlighting a potential risk, prompting further testing and, if necessary, decontamination.

Inactivation strategies, processing contamination and environmental controls: Once raw materials enter the production process, inactivation and preservation strategies come into play. While raw materials may initially introduce *Salmonella* into the processing facility, contamination of finished products is often linked to inadequate processing or poor hygiene on feed contact surfaces during operations, maintenance, transport or storage (Podolak et al., 2010).

Inadequate heat treatment during feed processing may fail to eliminate *Salmonella*, allowing it to persist in the product and contaminate feed contact surfaces. Thermal treatments such as roasting or pelleting under high-temperature steam conditioning significantly reduce bacterial loads in feed. However, while these methods are effective at inactivating most pathogens, like *Salmonella*, they do not entirely eliminate the risk, as finished feed remains susceptible to recontamination during cooling, bagging, and transportation (Boltz et al., 2021; Jones, 2011; Obe, 2023). The addition of organic acids during or after processing can provide long lasting antimicrobial protection, particularly against Gram-negative bacteria like *Salmonella* and other Enterobacteriaceae.

Salmonella can survive for extended periods in the processing environment, particularly in humid conditions with organic residue buildup (Gong et al., 2017). High levels of contamination have been observed in receiving areas of rendering plants, with *Salmonella* detected on employee shoes, gloves, and floor swabs (Morita et al., 2006). Dust accumulation in mill processing contributes significantly to the persistence of *Salmonella* (Butcher et al. 2019; Jones et al. 2004). In feed mills, airborne dust is a key factor in cross-contamination, facilitating the spread of the pathogen. This issue is particularly prevalent in facilities running continuous production cycles, where insufficient cleaning of mixing equipment and conveyors increases the risk of recontamination (Podolak et al., 2010).

Implementing effective biosecurity measures such as segregating "dirty" and "clean" zones, restricting employee movement between receiving and finished product areas, and treating incoming raw ingredients as potentially contaminated can reduce the risk of cross-contamination (Morita et al., 2006). Designing feed production facilities with hygiene in mind, using easy to clean equipment and minimizing dust accumulation further reduce contamination risks (Podolak et al., 2010). Proper sanitation and effective cleaning protocols are essential to preventing microbial persistence, biofilm formation and recurrent contamination (Gong et al., 2017). Routine environmental monitoring and product testing for key indicator microorganisms after each identified CCP can help detect or prevent recontamination, assess sanitation effectiveness, and identify potential bacterial growth niches.

Storage contamination and preservation strategies: Poor storage conditions, including high humidity and warm temperatures, create ideal conditions for bacterial growth, making stored feed vulnerable. Older ingredients stored for extended periods are particularly susceptible and can become contamination hotspots if not properly managed.

To preserve feed quality during storage, maintaining dryness, cleanliness, and effective pest control is crucial (Butcher et al., 1995). Sealed, waterproof silos protect against environmental contaminants, while treating raw materials with organic acids or other chemical additives can inhibit microbial growth. Routine inspections should be conducted to detect contaminants such as rodent droppings or mold. Implementing a "first-in, first-out" system prevents older, more vulnerable materials from becoming a potential source of contamination.

Salmonella Monitoring and Testing Methods

To ensure compliance, feed establishments are encouraged to implement an internal monitoring program for *Salmonella* (CFIA, 2024). While accredited laboratories and validated testing methods are required for reporting results to the CFIA, alternative rapid screening methods can be used in-house for internal process control of biological hazards allowing for more frequent monitoring and providing real-time insights. By complementing accredited laboratory testing, they help identify potential contamination earlier, enabling corrective actions to maintain environmental and product safety.

Salmonella testing in feed and environmental samples can be performed using various methods, which fall into three main categories: culture-based methods, rapid DNA screening, immunoassays. Culture-based agar methods, such as MFHPB-20, are considered the gold standard due to their ability to detect and biochemically confirm viable bacteria. However, they are often more labor intensive, requiring multiple incubation steps and skilled technicians, making them more suited for reference laboratories rather than routine in-house testing.

Rapid DNA screening methods, including PCR and LAMP-based assays, can provide results within 24 h by detecting *Salmonella*-specific genetic markers. Pre-prepared kits with minimal hands-on steps and technical expertise requirements simplify in-house implementation. These methods are fast, highly specific, and sensitive, making them also widely used in reference laboratories.

Immunoassays, such as lateral flow devices and ELISA, while often considered less specific and sensitive, still offer a very simple and cost-effective screening option.

For most feed and environmental monitoring programs, rapid DNA screening and immunoassays are the easiest to implement due to their speed, cost-effectiveness and minimal laboratory infrastructure requirements. While these pathogen testing methods are sensitive, they all require sample enrichment to improve detection probability. Proper biosafety measures, including biohazardous waste disposal, must be in place to handle enriched samples safely (PHAC, 2025). Also, although specific *Salmonella* testing is necessary to determine presence or absence, its detection in feed products can be challenging due to low contamination levels, sample heterogeneity, and large production volumes.

A Broader Approach to Feed Safety: Enterobacteriaceae as a Hygiene Indicator

Given the challenges associated with *Salmonella* testing, incorporating a broader testing strategy can enhance feed safety and environmental control. Rather than focusing solely on *Salmonella* detection, monitoring the overall hygienic status of the environment and products provides a more comprehensive evaluation of contamination risks. Testing for indicator microorganisms such as total aerobic bacteria, Enterobacteriaceae or coliforms can offer additional insights into sanitation effectiveness, process control and background microflora.

Enterobacteriaceae testing is increasingly preferred in the food industry due to its broad inclusivity, providing a comprehensive assessment of hygiene and microbial control. Enterobacteriaceae are a diverse group of Gram-negative bacteria including coliforms and pathogens like *Salmonella* (Figure 3). Commonly found in the intestinal tracts of humans and animals, as well as in environmental sources such as soil and water, they primarily serve as general hygiene indicators rather than a pathogen test. While their presence does not directly imply *Salmonella* contamination, high Enterobacteriaceae counts suggest increased risk of presence and indicate conditions conducive to *Salmonella* growth. A compact, sample-ready culture-based method requiring minimal expertise and laboratory equipment can be implemented for in-house surface and product testing, delivering results within 24h. Monitoring Enterobacteriaceae levels in feed and on environmental surfaces helps evaluate the effectiveness of control measures in minimizing *Salmonella* contamination. Lower Enterobacteriaceae counts correlate with well-controlled processes and improved hygiene (Neogen, 2023a). Quantitative results from indicator testing enable trend analysis, establishment of acceptance limits, and the detection of process deviations, prompting corrective actions and further pathogen testing when necessary (Figure 4).

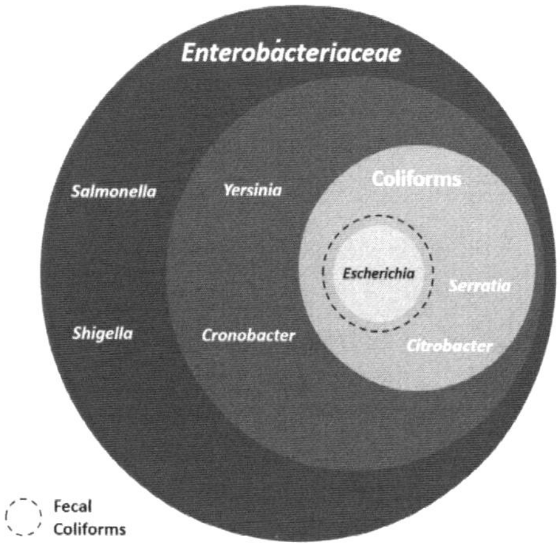

Figure 3. Diagram - Classification of Enterobacteriaceae and related bacteria. Adapted from Baylis et al. 2011.

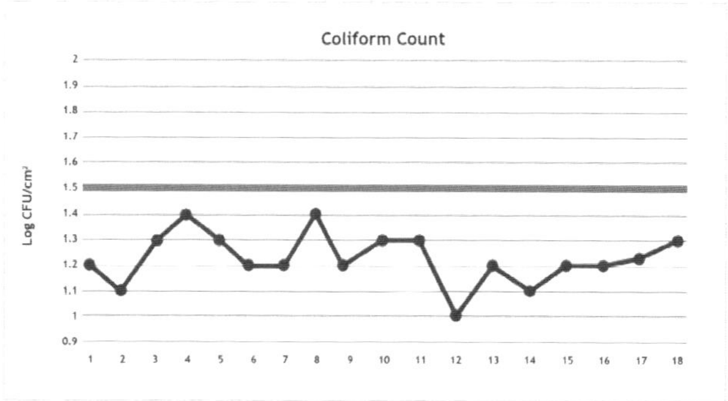

Figure 4. Example of coliform counts and baseline-level post-sanitation. Adapted from Neogen, 2023a.

Establishing an Effective Indicator Sampling Program

An effective indicator sampling program is essential for ensuring feed safety and preventing contamination in processing facilities. Such a program requires careful planning, execution, and continuous evaluation. A multidisciplinary team with expertise in processing systems and cleaning protocols should design the sampling plan. This team is responsible for creating a comprehensive plan that includes training for proper sampling techniques and site selection.

Sampling site selection begins with a detailed mapping of the production process. This involves identifying key processing steps, equipment, and surface types (e.g stainless steel or rubber), to determine the most appropriate sampling locations. Sampling is typically conducted using a zone-based approach, with Zone 1 focusing on product contact surfaces, which are considered the most critical. Zone 2 includes adjacent surfaces. Zone 3 further away but still in the production area is reserved for investigative purposes. Zone 4 outside the production facility are less frequently included. The choice of sampling tools is also important, with swabs being used for small difficult to reach areas such as crevices and sponges for larger readily accessible areas (Neogen, 2023a).

Sampling frequency and the number of samples should be determined using a risk-based approach. Factors such as product type, process risk, sanitation cycles, and facility conditions influence how often sampling should be conducted. The closer the sampling point is to the product, the greater the risk, and the harder it is to clean, the higher the probability for contamination. When both conditions are present, those test points should be monitored more closely and sampled more frequently. Routine verification is essential and should be conducted after every sanitation cycle. In some cases, equipment may need to run before sampling to dislodge and expose any hiding residual microbes. Testing frequency can be justified and adjusted accordingly based on historical data (Neogen, 2023a).

Data trending and the establishment of baselines are crucial for identifying patterns and potential contamination risks before they escalate. Regular review of sampling data helps detect trends, while long-term tracking provides insights into seasonal variations and operational impacts. Baselines are established overtime through consistent sanitation cycles. Standard operating procedures (SOPs) for indicator testing should clearly define the frequency of trending and the processes for data review to enable informed decision-making. Cut-off levels for indicator organisms should be based on historical data and baseline levels, with low microbial levels expected post-sanitation (Neogen, 2023a).

Corrective action is necessary, if indicator results fall outside acceptable parameters. Thorough documentation is essential. Significant deviations require re-evaluation and retraining may be needed if improper techniques are identified. Increasing sampling to identify the sources of indicator organisms is a key step in preventing contamination. Common sources include cross-contamination from external sources, water contamination, and raw materials, which are primary contributors to contamination in Zones 1 and 2. Other factors, such as facility changes, new sanitation chemicals or personnel, and sporadic equipment failures, can also play a role. By addressing these sources and continuously refining the sampling program, establishments can enhance their food safety protocols and reduce the risk of contamination (Neogen, 2023a).

Conclusion

Ensuring feed safety through effective *Salmonella* control is essential for protecting animal health, preventing economic losses, and mitigating public health risks. The new feed regulations provide flexibility in how safety outcomes are achieved, as long as feed remains safe. A proactive approach to *Salmonella* control, supported by comprehensive product and environment monitoring, data collection and analysis plays a critical role in early contamination risk detection and assessing the effectiveness of mitigation strategies. Routine Enterobacteriaceae testing offers valuable insights into feed hygiene, process control and sanitation effectiveness ultimately strengthening feed safety programs. By shifting from reactive to predictive control through continuous monitoring, data-driven decision-making and effective interventions, the feed industry can improve control over microbiological hazards, ensure compliance with Canadian standards and foster a positive feed safety culture.

References

Baylis, C., M. Uyttendaele, H. Joosten and A. Davies. 2011. The Enterobacteriaceae and their significance to the food industry. ILSI Europe Report Series.1-48.

Bhat, R.V. and J.D. Miller. 1991. *Mycotoxins and food supply. 1*, 27–31. Food, Nutrition and Agriculture FAO. Available at: https://www.fooddiagnostics.dk/wp-content/uploads/FAO-Mycotoxins_and_Food_Supply.pdf

Boltz, T.P. , J.S. Moritz, V.E. Ayres, C.L. Showman, J. Jaczynski and C. Shen. 2021. Modeling thermal inactivation of *Salmonella* Typhimurium in mash broiler feed. J Appl Poult Res. 30(4):100208.

Butcher, G.D. and R.D. Miles. 1995. Minimizing Microbial Contamination in Feed Mills Producing Poultry Feed. Available at: https://edis.ifas.ufl.edu/publication/VM054

CFIA - Canadian Food Inspection Agency. 2018. National feed inspection program: Results from 1989 to 2018 for the *Salmonella* monitoring program for livestock feeds. Available at: https://inspection.canada.ca/en/animal-health/livestock-feeds/inspection-program/salmonella-monitoring-program-livestock-feeds

CFIA - Canadian Food Inspection Agency. 2024. Feeds Regulations, 2024: SOR/2024-132 - Canada Gazette, Part II, Volume 158, Number 14. Available at: https://canadagazette.gc.ca/rp-pr/p2/2024/2024-07-03/html/sor-dors132-eng.html

CFIA - Canadian Food Inspection Agency. 2024. Preventive control plan guidance: General overview *Feeds Regulations, 2024.* Available at: https://inspection.canada.ca/en/preventive-control-plan-guidance

CFIA - Canadian Food Inspection Agency. 2024. RG-11 Regulatory Guidance: Prevention, control and mitigation of *Salmonella* in livestock feeds. Available at: https://inspection.canada.ca/en/animal-health/livestock-feeds/regulatory-guidance/rg-11

Coradi, P.C., A.F. de L. Filho, and J.B.P Chaves. 2013. Identification of the Critical Control Points in a Feed Mill: a case study. *Revista Brasileira de Tecnologia Agroindustrial*, *07*(1981-3686), 977–992.

Crump, J.A., P.M. Griffin and F.J. Angulo. 2002. Bacterial contamination of animal feed and its relationship to human foodborne illness. Clin Infect Dis. Oct 1;35(7):859-65.

Desjardins, A., C. Maragos, W. Norred, J. Pestka, T. Phillips, P. Vardon, T. Whitaker, G. Wood, and H. Van Egmond. 2003. *Mycotoxins: Risks in Plant, Animal, and Human Systems* (Vol. 139). Council for Agricultural Science and Technology.

Eeckhout, M., G. Haesaert, S. Landschoot, N. Deschuyffeleer, S. De Laethauwer. 2013. *Guidelines for prevention and control of mould growth and mycotoxin production in cereals.* Mytox. Available at: https://mytox.be/wp-content/uploads/Guidelines-for-prevention-and-control-of-mould-growth-and-mycotoxin-production-in-cereals.pdf

Erber, E., and C.H. Yun. 2004. Managing the risk of mycotoxins in modern feed production. In *Procedding of Korea Feed Ingredients Association Conference* (Vol. 5, pp. 19-56). Korea Feed Ingredients Association. Available at: https://koreascience.kr/article/JAKO200451922836738.pdf

Fumagalli, F., M. Ottoboni, L. Pinotti, and F. Cheli. 2021. Integrated Mycotoxin Management System in the Feed Supply Chain: Innovative Approaches. *Toxins*, *13*(8), 572.

Ge, B., P.C. LaFon, P.J. Carter, S.D. McDermott, J. Abbott, A. Glenn, S.L. Ayers, S.L. Friedman, J.C. Paige, D.D. Wagner, S. Zhao, P.F. McDermott and M.A. Rasmussen. 2013. Retrospective analysis of *Salmonella*, *Campylobacter*, *Escherichia coli*, and *Enterococcus* in animal feed ingredients. Foodborne Pathog Dis. Aug;10(8):684-91.

Gong, C. and X. Jiang. 2017. Characterizing *Salmonella* Contamination in Two Rendering Processing Plants. J Food Prot. Feb;80(2):265-270.

Government of Canada. 2016. Yearly food-borne illness estimates for Canada. Available at: https://www.canada.ca/en/public-health/services/food-borne-illness-canada/yearly-food-borne-illness-estimates-canada.html

Herrman, T.J., M.R. Langemeier, and M. Frederking. 2007. Development and Implementation of Hazard Analysis and Critical Control Point Plans by Several U.S. Feed Manufacturers. *Journal of Food Protection*, *70*(12), 2819–2823.

Humphrey, T. 2004. *Salmonella*, stress responses and food safety. Nat Rev Microbiol. Jun;2(6):504-9.

Hooker, D.C., and A.W. Schaafsma. 2005. Agronomic and environmental impacts on concentrations of deoxynivalenol and fumonisin B_1 in corn across Ontario. *Canadian Journal of Plant Pathology*, *27*(3), 347–356.

Ileleji, K.E., D.E. Maier, C. Bhat, and C.P. Woloshuk. 2006. Detection of a developing hot spot in stored corn with a CO2 sensor. *Applied Engineering in Agriculture*, *22*(2), 275–289.

Janik, E., M. Niemcewicz, M. Podogrocki, M. Ceremuga, M. Stela, and M. Bijak. 2021. T-2 Toxin—The Most Toxic Trichothecene Mycotoxin: Metabolism, Toxicity, and Decontamination Strategies. *Molecules*, *26*(22), 6868.

Jones, F.T., and K.E. Richardson. 2004. *Salmonella* in commercially manufactured feeds. Poult Sci. Mar;83(3):384-91.

Jones, F.T. 2011. A review of practical *Salmonella* control measures in animal feed. The Journal of Applied Poultry Research 20(1):102-113.

Jones, R.K., H.E. Duncan, and P.B. Hamilton. 1981. Planting date, harvest date, and irrigation effects on infection and aflatoxin production by Aspergillus flavus in field corn [Factors predisposing corn to fungal infection]. *Phytopathology*, *71*, 810–816.

Juven, B.J., N. A. Cox, J.S. Bailey, J.E. Thomson, O.W. Charles and J.V. Shutze. 1984. Survival of *Salmonella* in Dry Food and Feed. J Food Prot. Jun;47(6):445-448.

Kabak, B., A.D.W. Dobson, and I. Var. 2006. Strategies to Prevent Mycotoxin Contamination of Food and Animal Feed: A Review. *Critical Reviews in Food Science and Nutrition*, *46*(8), 593–619.

Kumar, A. 2010. Management of Mycotoxin Contamination in Preharvest and Post Harvest Crops: Present Status and Future Prospects. *The Journal of Phytology*, *2*(7), 37–52.

Lauren, D.R., W.A. Smith, and E. Di. 2007. Influence of harvest date and hybrid on the mycotoxin content of maize grain grown in New Zealand. *New Zealand Journal of Crop and Horticultural Science*, *35*(3), 331–340.

Leslie, J.F., and A. Logrieco. 2014. *Mycotoxin reduction in grain chains*. Wiley Blackwell. Available at: https://www.wiley.com/en-us/Mycotoxin+Reduction+in+Grain+Chains-p-9780813820835

Lin, Y.A. 2024. Impact of feed quality on animal health and productivity. Feed additive. Available at: https://www.feedandadditive.com/impact-of-feed-quality-on-animal-health-and-productivity/

Lynch, R.E., and D.M. Wilson. 1991. Enhanced Infection of Peanut, Arachis hypogaea L., Seeds with Aspergillus flavus Group Fungi Due to External Scarification of Peanut Pods by the Lesser Cornstalk Borer, Elasmopalpus lignosellus (Zeller)1. *Peanut Science*, *18*(2), 110–116.

Magan, N., and D. Aldred. 2007. Post-harvest control strategies: Minimizing mycotoxins in the food chain. *International Journal of Food Microbiology*, *119*(1-2), 131–139.

Milani, J. 2013. Ecological conditions affecting mycotoxin production in cereals: a review. *Veterinární Medicína*, *58*(No. 8), 405–411.

Morita, T., H. Kitazawa, T. Iida and S. Kamata. 2006. Prevention of *Salmonella* cross-contamination in an oilmeal manufacturing plant. J Appl Microbiol. Aug;101(2):464-73.

Neogen. 2023a. Food and Beverage Environmental Monitoring Handbook. Available at: https://www.neogen.com/neocenter/resources/food-beverage-environmental-monitoring-handbook/

Neogen. 2023b. *Mycotoxin Handbook: Navigate Testing Methods*. Available at https://www.neogen.com/neocenter/resources/mycotoxin-handbook/

Nolan, P., S. Auer, A. Spehar, C.T. Elliott, and K. Campbell. 2019. Current trends in rapid tests for mycotoxins. *Food Additives & Contaminants: Part A*, *36*(5), 800–814.

Obe, T., T. Boltz, M. Kogut, S.C. Ricke, L.A. Brooks, K. Macklin and A. Peterson. 2023. Controlling *Salmonella*: strategies for feed, the farm, and the processing plant. Poult Sci. 2023 Dec;102(12):103086.

Parker, E.M., A.J. Parker, G. Short, A.M. O'Connor and T.E. Wittum. 2022. *Salmonella* detection in commercially prepared livestock feed and the raw ingredients and equipment used to manufacture the feed: A systematic review and meta-analysis. Prev Vet Med. Jan;198:105546.

Pestka, J., and L. Casale, L. 1989. Naturally occurring fungal toxins. *Advances in Environmental Science and Technology*, *23*(0065-2563), 613–638.

Podolak, R., E. Enache, W. Stone, D.G. Black and P.H. Elliott. 2010. Sources and risk factors for contamination, survival, persistence, and heat resistance of *Salmonella* in low-moisture foods. J Food Prot. 2010 Oct;73(10):1919-36.

PHAC - Public Health Agency of Canada. 2025. Canadian Biosafety Standards and Guidelines. Available at: https://www.canada.ca/en/public-health/services/canadian-biosafety-standards-guidelines.html

Rukambile, E., V. Sintchenko, G. Muscatello, R. Kock and R. Alders. 2019. Infection, colonization and shedding of *Campylobacter* and *Salmonella* in animals and their contribution to human disease: A review. Zoonoses Public Health. Sep;66(6):562-578.

Sanchez, S., C.L. Hofacre, M.D. Lee, J.J. Maurer and M.P. Doyle. 2002. Animal sources of salmonellosis in humans. J Am Vet Med Assoc. Aug 15;221(4):492-7.

Sargeant, J.M., S.C. Totton, M. Plishka and E.R. Vriezen. 2021. *Salmonella* in Animal Feeds: A Scoping Review. Front Vet Sci. Nov 4;8:727495.

Scarpino, V., A. Reyneri, M. Sulyok, R. Krska, and M. Blandino. 2015. Effect of fungicide application to control Fusarium head blight and 20 Fusarium and Alternaria mycotoxins in winter wheat (Triticum aestivum L.). *World Mycotoxin Journal*, *8*(4), 499–510.

Stewart, S.C., S.S. Dritz, J.C. Woodworth, C. Paulk and C.K. Jones. 2020. A review of strategies to impact swine feed biosecurity. Animal Health Research Reviews. 21(1):61-68.

Streit, E., K. Naehrer, I. Rodrigues, and G. Schatzmayr. 2013. Mycotoxin occurrence in feed and feed raw materials worldwide: long-term analysis with special focus on Europe and Asia. *J Sci Food Agric*, *93*(12).

Trenholm, H.L., L.L. Charmley, D.B Prelusky, and R.M. Warner. 1991. Two physical methods for the decontamination of four cereals contaminated with deoxynivalenol and zearalenone. *Journal of Agricultural and Food Chemistry*, *39*(2), 356–360.

Trenholm, H.L., D.B. Prelusky, J.C. Young, and J.D. Miller. 1989. A practical guide to the prevention of Fusarium mycotoxins in grain and animal feedstuffs. *Archives of Environmental Contamination and Toxicology*, *18*, 443–451.

Zheng, Z., C.W. Humphrey, R.S. King, and J.L. Richard. 2005. Validation of an ELISA test kit for the detection of total aflatoxins in grain and grain products by comparison with HPLC. *Mycopathologia*, *159*(2), 255–263.

Poultry Ca and P nutrition as influenced by vitamin D_3 and phytase application

Influence de la vitamine D_3 et de la phytase sur l'apport en Ca et P dans l'alimentation des volailles

D. R. Korver[1,2]
[1]*Alpine Poultry Nutrition, Edmonton, AB*
[2]*University of Alberta, Edmonton, AB*
dkorver@alpinepn.com

PRE-CONFERENCE SPONSORED BY / PRÉ-COLLOQUE COMMANDITÉ PAR

HUVEPHARMA®

Abstract

Vitamin D has been added to poultry diets for as long as poultry have been raised in confinement, because the birds are unable to synthesize sufficient levels of vitamin D in the absence of sufficient exposure to sunlight. Initially, vitamin D was known primarily to be involved in the regulation of absorption of calcium and phosphorus from the diet, particularly when dietary levels were low relative to the birds' requirements. Since that time, it has become clear that vitamin D plays important roles in other physiological functions such as immunity, muscle growth and cell growth regulation. The form of vitamin D traditionally supplemented to poultry diets (cholecalciferol) is not the active form in the body. Additional isomers of vitamin D have also been developed as effective dietary supplements to meet the vitamin D requirements of commercial poultry.

Phytase is an enzyme that liberates phosphate molecules from phytate, the storage form of P in plant-based ingredients. Phytate-bound P is poorly available to poultry, and additional P must be provided, either as supplemental P, or by increasing the availability of the intrinsic P in the plant ingredients. The use of phytase has allowed the poultry industry to reduce the use of inorganic mineral phosphate supplementation. This reduces diet costs and environmental pollution caused by excretion of the undigested P. Degradation of the phytate molecule also reduces interference in availability of other nutrients including calcium, and the myo-inositol ring of phytate can be used in several metabolic and regulatory processes.

Primarily because of their independent effects on calcium metabolism, numerous studies have been conducted to investigate the potential interactive effects of phytase and vitamin D supplementation. Effective use of these supplements can reduce the need for phosphorous supplementation, detrimental interactions between calcium and phosphorus in the diet, and increase performance, skeletal integrity, meat yield (broilers) and shell quality (laying hens).

Keywords: broiler chicken, laying hen, vitamin D, phytase

Résumé

La vitamine D a été ajoutée au régime alimentaire des volailles depuis qu'elles sont élevées en milieu confiné. En effet, les oiseaux ne synthétisent pas assez de vitamine D pour combler leurs besoins sans une exposition suffisante à la lumière du jour. On sait depuis longtemps que la vitamine D intervient principalement dans la régulation de l'absorption du calcium et du phosphore, en particulier lorsque leur teneur dans les aliments est faible par rapport aux besoins des oiseaux. Il est maintenant clair que la vitamine D joue un rôle important dans plusieurs autres fonctions physiologiques, notamment l'immunité, la croissance musculaire et la régulation de la croissance cellulaire. La forme de vitamine D couramment ajoutée à l'alimentation des volailles (cholécalciférol) ne correspond pas à sa forme active dans l'organisme. D'autres isomères de la vitamine D ont été développés comme compléments alimentaires efficaces pour répondre aux besoins en vitamine D des volailles commerciales.

La phytase est une enzyme qui libère les molécules de phosphate des phytates, la principale forme de stockage du phosphore dans les ingrédients d'origine végétale. Le phosphore lié aux phytates est peu assimilable par les volailles. Par conséquent, il est nécessaire de fournir aux oiseaux une source additionnelle de phosphore, que ce soit sous forme de complément ou en augmentant la disponibilité du phosphore des ingrédients végétaux. Grâce à l'utilisation de la phytase, l'industrie avicole a pu diminuer le recours aux suppléments de phosphate inorganique. Cette pratique permet de réduire les coûts de l'alimentation et la pollution de l'environnement due à l'excrétion du phosphore non digéré. Une fois dégradée, la molécule de phytate améliore aussi la disponibilité d'autres nutriments, y compris le calcium, et le groupement *myo*-inositol du phytate peut être utilisé dans plusieurs processus métaboliques et de régulation.

De nombreuses études ont été menées pour examiner les effets interactifs potentiels de la phytase et de la supplémentation en vitamine D, surtout en raison de leurs effets indépendants sur le métabolisme du calcium. L'utilisation optimale de ces compléments permet de diminuer les apports en phosphore supplémentaire, d'atténuer les effets indésirables des interactions entre le calcium et le phosphore dans la ration et d'améliorer les performances, l'intégrité du squelette, le rendement en viande (poulets de chair) et la qualité de la coquille (pondeuses).

Introduction

Calcium and phosphorus are primarily found in the bones of poultry, and their requirements in poultry largely reflect the proportion of each mineral in bone, although a biologically notable exception exists in laying hens, which have a large additional Ca demand to support eggshell formation. However, each of these nutrients play additional roles in metabolism and various bodily functions. Bone formation is critical to support the rapid growth rate of modern broilers and turkeys. In layer and breeder pullets, proper development of the skeletal system is essential to support the long laying cycles of which modern laying hens are capable. During the laying phase, bone Ca is mobilized at night, when the amount of Ca absorbed from the gut is insufficient to support eggshell formation, and when a shell is not being formed, hens can re-deposit medullary (but not structural) bone on a daily basis (Kerschnitzki et al., 2014; Sinclair-Black et al., 2024).

Calcium and Phosphorus Metabolism in Poultry

Dietary Ca absorption involves two distinct mechanisms. At low levels of dietary intake relative to the requirement, Ca is primarily absorbed via a saturable, vitamin D-dependent receptor on the enterocyte surface. At high levels of dietary Ca, a non-saturable, paracellular mechanism is involved. Thus, excessive intakes of dietary Ca relative to requirement lead to high levels of Ca absorption. Similarly, P is absorbed via a transcellular, active transport process at low to adequate intakes. It appears that there is also a paracellular absorption pathway at high intakes (Fishbane and Nigwekar, 2021), although this is less well understood.

The metabolic regulation of Ca and P homeostasis is closely related, and mainly involves two overlapping endocrine pathways – parathyroid hormone (PTH) and vitamin D, and three main target tissues – the intestine, bones, and the kidneys. Blood Ca levels are tightly regulated via alterations in the level of absorption from the intestine, mobilization or deposition in the bone, and resorption from the kidney. Phosphorus homeostasis primarily involves regulation at the level of the kidney. When plasma P is high, bone tissue produces fibroblast growth factor-23 (FGF-23), which increases P excretion by the kidney, and inhibits production of PTH and 1α-hydroxylase (Sinclair-Black et al., 2023).

When plasma Ca or P are low, PTH is released by the parathyroid. PTH stimulates bone resorption and the release of both Ca and P to the blood. At the kidney, PTH decreases Ca excretion but increases P excretion PTH also stimulates the kidneys to convert 25-hydroxyvitamin D_3 ($25OHD_3$) to active vitamin D, $1{,}25(OH)_2$ vitamin D_3 ($1{,}25OH_2D_3$). The $1{,}25OH_2D_3$ acts on the intestine to increase the active absorption of Ca and P from the diet. Simultaneously, bone resorption is increased by $1{,}25OH_2D_3$, resulting in a release of both Ca and P to the plasma.

Vitamin D Metabolism in Poultry

Modern commercial poultry are typically supplemented vitamin D at many times the minimum requirement for growth, bone formation and eggshell formation (National Research Council,

1994). This over-supplementation reflects the approach of nutritionists recognizing that losses may occur during feed production and storage, the desire for substantial safety margins, and the relatively small contribution of vitamin D to the overall cost of the feed. Vitamin D activity can be supplemented in a number of different forms, including vitamin D_3 (Bassi et al., 2025; Saunders-Blades and Korver, 2015), $25OHD_3$ (Bassi et al., 2025; Saunders-Blades and Korver, 2015), 1α-vitamin D_3 (Warren et al., 2020), and $1{,}25OH_2D_3$ glycoside derived from the plant *Solanum glaucophyllum* (Alharbi et al., 2025). Interestingly, although each of the aforementioned products provide vitamin D activity, the supplementation of some of these metabolites in addition to or in place of vitamin D_3 has beneficial effects on functions beyond bone metabolism and shell quality.

Vitamin D can be produced naturally by the chicken with exposure of the skin to ultraviolet light (Rana and Campbell, 2021). However, commercial poultry raised in confinement do not receive exposure to sunlight, and even birds raised in open-sided houses in hot climates do not receive direct sunlight. Whether produced in the skin or provided in the diet, vitamin D_3 is efficiently converted to $25OHD_3$ by 25-hydroxylase, although this conversion is limited in young chicks (Saunders-Blades and Korver, 2014), suggesting impaired metabolism in very young birds. The main form of vitamin D in the body is $25OHD_3$. When required, the $25OHD_3$ is converted in the kidney by 1α-hydroxylase to the active hormone, $1{,}25OH_2D_3$. At excessive levels of $25OHD_3$, 24-hydroxylase produces inactive 24,2-5OH_2 vitamin D_3 ($24{,}25OH_2D_3$) (Warren, 2024). When no longer needed, or when the active hormone is produced in excess of requirement, 24 hydroxylase converts the active hormone to $1{,}24{,}25(OH_3)D_3$, a water-soluble metabolite for excretion. Interestingly, older birds such as laying hens at the end of the production cycle have greatly reduced levels of 1α-hydroxylase, and increased levels of 24-hydroxylase, suggesting impaired vitamin D metabolism with advanced age (Abe et al., 1982; Joyner et al., 1987; Kuwata et al., 2024). Although the endocrine functions of vitamin D are regulated via this mechanism, multiple extrarenal tissues also express 1α-hydroxylase (Shanmugasundaram and Selvaraj, 2012) and produce active vitamin D for autocrine and paracrine roles in immune function, muscle development, gut function, reproduction and angiogenesis.

Phytate and Phytase in Poultry Diets

Much of the P found in plant-based feed ingredients is in the form of phytate (Humer et al., 2015), which is generally poorly available to poultry (Venter et al., 2024), but is surprisingly more available to broiler chicks than to laying hens (Ma et al., 2019). Phytate (inositol hexaphosphate, IP_6) has a high affinity for P and other nutrients, particularly at pH approaching neutral. Sequential removal of phosphate groups reduces the binding affinity of phytate for other nutrients. For many types of poultry, typical diets contain adequate total P, but because of the poor availability, P is supplemented in the form of (inorganic) rock phosphates. This creates several problems. First, as a limited resource (Cordell and White, 2011), such supplementation is expensive. Second, because the diet contains a substantial amount of undigestible P, that P is excreted into the environment, which can lead to P runoff into surface waters, leading to eutrophication and the killing of aquatic life. Third, even in the presence of adequate available P, high levels of phytate are negatively correlated with functional properties such as bone breaking strength (Figure 1; Bello et al., 2019).

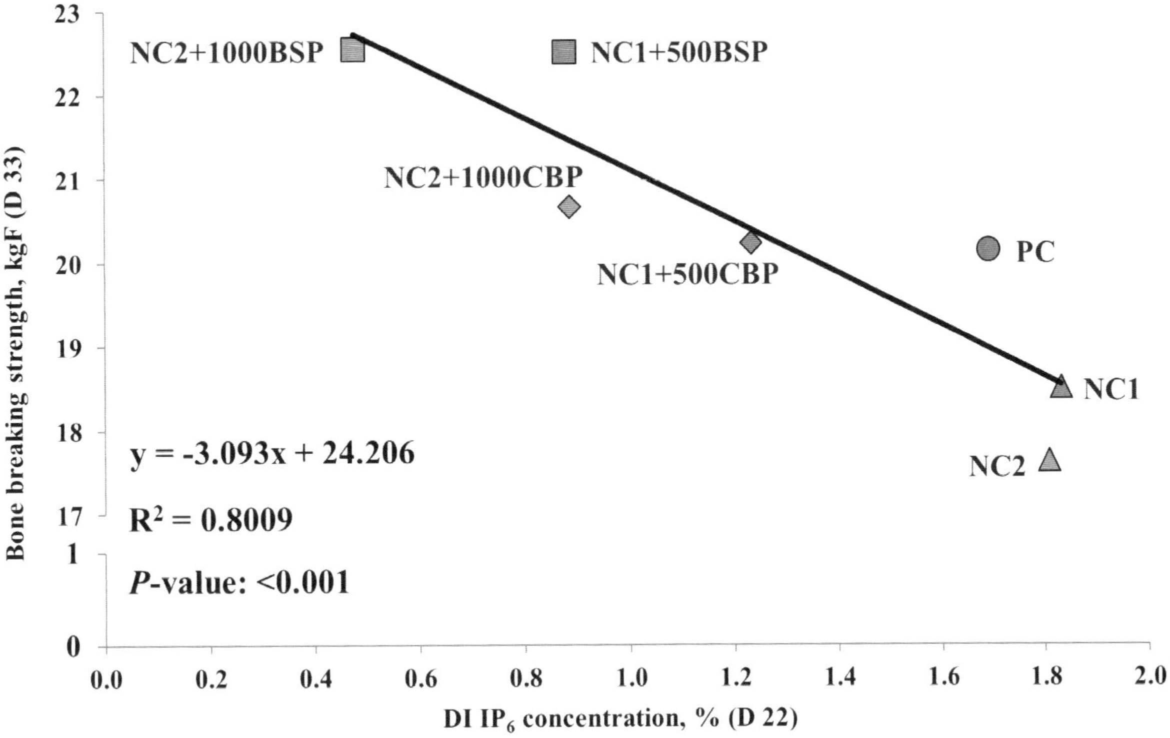

Figure 1. Relationship between dietary inositol hexaphosphate (IP_6; phytate) and bone breaking strength at 33 days of age in broiler chickens. PC = positive control; diet containing 0.711% Ca and 0.395% available P; NC = Negative control, diet contained 0.656% Ca and 0.249% available P; CBP = *Citrobacter braakii* phytase; BSP = *Buttiauxella* sp. Phytase; 500 and 1000 refer to the FTU/g of phytase activity added. Data are from Bello et al. (2019).

In addition to P, phytate can also bind other nutrients including divalent metal cations such as Ca, Zn, Cu, Zn and others, as well as proteins and carbohydrates, thus reducing nutrient availability beyond P (Selle et al., 2023).

In order to make a greater proportion of the intrinsic P available to the bird and therefore reduce the need for inorganic P supplementation, phytase is routinely added to poultry diets. Phytase sequentially removes phosphate groups from the molecule, thus reducing the anti-nutritive effect of phytate, and increasing availability of P and other nutrients in the diet. Removal of all 6 phosphate groups is not necessary for a substantial increase in P and other nutrient digestibility, however the remaining inositol ring can be absorbed from the intestine and used by the bird (Dersjant-Li et al., 2022). Supplementation of very high levels (relative to typical commercial levels) of phytase, or "superdosing" may have additional benefits (David et al., 2021; Woyengo and Wilson, 2019).

In order to most effectively use phytase in poultry diets, several considerations must be kept in mind.

1) The diet must contain less available P than is required by the bird. If the diet already contains an adequate amount of available P, there is no benefit (but potential detriments) to providing more available P. Increasing dietary P intake from 0.4 g to 2.4 g per hen per day resulted in a substantial decrease in egg specific gravity (Miles et al., 1983). Therefore, adding phytase to a diet already adequate in available P could reduce shell quality in laying hens (Javadi et al., 2021).
2) The diet must contain phytate P that can be liberated at quantities sufficient to meet the shortfall described in point 1 above. Phytate levels vary among plant types (Humer et al., 2015), and even from year to year (Korge et al., 2025).
3) Inorganic phosphate sources such as rock phosphates, and animal-derived sources such as meat and bone meal do not contain phytase, and will not be affected by phytase addition.
4) The phytase must be active. Most modern commercial phytases are generally heat-stable and can survive the pelleting process, but excessive temperatures can destroy activity, as can improper storage (De Jong et al., 2017).

Interactions of Vitamin D and Phytase in Poultry

Vitamin D is critical to Ca and P homeostasis, and supplementing the diet with vitamin D activity in various forms can reduce the effects of moderate reductions in dietary Ca and available P (Lopes et al., 2024; Rao et al., 2009; Vieites et al., 2018) by influencing Ca and P metabolism and homeostasis. Dietary phytase dephosphorylates the phytate molecule and increases availability of P and other nutrients, thus allowing a reduction in supplementation of those nutrients (Gautier et al., 2018; Scholey et al., 2018). Because these two types of feed additives have different modes of action it is intriguing to speculate that there may be interactive effects when both are used strategically in the diet.

Most studies investigating the use of these two categories of feed supplements have found each to be effective in reducing the effects of Ca- and P-reduced diets, but the effects are not generally synergistic (Bassi et al., 2025; Bilal et al., 2010; Delezie et al., 2012; Han et al., 2009; Shi et al., 2024a; Shi et al., 2024b; Snow et al., 2004). A similar lack of synergy has been reported in pigs (O'Doherty et al., 2010). In spite of the overall lack of synergy in the published literature, it has been demonstrated for many years that both vitamin D (or its metabolites) and phytase are important additions to the diet of poultry.

References

Abe, E., H. Horikawa, T. Masumura, M. Sugahara, M. Kubota, and T. Suda. 1982. Disorders of cholecalciferol metabolism in old egg-laying hens. J. Nutr. 112:436-446.

Alharbi, K., A. Asnayanti, A. Hasan, W. J. Vaught, K. Buehler, J. D. Van der Klis, J. Gonzalez, M. T. Kidd, and A. Alrubaye. 2025. Investigating the effect of 1,25 dihydroxycholecalciferol-glycosides and phytogenic antioxidants against bacterial chondronecrosis induced by aerosol transmission model. J. Appl. Poult. Res. 34:100507.

Bassi, L. S., C. C. S. Martins, C. A. Lozano-Poveda, I. C. Dias, S. G. Oliveira, and A.

Maiorka. 2025. Effect of 25-hydroxycholecalciferol and high phytase doses on performance, vitamin d status, bone mineralization, and mechanistic target of rapamycin gene expression of broilers. Animal 19:101353.

Bello, A., Y. Dersjant-Li, and D. R. Korver. 2019. The efficacy of 2 phytases on inositol phosphate degradation in different segments of the gastrointestinal tract, calcium and phosphorus digestibility, and bone quality of broilers. Poult. Sci. 98:5789-5800.

Bilal, T., H. Eseceli, A. Altiner, and O. Keser. 2010. Effect of vitamin d3 and/or zeolite supplementation to laying hen rations added microbial phytase on some blood indices. 1. Calcium and inorganic phosphorus levels and alkaline phosphatase activity. J. Anim. Vet. Adv. 9:1458-1462.

Cordell, D., and S. White. 2011. Peak phosphorus: Clarifying the key issues of a vigorous debate about long-term phosphorus security. Sustainability 3:2027-2049.

David, L. S., M. R. Abdollahi, M. R. Bedford, and V. Ravindran. 2021. True ileal calcium digestibility in soybean meal and canola meal, and true ileal phosphorous digestibility in maize-soybean meal and maize-canola meal diets, without and with microbial phytase, for broiler growers and finishers. Br. Poult. Sci. 62:293-303.

De Jong, J. A., J. C. Woodworth, J. M. DeRouchey, R. D. Goodband, M. D. Tokach, S. S.

Dritz, C. R. Stark, and C. K. Jones. 2017. Stability of four commercial phytase products under increasing thermal conditioning temperatures. Transl. Anim. Sci. 1:255-260.

Delezie, E., L. Maertens, and G. Huyghebaert. 2012. Consequences of phosphorus interactions with calcium, phytase, and cholecalciferol on zootechnical performance and mineral retention in broiler chickens. Poult. Sci. 91:2523-2531.

Dersjant-Li, Y., T. Christensen, S. Knudsen, A. Bello, M. Toghyani, S. Liu, P. Selle, and L. Marchal. 2022. Effect of increasing dose level of a novel consensus bacterial 6-phytase variant on phytate degradation in broilers fed diets containing varied phytate levels. British Poult. Sci. 63:395-405.

Fishbane, S. N., and S. Nigwekar. 2021. Phosphate absorption and hyperphosphatemia management in kidney disease: A physiology-based review. Kidney Med 3:1057-1064.

Gautier, A. E., C. L. Walk, and R. N. Dilger. 2018. Effects of a high level of phytase on broiler performance, bone ash, phosphorus utilization, and phytate dephosphorylation to inositol. Poult. Sci. 97:211-218.

Han, J. C., X. D. Yang, T. Zhang, H. Li, W. L. Li, Z. Y. Zhang, and J. H. Yao. 2009. Effects of 1 alpha-hydroxycholecalciferol on growth performance, parameters of tibia and plasma, meat quality, and type iib sodium phosphate cotransporter gene expression of one- to twenty-one-day-old broilers. Poult. Sci. 88:323-329.

Humer, E., C. Schwarz, and K. Schedle. 2015. Phytate in pig and poultry nutrition. J. Anim. Physiol. Anim. Nutr. 99:605-625.

Javadi, M., J. J. Pascual, M. Cambra-López, J. Macías-Vidal, A. Donadeu, J. Dupuy, L. Carpintero, P. Ferrer, and A. Cerisuelo. 2021. Effect of dietary mineral content and phytase dose on nutrient utilization, performance, egg traits and bone mineralization in laying hens from 22 to 31 weeks of age. Animals 11:14.

Joyner, C. J., M. J. Peddie, and T. G. Taylor. 1987. The effect of age on egg production in the domestic hen. Gen. Comp. Endocrinol. 65:331-336.

Kerschnitzki, M., T. Zander, P. Zaslansky, P. Fratzl, R. Shahar, and W. Wagermaier. 2014. Rapid alterations of avian medullary bone material during the daily egg-laying cycle. Bone 69:109-117.

Korge, M., M. Alaru, I. Keres, K. Möll, L. Talgre, I. Voor, I. Altosaar, and E. Loit-Harro. 2025. Phytate content in cereals impacted by cropping system and harvest year. Foods 14:446.

Kuwata, N., H. Mukohda, H. Uchida, R. Takamatsu, M. M. Binici, T. Yamada, and T. Sugiyama. 2024. Renal endocytic regulation of vitamin D metabolism during maturation and aging in laying hens. Animals 14:12.

Lopes, T. S. B., H. Shi, D. White, I. C. S. Araujo, and W. K. Kim. 2024. Effects of 25-hydroxycholecalciferol on performance, gut health, and bone quality of broilers fed with reduced calcium and phosphorus diet during *Eimeria* challenge. Poult. Sci. 103:13.

Ma, Q. G., M. Rodehutscord, M. Novotny, L. Li, and L. Q. Yang. 2019. Phytate and phosphorus utilization by broiler chickens and laying hens fed maize-based diets. Front. Agric. Sci. Eng. 6:380-387.

Miles, R., P. Costa, and R. Harms. 1983. The influence of dietary phosphorus level on laying hen performance, egg shell quality, and various blood parameters. Poult. Sci. 62:1033-1037.

National Research Council. 1994. Nutrient requirements of poultry, 9th revised ed.

O'Doherty, J. V., D. A. Gahan, C. O'Shea, J. J. Callan, and K. M. Pierce. 2010. Effects of phytase and 25-hydroxyvitamin D_3 inclusions on the performance, mineral balance and bone parameters of grower–finisher pigs fed low-phosphorus diets. Animal 4:1634-1640.

Rana, M. S., and D. L. M. Campbell. 2021. Application of ultraviolet light for poultry production: A review of impacts on behavior, physiology, and production. Front. Anim. Sci. 2.

Rao, S. V. R., M. Raju, A. K. Panda, G. S. Sunder, and R. P. Sharma. 2009. Performance and bone mineralisation in broiler chicks fed on diets with different concentrations of cholecalciferol at a constant ratio of calcium to non-phytate phosphorus. Br. Poult. Sci. 50:528-535.

Saunders-Blades, J. L., and D. R. Korver. 2014. The effect of maternal vitamin D source on broiler hatching egg quality, hatchability, and progeny bone mineral density and performance. J. Appl. Poult. Res. 23:773-783.

Saunders-Blades, J. L., and D. R. Korver. 2015. Effect of hen age and maternal vitamin D source on performance, hatchability, bone mineral density, and progeny in vitro early innate immune function. Poult Sci 94:1233-1246.

Scholey, D. V., N. K. Morgan, A. Riemensperger, R. Hardy, and E. J. Burton. 2018. Effect of supplementation of phytase to diets low in inorganic phosphorus on growth performance and mineralization of broilers. Poult. Sci. 97:2435-2440.

Selle, P. H., S. P. Macelline, P. V. Chrystal, and S. Y. Liu. 2023. The contribution of phytate-degrading enzymes to chicken-meat production. Animals 13:20.

Shanmugasundaram, R., and R. K. Selvaraj. 2012. Vitamin D-1α-hydroxylase and vitamin D-24-hydroxylase mRNA studies in chickens. Poult. Sci. 91:1819-1824.

Shi, H. Y., V. S. R. Choppa, D. Paneru, and W. K. Kim. 2024a. Effects of phytase and 25-hydroxycholecalciferol supplementation in broilers fed calcium-phosphorous deficient diets, with or without E*imeria* challenge, on growth performance, body composition, bone development, and gut health. Anim. Nutr. 19:411-428.

Shi, H. Y., D. Paneru, M. K. Sharma, J. H. Lee, H. R. R. Naeini, F. S. Mahdavi, and W. K. Kim. 2024b. Phytase, 25-hydroxyvitamin D_3 and cocci vaccination to broilers fed a calcium and phosphorus-reduced diet under *Eimeria* spp. challenge: Effects on growth performance and intestinal health. Poult. Sci. 103: 104326.

Sinclair-Black, M., R. A. Garcia-Mejia, L. R. Blair, R. Angel, X. Arbe, D. Cavero, and L. E. Ellestad. 2024. Circadian regulation of calcium and phosphorus homeostasis during the oviposition cycle in laying hens. Poult. Sci. 103:103209.

Sinclair-Black, M., R. A. Garcia, and L. E. Ellestad. 2023. Physiological regulation of calcium and phosphorus utilization in laying hens. Front. Physiol. 14:1112499.

Snow, J., D. Baker, and C. Parsons. 2004. Phytase, citric acid, and 1α-hydroxycholecalciferol improve phytate phosphorus utilization in chicks fed a corn-soybean meal diet. Poult. Sci. 83:1187-1192.

Venter, K., W. Li, R. Angel, P. W. Plumstead, M. Proszkowiec-Weglarz, H. Enting, and L. E. Ellestad. 2024. Calcium and phosphorus digestibility in broilers as affected by varying phytate concentrations from corn. Poult. Sci. 103:9.

Vieites, F. M., E. Brusamarelo, B. S. Vieira, F. G. da Silva, C. S. Souza, G. D. S. Corrêa, J. G. Caramori, and G. H. K. Moraes. 2018. 1,25-dihydroxycholecalciferol from *Solanum glaucophyllum* supports normal growth and reduces the negative effects of calcium and phosphorus restriction on broilers' bone tissue. Semin.-Cienc. Agrar. 39:2205-2214.

Warren, M. F., and K. Livingston. 2024. Implications of vitamin D research in chickens can advance human nutrition and perspectives for the future. Curr. Dev. Nutr. 5:nzab018.

Warren, M. F., T. C. Vu, O. T. Toomer, J. D. Fernandez, and K. A. Livingston. 2020. Efficacy of 1-α-hydroxycholecalciferol supplementation in young broiler feed suggests reducing calcium levels at grower phase. Front. Vet. Sci. 7:245.

Woyengo, T. A., and J. Wilson. 2019. Enhancing nutrient utilization in maize for broiler chickens by superdosing phytase. Anim. Feed Sci. Technol. 252:34-40.

Recent Evidence to Support Modernized, Optimum Vitamin D Nutrition for Efficient Swine Production

Données récentes en faveur de l'actualisation et de l'optimisation de la supplémentation en vitamine D pour favoriser l'efficacité de la production porcine

J.R. Bergstrom, Ph.D.[1]

[1]*dsm-firmenich, Animal Nutrition & Health, Plainsboro, NJ, USA*

jon.bergstrom@dsm-firmenich.com

PRE-CONFERENCE SPONSORED BY / PRÉ-COLLOQUE COMMANDITÉ PAR

dsm-firmenich

Abstract

Vitamin D_3 is an essential nutrient required for normal growth and development of pigs and other mammals, with emphasis classically placed on its importance for maintenance of Ca and P homeostasis and bone mineralization. With the continuing recognition that there are economic losses and welfare concerns resulting from cases of lameness and metabolic bone diseases in swine, there has been increased research to improve the understanding of requirements for vitamin D, Ca, and P to optimize the bone health and performance of modern genotypes. This is particularly important because either deficiencies or over-feeding of certain nutrients (i.e., Phosphorus) can be costly, especially when considering the implications for efficiency and sustainability.

Research in the last decades has revealed that vitamin D is essential for physiological functions "beyond bone" and displays important functions for reproduction, cellular development, and immune system function, to name a few. From an evolutionary perspective, significant amounts of vitamin D are obtained from exposure to sunlight and vitamin D photosynthesis in the skin. Pigs that are housed indoors, especially younger animals, are frequently found to have a low vitamin D status when compared to pigs housed outdoors during summer conditions, even when receiving "typical" levels of vitamin D_3 supplementation in the diet.

In this brief review, recent research is provided to support recommendations for achieving an optimum vitamin D status for modern, lean genotypes housed indoors; and the benefits of dietary supplementation using 25-OH-D_3 to safely achieve healthy plasma levels of 60 – 90 ng/ml of 25-OH-D_3 are demonstrated.

Keywords: performance, swine, 25-OH-D_3, vitamin D

Résumé

La vitamine D3 est un nutriment essentiel pour la croissance et le développement normaux des porcs et d'autres mammifères. Depuis longtemps, l'accent est d'abord placé sur l'importance de cet élément pour le maintien de l'homéostasie du Ca et du P et la minéralisation osseuse. Parce que l'on reconnaît de plus en plus les pertes économiques et les problèmes de bien-être associés à la boiterie et aux maladies osseuses métaboliques chez les porcs, la recherche a intensifié ses efforts pour mieux comprendre les besoins en vitamine D, calcium (Ca) et phosphore (P), l'objectif étant d'optimiser la santé osseuse et les performances des génotypes modernes. Ceci est particulièrement important, car les carences ou les excès en certains nutriments (par exemple, le P) peuvent être coûteux, surtout si l'on considère les implications en termes d'efficacité et de durabilité.

Les recherches des dernières décennies ont montré que la vitamine D est importante non seulement pour les os, mais aussi pour d'autres fonctions physiologiques. Elle intervient dans la reproduction, le développement cellulaire et le fonctionnement du système immunitaire, entre autres. Du point de vue de l'évolution, des quantités importantes de vitamine D sont obtenues par l'exposition au soleil et la synthèse dans la peau. Les porcs gardés à l'intérieur, en particulier les jeunes animaux, présentent souvent un bilan pauvre en vitamine D par rapport aux porcs élevés à l'extérieur en été, même lorsqu'ils reçoivent les concentrations « habituelles » de supplément en vitamine D3 dans leur alimentation.

Dans ce bref exposé, des recherches récentes viennent étayer les recommandations visant à optimiser le bilan en vitamine D pour les animaux modernes à génotype maigres gardés à l'intérieur. Les avantages d'une supplémentation en 25-OH-D3 pour atteindre en toute sécurité des concentrations plasmatiques suffisantes de 60 à 90 ng/ml de 25-OH-D3 sont démontrés.

Introduction

In humans, pigs, and other terrestrial vertebrate species, sunlight-dependent photosynthesis of vitamin D_3 in the skin evolved to fulfill a critical physiological role in maintaining calcium homeostasis (Lucock et al., 2015). Consequently, vitamin D has become known as the "sunshine vitamin", and many species evolved with a dependency for sunlight as a primary source to fulfill their requirements for Vitamin D. Indeed, the rickets afflictions of children that became epidemic in the industrialized urban areas of 19th- and early 20th-century northern Europe and the U.S. were attributed to the lack of adequate exposure to sunlight. With the recognition that both people and animals benefited from exposure to ultraviolet radiation to prevent and treat rickets, recommendations for sensible and adequate sunlight exposure (without causing sunburn) were developed and promoted in the 1930s to prevent rickets in children. As diets improved and some foods and supplements became fortified with vitamin D, the importance of sunlight exposure had fallen out of favor as concerns about the risks for developing cancer from excessive sunlight were growing. Collectively, recommendations to "cover up", the use of sunscreens, and a shift to more sedentary and indoor lifestyles have contributed to a continuing prevalence of vitamin D insufficiency because dietary sources are largely limited or inadequate for achieving a sufficient vitamin D status (DeLuca, 2014; Holick, 2008; Holick, 2007; Macdonald, 2013; Uçar and Holick, 2025).

Pigs are like humans in their needs and utilization of vitamin D and have sometimes been utilized in research to obtain a better understanding of vitamin D supplementation and its effects in the body (Heaney et al., 2009). Beginning in the U.S. during the 1980s, a dramatic shift to indoor and specialized pig production occurred as farms developed and adapted improvements in technology and management systems while becoming larger in scale (Zering, 2019). Nowadays, most pigs in the commercial farms of North America are housed almost entirely indoors. Consequently, modern, commercially raised pigs are commonly found to have an inadequate vitamin D status (based on their serum or plasma levels of 25-hydroxy-vitamin D_3 [25-OH-D_3]; Table 1.), especially younger growing pigs, despite typically having dietary vitamin D_3 supplementation levels that are greater than the estimated dietary requirements suggested by the NRC (2012). An inadequate or reduced vitamin D status is often identified in cases of hypocalcemia, metabolic bone diseases, and non-infectious lameness (Arnold et al., 2015; Faccin et al., 2023; Madson et al., 2012).

Table 1. Age-based serum 25-OH-D_3 concentrations that are normally adequate in swine.

Pig age	Serum 25-OH-D_3 (ng/mL)
Neonate	5 – 15
10 days	8 – 23
Growing pigs (post-weaning)	18 – 30
Mature	35 – 70
Parturition	35 – 100

From Madson et al., 2012

Many indoor-reared pigs that are healthy can be found to have normal bone mineralization and growth or development while arguably having a vitamin D status that is deemed insufficient or suboptimal ($\leq$18-30 ng/mL serum or plasma 25-OH-D_3). This is generally the case when the dietary supply of vitamin D_3 is sufficient to meet or exceed the NRC (2012) requirement estimate while the supply of digestible Ca and P are generously sufficient to meet the requirements for optimal growth or bone mineralization in healthy pigs. Nevertheless, subclinical hypovitaminosis D is common, especially in younger and growing pigs housed indoors, and unhealthy pigs can be found to have a lower vitamin D status with reduced serum Ca and P and bone mineral density when compared to their healthy contemporaries (Arnold et al., 2015; Williams et al., 2024). Importantly, the absence of dietary vitamin D or feeding very low levels (i.e., below the requirement estimate; NRC, 2012) for up to 3 weeks or longer will result in a significant reduction in vitamin D status that is associated with long-term compromises in growth performance and bone formation, even when there may be a sufficient supply of dietary Ca and P (Amundson et al., 2016; Kokinos et al., 2023; McCue, et al., 2019; Phiri, et al. 2024; Rortvedt and Crenshaw, 2012).

Beyond the classically recognized functions of vitamin D for active Ca/P absorption and sustaining proper bone mineralization, much of the research within the last decades has identified other important functions of vitamin D and that the requirements and optimal status are higher for improving various measures of health outcomes while reducing the risks for diseases and mortalities (Grant et al., 2025; Heaney et al., 2009; Figure 1.). Additionally, there is evidence to suggest that dietary or orally supplemented vitamin D_3 may be an inferior source of vitamin D_3 compared to that produced by photosynthesis in the skin. Vitamin D_3 produced in the skin has a longer half-life because nearly 100% becomes associated with vitamin D binding protein (DBP) while as much as 40% of ingested vitamin D_3 can be bound in the lipoprotein fraction and become rapidly removed from the circulation. Another relevant difference is that the synthesis of vitamin D_3 in the skin does not cause vitamin D toxicity, although the risk for ingesting vitamin D_3 at levels associated with vitamin D toxicity or hypercalcemia is extremely low (Holick, 2008; Lucock et al., 2015; Macdonald, 2013; Uçar and Holick, 2025). In swine, there is substantial and increasing evidence to indicate that the NRC (2012) requirement estimates for dietary vitamin D supplementation should be updated. However, with regulatory limitations for the supplementation of vitamin D in swine diets within some countries, it can be difficult to achieve an optimal vitamin D status in pigs reared indoors. Fortunately, feeding a pure source of 25-OH-D_3 (Hy·D®) to swine has gained acceptance in many countries, and supplementation using 25-OH-D_3 has been demonstrated to be more effective than feeding higher levels of vitamin D_3 alone for safely improving vitamin D nutrition and associated physiological outcomes in swine (Hasan et al., 2023).

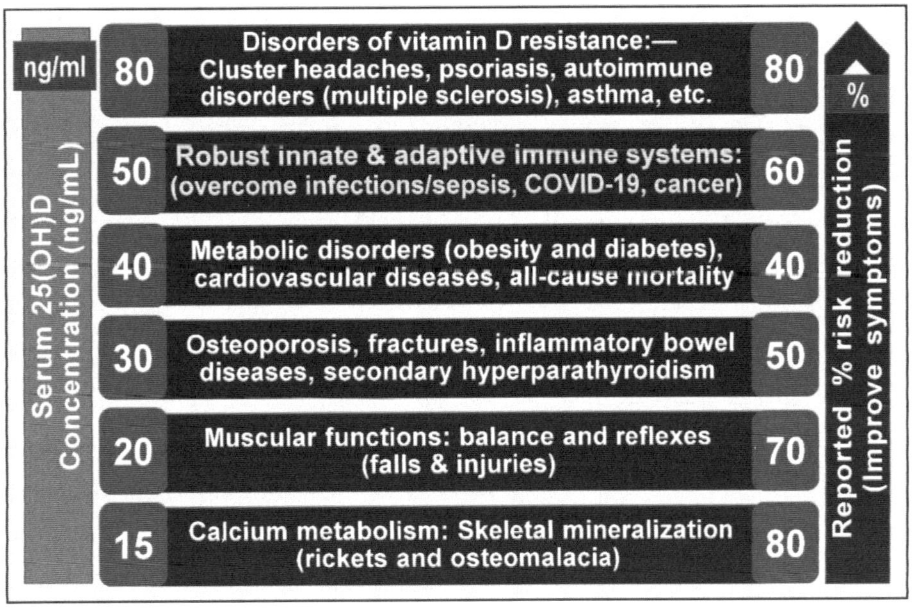

Figure 1. Serum 25-OH-D concentrations needed to overcome various conditions and disorders in humans, and the reported average (%) improvements in primary clinical outcomes. From Grant et al., 2025; and Wimalawansa et al., 2024. This figure is from an open-access article distributed

The Importance of Vitamin D for Maternal Nutrition in Swine

Since publication of the 2012 Nutrient Requirements of Swine (NRC), there has been continuing genetic advancements that have resulted in significant increases in litter sizes and lean growth rates and efficiencies. These changes have coincidentally occurred with increasing concerns related to sow mortalities, pre-weaning mortalities, and post-weaning mortalities on farms. Aside from farm-specific health and management factors, increased litter sizes are known to result in a reduction in the mean birth weights of pigs, with more low birth weight pigs contributing to an increased risk for pre-weaning mortality. Additionally, sows giving birth to larger litters may have a longer farrowing duration with increased stress and risk for mortality (Oliviero et al., 2019). Sudden death, lameness, and prolapses are among the most common reasons given for sow mortalities, with mortalities reported to occur most frequently during the peripartum or post-partum periods. Of particular concern is that breeding gilts and young sows make up a significant number of the sow farm mortalities or removals (Kikuti et al., 2022; Kongsted et al., 2021; Paiva et al., 2023). Interestingly, recent research indicates that post-weaning performance and mortality for groups of pigs from weaning-to-finish can be associated with some corresponding sow farm performance variables, such as piglet birth weights and/or pre-weaning mortalities (Gebhardt et al., 2020; Magalhães et al., 2022). Collectively, these challenges have generated a significant amount of continued interest and increased research for improving our understanding of the nutritional requirements for modern sows, including the importance of vitamin D.

In 2010, original research to investigate the cause of kyphosis ("humpy-back", abnormal spinal curvature) in young growing pigs at the University of Wisconsin Swine Research Farm provided a fresh perspective on the importance of Vitamin D for maternal nutrition and progeny outcomes, and this stimulated a lot of interest in determining the dietary needs for vitamin D more precisely (Amundson and Crenshaw, 2020; Rortvedt and Crenshaw, 2012; Rortvedt et al., 2010). Briefly, although there were not obvious effects on the performance or health of the sows or their progeny during gestation and lactation, an outbreak of kyphosis in pigs during the post-weaning period led to the discovery that an accidental omission of vitamin D in the premix had occurred, and that the sows had not been receiving the expected supplementation of 325 IU vitamin D_3/kg diet. Subsequent research at the University of Wisconsin led to the development of a useful research model for understanding the importance of maternal vitamin D and other nutrients, such as Ca and P, for pig growth and skeletal development. In the early studies, when compared to feeding a diet with 0 or 45 IU vitamin D3/kg diet, a restored level of maternal supplementation to achieve 280 or 325 IU vitamin D_3/kg (like the requirement estimate of the 1998 NRC) appeared to be sufficient for reducing kyphosis and improving/restoring the growth and bone mineralization in the progeny post-weaning, especially when there was sufficient Ca and P fed in the post-weaning period. In follow-up studies that included a higher level of maternal vitamin D_3 to represent the average level being fed on farms at that time, pigs born to sows fed 1,750 IU vitamin D_3/kg diet had increased growth and improved bone characteristics from 4 to 8 weeks of age when compared to the progeny from sows fed either 0 or 325 IU/kg, especially when vitamin D_3 was absent from both the maternal and nursery diets (Amundson et al., 2016; Amundson et al., 2017; Flohr et al., 2016[a]; Kokinos et al., 2023).

As previously mentioned, the serum/plasma level of 25-OH-D_3 is the standard biomarker that is normally used for determining vitamin D sufficiency in pigs, humans, and other species. The serum status for 25-OH-D_3 was evaluated and reported in one of the previous studies at the University of Wisconsin, with the status of the sows at the time of farrowing reported to be not detectable (ND), 10.6 ng/mL, and 22.9 ng/mL for the 0, 325, and 1,750 IU of vitamin D_3/kg maternal diet treatments, respectively (Amundson et al., 2017). However, serum or plasma levels of 25-OH-D_3 typically become lower in sows around the time of parturition, so the life-stage and other factors at the time of sampling should be carefully considered when assessing vitamin D status (Flohr et al., 2016[b]; Flohr et al., 2014[a]; Thayer et al., 2019; Weber et al., 2014).

Although differences in the reproductive and lactational litter performance of sows receiving varying levels of vitamin D_3 have not always been obvious or consistent in the research, differences in bone characteristics of the progeny are readily apparent when there are comparisons including sows with insufficient levels of vitamin D, even with an older weaning age. Witschi et al. (2011), when comparing the progeny of sows fed either 200 (NRC, 1998) or 2,000 IU vitamin D_3/kg diet, found that the progeny of sows fed the low level of vitamin D had reduced tibial and metatarsal bone weight, ash weight, total Ca, breaking strength, mineral content (BMC), and mineral density (BMD) at a weaning age of 35 days. Like the findings of others (i.e., Amundson et al., 2017; Thayer et al., 2019; and Weber et al., 2014), the levels of 25-OH-D_3 in the blood, colostrum/milk, and piglets' blood from sows fed higher levels of vitamin D were increased, but the colostrum/milk and piglet serum/plasma concentrations were all relatively low during the suckling period, and not proportional to the differences in sow levels of serum 25-OH-D_3 that were observed across the maternal dietary treatment levels.

Regardless of the dietary level or source of maternal vitamin D and the potential for modest differences in serum 25-OH-D_3 of the piglets at weaning, few piglets (reared indoors) have achieved a serum 25-OH-D_3 status that is regarded as sufficient by the time they're weaned, even when weaned at 35 days of age. Importantly, concentrations of Ca and P increase in milk during lactation, and it's likely that the sensitivity and importance of comparing serum/plasma levels of 25-OH-D_3 in pigs at typical weaning ages is less relevant, by itself, for accurately determining status and the risk for future compromises in growth or bone development (Amundson and Crenshaw, 2020). With increasing concentrations of Ca and P in milk during lactation, the importance of vitamin D status for assessing the risk of insufficiency in piglets might be delayed until sometime after sows' milk is no longer available or sufficient to provide these essential nutrients. Interestingly, in the study by Witschi et al. (2011), the post-weaning growth from weaning (d 35) until 70 days of age tended to be reduced for the progeny of sows fed the low (200 IU/kg diet) level of vitamin D. It's unclear, however, if the differences in post-weaning growth were due to either the maternal levels of vitamin D, the provision of the same vitamin D levels in the creep feeding diets that were offered to piglets from 3 weeks of age until 5 weeks post-weaning, or both. Nevertheless, the serum 25-OH-D_3 status was substantially increased at 77 days of age for pigs with 2,000 IU vitamin D/kg in the maternal and creep diets, especially when 25-OH-D_3 was the source of supplemented vitamin D. If weaning age is to be increased on farms in the future, the appropriate supplementation of vitamin D in both the maternal and creep feeding diets should be important considerations for ensuring optimal growth and bone development.

Since 2010, there's also been an increased amount of research to evaluate the importance of vitamin D nutrition in swine "beyond bone". Although increasing vitamin D_3 supplementation to sows at levels well above the 2012 NRC requirement estimate of 800 IU vitamin D_3/kg diet (in gestation and lactation) has been demonstrated to increase serum 25-OH-D_3 in sows and piglets, as well as the vitamin D_3 concentrations in milk and in neonatal tissues, there have not been any further improvements in bone mineralization of the progeny or the performance of sows and their piglets with maternal dietary levels up to 9,600 IU vitamin D_3/kg (Flohr et al., 2014[a], 2016[b]). Other research, however, has demonstrated that supplementation using a pure source of 25-OH-D_3 (Hy·D®) at 50 µg/kg[1] diet, either as a full replacement for supplemental vitamin D_3 or in addition to some level of vitamin D_3 supplementation during gestation and lactation, can improve the reproductive performance of gilts/sows and the muscle fiber development and post-weaning performance of piglets.

The earliest North American research to evaluate feeding 25-OH-D_3 (Hy·D®) in comparison to vitamin D_3 for breeding females was conducted by Coffey et al. (2012; Hines et al., 2013). When compared to supplementing 2,500 IU vitamin D_3/kg diet, gilts supplemented with 500 IU vitamin D_3 in combination with 2,000 IU (50 µg) 25-OH-D_3/kg beginning at 43 days prior to breeding had an increased pregnancy rate and litter size at gestational day 90, without any reduction or difference in the mean fetal weight. When evaluating development of the Longissimus Muscle (LM) and semitendinosus muscle in fetuses at gestational day 90, the fetuses from 25-OH-D_3-supplemented gilts had an increased number of LM fibers and myoblasts that displayed a prolonged proliferative capacity, suggesting a potential for greater postnatal muscle growth of fetuses when feeding 25-OH-D_3 rather than only vitamin D_3 to pregnant gilts. Subsequent research with reproducing gilts/sows and their progeny confirmed that feeding 50 µg 25-OH-D_3/kg diet, whether alone or in combination with some level of vitamin D_3 ($\leq$2,000 IU/kg), can improve the progeny outcomes, including improvements in pre- and post-weaning pig growth and bone development when compared to feeding high levels of vitamin D_3 alone (Flohr et al., 2016[b,c], Zhou et al., 2016 & 2017). Tsai et al. (2023), when evaluating the performance of progeny weaned from gilts and sows fed either 1,653 IU vitamin D3/kg or 500 IU vitamin D_3 in combination with 50 µg 25-OH-D_3/kg throughout their development and subsequent parities (including pigs born to parity 1 to parity 4 sows), found that pigs born from gilts and sows fed 25-OH-D_3 (Hy·D®) had greater average daily gain (ADG) and tended to have better feed efficiency (G:F) throughout the growing-finishing period, with associated increases in final bodyweight (BW, +3.19 kg), hot carcass weight (HCW, +3.56 kg), and yield percentage (+0.11%) when compared to the progeny of gilts and sows supplemented only with vitamin D_3.

[1] 1 µg vitamin D activity = 40 IU

Although litter sizes at birth, birth weights, number of pigs weaned per litter, and weaning weights have not always been found to be improved for gestating and lactating sows fed 50 μg 25-OH-D_3/kg diet, a collective assessment of the existing research reveals there are various potential benefits for sows with an improved maternal vitamin D status from feeding 25-OH-D_3. Other benefits that have occasionally been reported for sows fed 25-OH-D_3 include: fewer farrowing complications, a shorter farrowing duration, fewer stillborn pigs or neonatal mortalities, improved milk composition, and improved vitality of the newborn piglets (Meuter et al., 2016; Planchenault et al., 2018; Roger et al., 2023; Zhou et al., 2017).

Vitamin D Nutrition for Starting, Growing / Developing, and Finishing Swine

As mentioned previously, pigs that are managed and raised indoors during the suckling period are frequently found to have a low vitamin D (25-OH-D_3) status at weaning, regardless of the dietary level or source of vitamin D in the maternal diet. The common association of a low vitamin D status in cases of pigs experiencing metabolic bone disease, rickets, or lameness has led to increased research for improving the vitamin D status of pigs, especially at weaning. With the recognition that the dietary level and source of vitamin D in the maternal diet does not substantially increase the status of the piglets (without risking over-supplementation of the gilts/sows), there's been a considerable amount of research to evaluate the effectiveness of individual piglet supplementation using a mega-dose of either oral or injectable vitamin D_3 within the first days after birth. In all these studies, an individual mega-dose of 40,000 or 80,000 IU vitamin D_3 resulted in substantial increases in the vitamin D (25-OH-D_3) status at weaning, but there were not any differences in serum 25-OH-D_3 status, bone mineralization, and post-weaning growth and health at the end of the nursery period when compared to pigs not receiving a bolus of vitamin D_3 after birth (Flohr et al., 2014[b]; Jang et al., 2014; Rortvedt et al., 2012). In all these studies, however, the maternal diets contained substantial levels of vitamin D_3, and both maternal and nursery diets were considered adequate in Ca and P. Also, the actual levels of vitamin D_3 (in addition to 25-OH-D_3) in the piglet serum and tissues was not determined in these studies, and it's unknown whether the bolus of vitamin D_3 resulted in supraphysiological levels of vitamin D_3 that could have been antagonistic to the levels of 25-OH-D_3 in the pigs. Without more extensive research that includes the pre-weaning supplementation of individual piglets with reasonable doses of 25-OH-D_3, current practices should focus on providing good vitamin D nutrition for the sows, creep feeds and post-weaning diets.

Although there was little research on vitamin D for post-weaning pigs prior to 2012, producers in the U.S. typically supplement vitamin D at substantially higher levels than the estimated requirements suggested by the 2012 NRC. Supplementation of vitamin D at levels greater than that suggested by the NRC for growing pigs occurs for several reasons, including 1) the estimates for average daily feed intake (ADFI) that were used by the NRC are higher than that occurring on today's farms, especially in the first weeks after weaning, so greater dietary levels are needed; 2) the NRC requirement estimates for vitamin D are generally based on research from more than 30 years ago and are likely insufficient for modern pigs; 3) nutritionists, genetic suppliers, and veterinarians often assume that a higher level of supplementation is required to optimize performance and health; and 4) higher levels ensure that possible losses of vitamin activity or bioavailability during the storage of premixes and feed manufacturing do not result in a deficiency (Faccin et al., 2023). While there is research supporting a higher requirement for vitamin D in nursery pigs than that suggested by the 2012 NRC (220 IU vitamin D_3/kg for pigs from weaning to 11 kg BW), there's limited research for establishing a single, improved requirement estimate. Current recommendations for pigs from weaning to 11 kg BW differ according to the genetic supplier, ranging from 960 IU/kg for DanBred to a minimum of 1,800 IU/kg for Topigs-Norsvin (Oviedo-Rondon et al., 2023). Although much of the existing research in nursery pigs does not show any benefits for growth or bone mineralization when providing more than 1,378 IU vitamin D_3/kg diet or additional vitamin D_3 through the drinking water, some of the most recent commercial research has shown that ADG and ADFI may be improved in the first 3 weeks after weaning when pigs are fed 3,500 IU vitamin D/kg compared to 1,500 IU vitamin D_3/kg (Bergstrom et al., 2023 & 2024; Flohr et al., 2014[b]). Importantly, the combination of 1,500 IU of vitamin D_3 with 50 µg of 25-OH-D_3 has resulted in substantial improvements in the serum 25-OH-D_3 status when compared to feeding 3,500 IU/kg diet of only vitamin D_3.

In nursery pigs, growing-finishing pigs, or developing gilts, research within the last has 15 years has demonstrated the advantages of an improved 25-OH-D_3 status from feeding 50 µg of 25-OH-D_3/kg when compared to only feeding vitamin D_3. Konowalchuk et al. (2013) reported improvements in the immunomodulatory responses of nursery pigs that were fed the combination of 1,500 IU of vitamin D_3 with 50 µg of 25-OH-D_3 when compared to feeding only 1,500 or 3,500 IU vitamin D_3/kg. Other research has shown a variety of other benefits from dietary supplementation with 50 µg of 25-OH-D_3/kg when added to some existing level of vitamin D_3, including improvements in bone ash and calcium and phosphorus digestibility (especially when these minerals are marginal or deficient), reductions in the incidence of osteochondrosis or improved soundness scores, and reduced drip loss from carcasses when the levels of some other vitamins are also increased (Becker et al., 2024; Brana et al., 2012; Huerta et al., 2025; Jaramillo et al., 2025; Regassa et al., 2013; Sugiyama et al., 2013).

Insights Regarding the Differences in Supplementation with Vitamin D_3 and/or 25-OH-D_3 and Vitamin D Metabolism

The reasons for differences in outcomes when comparing dietary supplementation using vitamin D_3 with supplementation using 25-OH-D_3 are largely due to the greater bio-efficacy of ingested 25-OH-D_3. Ingested vitamin D_3 is more dependent on the presence of dietary fat for intestinal absorption, whereas 25-OH-D_3 is readily absorbed through a chylomicron-independent mechanism and rapidly increases or sustains serum/plasma levels of 25-OH-D_3. Whether produced endogenously in the skin or by ingestion and absorption, vitamin D_3 is the predominant form found in tissue reserves of fat, muscle, and liver. Nevertheless, to fulfill the physiological needs for vitamin D, vitamin D_3 must be mobilized to the liver via vitamin D-binding protein (DBP) for hydroxylation by the vitamin D-25-hydroxylase enzyme (CYP2R1) to become 25-OH-D_3 (Figure 2.). 25-Hydroxyvitamin D_3 is the major circulating form of vitamin D that is bound to DBP in plasma and provides a pool for sustaining the needs for this metabolite in tissues throughout the body (e.g., bone cells, keratinocytes, liver, muscle, placenta, etc.), especially the kidney. In the kidney, the classically recognized homeostatic regulation and tight control of 25-OH-D_3 hydroxylation to the metabolically active 1-25-$(OH)_2$-D_3 occurs to sustain calcium metabolism and skeletal health. However, many other body tissues are now known to have vitamin D receptors (VDRs) and contain vitamin D-1-hydroxylase enzyme (CYP27B1) for converting 25-OH-D_3 to active 1-25-$(OH)_2$-D_3. The regulation of this conversion in extra-renal tissues is more substrate dependent and therefore susceptible to deficiency of 25-OH-D_3 (Nair et al., 2018). Consequently, the circulating level of 25-OH-D_3 is recognized to be the best single measurement for assessing vitamin D status, and directly providing 25-OH-D_3 in the diet is an effective way to improve vitamin D status (Combs, 2012; Fleet, 2025).

Because greater levels of supplemental vitamin D_3 are required to increase the circulating levels of 25-OH-D_3 in pigs housed exclusively indoors, another reason for differences in the responses to increasing dietary supplementation of vitamin D_3 when compared to the benefits observed with feeding 25-OH-D_3 could be related to an accumulation of vitamin D_3 in the liver, adipose tissue, and circulation. Ingesting high concentrations of vitamin D_3 leads to its increased storage and saturation of the VDRs, especially with vitamin D_3 itself, 25-OH-D_3 and other "downstream" vitamin D metabolites. However, even in cases of toxicity due to supraphysiological doses or intakes of vitamin D_3, there's rarely an elevation in the total amount of active 1-25-$(OH)_2$-D_3. When there is toxicity, it's believed that the excess concentrations of the inactive forms of vitamin D exceed the DBP capacity in circulation and cause increased amounts of "free" 1-25-$(OH)_2$-D_3 to be taken up by target cells (Asif and Farooq, 2023; Jones, 2008). Although vitamin D toxicity is extremely rare, pigs and humans have a similar maximum threshold of approximately 120 – 150 ng/mL for serum/plasma 25-OH-D_3 to prevent toxicity and hypercalcemia, but this safety limit is entirely derived from cases involving toxic intakes of vitamin D_3 (Wimsatt et al., 1998). In a study of tolerance to dietary levels of 25-OH-D_3 in growing pigs, Rosenberg et al. (2016) observed normal growth and health in pigs fed 500 μg (20,000 IU) 25-OH-D_3 (Hy·D®)/kg diet for 42 days, even though plasma levels of 25-OH-D_3 in those pigs averaged 360 ng/mL at the end of 42 days. Similarly, Shepard and DeLuca (1980) reported a remarkable finding that rats given 460 nmol (184 μg) 25-OH-D_3/day for 14 days had an average plasma concentration of 436 ng/mL without adverse effects, while rats given 650 nmol (260 μg) vitamin D_3/day had an average plasma concentration of 643 ng/mL of 25-OH-D_3 and demonstrated signs of toxicity. Of particular interest in the research by Shepard and DeLuca is the fact that the rats given 25-OH-D_3 had no detectable levels of vitamin D_3 in plasma, whereas plasma concentrations of vitamin D_3 were substantially increased with greater doses of vitamin D_3. More recent evidence suggests that the dosing or administration of supraphysiological amounts of vitamin D_3 can be antagonistic to 25-OH-D_3, especially while circulating levels of 25-OH-D_3 are low (Vasquez, 2021). Although the serum/plasma level of 25-OH-D_3 is a useful measure for diagnostic cases and health assessments, a broader understanding of other physiologically relevant vitamin D metabolites is likely to become more important for improving animal and human health in the future (Tang et al., 2019; Tuckey et al., 2019).

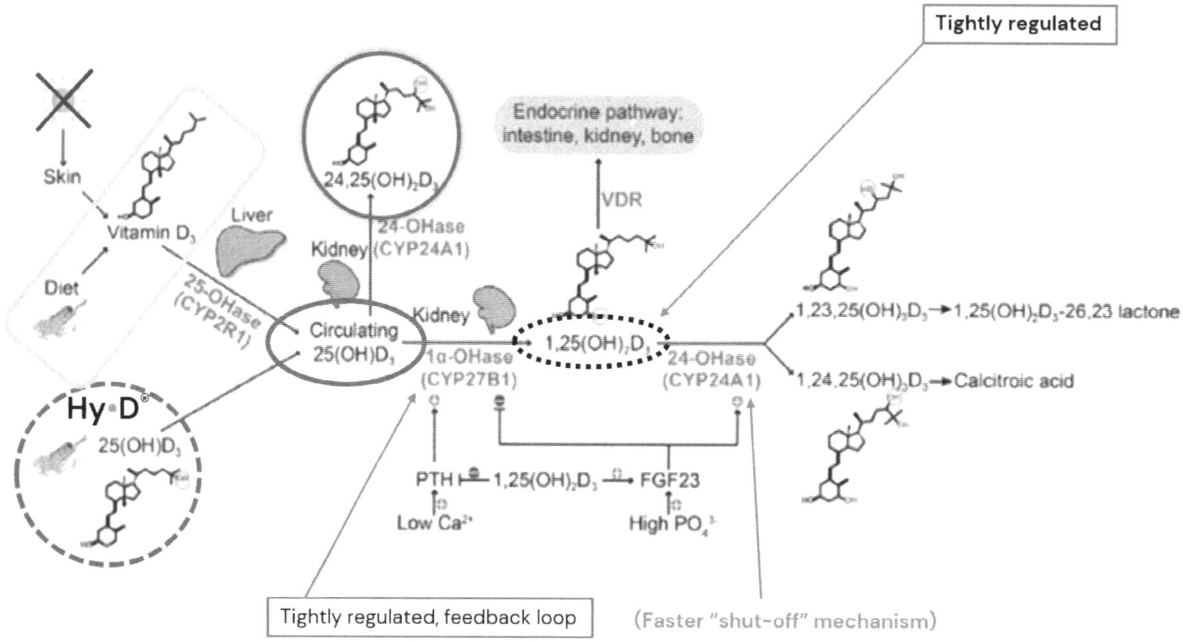

Figure 2. Simplified diagram of the exogenous sources and endogenous production of biologically relevant vitamin D metabolites, including some of the reactions that affect or regulate the metabolically active forms.

Summary and Recommendations

Interest in vitamin D nutrition has grown substantially in the last decades, especially with the awareness that vitamin D is important for physiological functions "beyond bone". Although a serum/plasma vitamin D status of at least 15 ng/mL of 25-OH-D_3 is generally recognized to be adequate for sustaining normal bone mineralization, higher levels are required to ensure adequacy for optimizing other important functions, including the needs for processes in cellular development, immune system function, and tissue maintenance or repair. Without exposure to sufficient sunlight, especially at latitudes greater than 37 degrees north or south during winter, the needs for dietary or supplemental sources of vitamin D are an important consideration. Since pigs are primarily housed indoors and have become leaner with a vastly improved productive efficiency nowadays, the efforts to improve our understanding of the pigs' requirements for vitamin D have increased substantially. The requirements to optimize productive outcomes in swine are greater than those established by the 2012 NRC, and an improved vitamin D status is becoming more important as pork producers strive to minimize nutrient excretion and utilize less dietary calcium and phosphorus. Although an optimal serum/plasma vitamin D status for swine is generally within 60 – 90 ng/ml of 25-OH-D_3, the use of greater levels of supplementation with vitamin D_3 alone is not effective for achieving the desired outcomes. Supplementing swine feeds with 50 µg of 25-OH-D_3/kg diet has been demonstrated to be an effective approach for improving the vitamin D status and the expected outcomes in swine. For swine housed exclusively indoors, attempts to improve the vitamin D status by only using high dietary levels of vitamin D_3 are less effective and may be harmful. With a maximum limit of 3,500 IU vitamin D/kg feed for swine in Canada, a combination of 1,500 IU of vitamin D_3 with 50 μg of 25-OH-D_3 is recommended for improving vitamin D related outcomes.

References

Amundson, L.A., and T.D. Crenshaw. 2020. Lessons learned from the hypovitaminosis D kyphotic pig model. J. Anim. Sci. 98(Suppl. 1):S52-S57.

Amundson, L.A., L.L. Hernandez, and T.D. Crenshaw. 2017. Serum and tissue 25-OH vitamin D_3 concentrations do not predict bone abnormalities and molecular markers of vitamin D metabolism in the hypovitaminosis D kyphotic pig model. Brit. J. Nutr. 118(1):30-40.

Amundson, L.A., L.L. Hernandez, J. Laporta, and T.D. Crenshaw. 2016. Maternal dietary vitamin D carry-over alters offspring growth, skeletal mineralization and tissue mRNA expressions of genes related to vitamin D, calcium and phosphorus homeostasis in swine. Brit. J. Nutr. 116:774-787.

Arnold, J., D.M. Madson, S.M. Ensley, J.P. Goff, C. Sparks, G.W. Stevenson, T. Crenshaw, C. Wang and R. Horst. 2015. Survey of serum vitamin D status across stages of swine production and evaluation of supplemental bulk vitamin D premixes used in swine diets. J. Swine Health Prod. 23(1):28-34.

Asif, A, N. Farooq. 2023. Vitamin D Toxicity. In: StatPearls [Internet]. Treasure Island (FL): StatPearls Publishing.

Becker, L.L., J.T. Gebhardt, M.D. Tokach, J.C. Woodworth, R.D. Goodband, J.M. DeRouchey, J.R. Bergstrom, and C.L. Siepker. 2024. Effects of added 25(OH)D_3 with varying standardized total tract digestible phosphorus concentrations on nursery pig performance, bone characteristics, and serum vitamin D status. J. Anim. Sci. 102:skae254.

Bergstrom, J., S. Hough, and K. Herkelman. 2023. The effects of different dietary levels of supplemental vitamin D, 25-OH-D3, and vitamin E on nursery pig performance and vitamin status. Proc. 54th AASV Annual Meeting.

Bergstrom, J.R., J. Landero, M. Young, and S. Hough. 2024. The effects of supplementing diets with 25-OH-D_3, with or without increased vitamin E, on the performance of nursery pigs. J. Anim. Sci. 102(Suppl. 2):162-163.

Brana, D., J. Gabriel-Landon, J. Cervantes-Lopez, and J.A. Cuaron. 2012. Use of 25OHD_3 favors opportune and sound bone maturation. J. Anim. Sci. 90(Suppl. 2):114.

Coffey, J.D., E.A. Hines, J.D. Starkey, C.W. Starkey, and T.K. Chung. 2012. Feeding 25-hydroxycholecalciferol improves gilt reproductive performance and fetal vitamin D status. J. Anim. Sci. 90:3783-3788.

Combs, Jr., G.F. 2012. The Vitamins (4th Edition). Academic Press, London.

DeLuca, H.F. 2014. Vitamin D: Bones and Beyond. J. Anim. Sci. 92(3):917-929.

Faccin, J.E.G., M.D. Tokach, R.D. Goodband, J.M. DeRouchey, J.C. Woodworth, and J.T. Gebhardt. 2023. Industry survey of added vitamins and trace minerals in U.S. swine diets. Transl. Anim. Sci. 7(1):txad035.

Fleet, J.C. 2025. Differences in the absorption and metabolism of vitamin D_2, vitamin D_3, and 25 hydroxyvitamin D. J. Steroid Biochem. Mol. Biol. 249:106718.

Flohr, J.R., J.M. DeRouchey, J.C. Woodworth, M.D. Tokach, R.D. Goodband, and S.S. Dritz. 2016[a]. A survey of current feeding regimens for vitamins and trace minerals in the US swine industry. J. Swine Health Prod. 24(6):290-303.

Flohr, J.R., M.D. Tokach, S.S. Dritz, J.M. DeRouchey, R.D. Goodband, J.L. Nelssen, and J.R. Bergstrom. 2014[a]. An evaluation of the effects of added vitamin D_3 in maternal diets on sow and pig performance. J. Anim. Sci. 92:594-603.

Flohr, J.R., M.D. Tokach, S.S. Dritz, J.M. DeRouchey, R.D. Goodband, J.L. Nelssen, S.C. Henry, L.M. Tokach, M.L. Potter, J.P. Goff, N.J. Koszewski, R.L. Horst, E.L. Hansen, and E.D. Fruge. 2014[b]. Effects of supplemental vitamin D_3 on serum 25-hydroxycholecalciferol and growth of preweaning and nursery pigs. J. Anim. Sci. 92:152-163.

Flohr, J.R., J.C. Woodworth, J.R. Bergstrom, M.D. Tokach, S.S. Dritz, R.D. Goodband, and J.M. DeRouchey. 2016[b]. Evaluating the impact of maternal vitamin D supplementation: I. Sow performance, serum vitamin metabolites, and neonatal muscle characteristics. J. Anim. Sci. 94:4629-4642.

Flohr, J.R., J.C. Woodworth, J.R. Bergstrom, M.D. Tokach, S.S. Dritz, R.D. Goodband, and J.M. DeRouchey. 2016[c]. Evaluating the impact of maternal vitamin D supplementation on sow performance: II. Subsequent growth performance and carcass characteristics of growing pigs. J. Anim. Sci. 94:4643-4653.

Gebhardt, J.T., M.D. Tokach, S.S. Dritz, J.M. DeRouchey, J.C. Woodworth, R.D. Goodband, and S.C. Henry. 2020. Postweaning mortality in commercial swine production. I: review of non-infectious contributing factors. Transl. Anim. Sci. 4(2):462-484.

Grant, W.B., S.J. Wimalawansa, P. Pludowski, and R. Z. Cheng. 2025. Vitamin D: Evidence-Based Health Benefits and Recommendations for Population Guidelines. Nutrients 17:277.

Hasan, M., M. Oster, H. Reyer, K. Wimmers, and D.C. Fischer. 2023. Efficacy of dietary vitamin D_3 and 25(OH)D_3 on reproductive capacities, growth performance, immunity and bone development in pigs. Br. J. Nutr. 130(8):1298-1307.

Heaney, R.P., R.L. Horst, D.M. Cullen and L.A.G. Armas. 2009. Vitamin D_3 Distribution and Status in the Body. J. Am. Coll. Nutr. 28(3):252-256.

Hines, E.A., J.D. Coffey, C.W. Starkey, T.K. Chung, and J.D. Starkey. 2013. Improvement of maternal vitamin D status with 25-hydroxycholecalciferol positively impacts porcine fetal skeletal muscle development and myoblast activity. J. Anim. Sci. 91:4116-4122.

Holick, M.F. 2008. Sunlight, UV-Radiation, Vitamin D and Skin Cancer: How Much Sunlight Do We Need? Adv. Exp. Med. Biol. 624:1-15.

Holick, M.F. 2007. Vitamin D Deficiency. N. Engl. J. Med. 357:266-281.

Huerta, D.G., M.A.P. Alvarado, L.H.L. Hernandez, I.E.C. Aranda, J.A.C. Ibarguengoytia, and H.H. Stein. 2025 (in press). Influence of dietary vitamin concentrations and of 25(OH)D_3 in diets for growing-finishing pigs. Abstract at the Annual Midwest Section Meeting of ASAS.

Jang, Y.D., M.D. Lindemann, H.J. Monegue, and R.L. Stuart. 2014. The Effects of Fat-soluble Vitamin Administration on Plasma Vitamin Status of Nursing Pigs Differ When Provided by Oral Administration or Injection. Asian Australas. J. Anim. Sci. 27(5):674-682.

Jaramillo, B.A., S.A. Lee, J.P. Acosta, and H.H. Stein. 2025 (in press). Additive effects of 25-hydroxy-vitamin D_3 and microbial phytase on digestibility of calcium and phosphorus, blood vitamin D_3 metabolites, and bone biomarkers in growing pigs. Abstract at the Annual Midwest Section Meeting of ASAS.

Jones, G. 2008. Pharmacokinetics of vitamin D toxicity. Am. J. Clin. Nutr. 88(suppl):582S-586S.

Kikuti, M., G.M. Preis, J. Deen, J.C. Pinilla, and C.A. Corzo. 2022. Sow mortality in a pig production system in the midwestern USA: Reasons for removal and factors associated with increased mortality. Vet. Rec. e2539.

Kokinos, B.P., L.A. Amundson, T.D. Crenshaw, and P.J. Lang. 2023. Maternal Vitamin D Deficiency Increases the Incidence of Osteochondrotic-Like Lesions in the Distal Femur of Growing Pigs. J. Anim. Sci. 101(Suppl. 2):271-272.

Kongsted, H., S. Haugegaard, A.S. Juel, C.M. Salomonsen, and T.K. Jensen. 2021. Causes of spontaneous sow deaths in the farrowing units of 10 Danish sow herds. Res. Vet. Sci. 139:127-132.

Konowalchuk, J.D., A. M. Rieger, M.D. Kiemele, D.C. Ayres, and D.R. Barreda. 2013. Modulation of weanling pig cellular immunity in response to diet supplementation with 25-hydroxyvitamin D_3. Vet. Immunol. Immunopathol. 155(1-2):57-66.

Lucock, M., P. Jones, C. Martin, E. Beckett, Z. Yates, J. Furst and M. Veysey. 2015. Vitamin D: Beyond Metabolism. J. Evid. Based Complementary Altern. Med. 20(4):310-322.

Macdonald, H.M. 2013. Contributions of Sunlight and Diet to Vitamin D Status. Calcif. Tissue Int. 92:163-176.

Madson, D.M., S.M. Ensley, P.C. Gauger, K.J. Schwartz, G.W. Stevenson, V.L. Cooper, B.H. Janke, E.R. Burrough, J.P. Goff and R.L. Horst. 2012. Rickets: case series and diagnostic review of hypovitaminosis D in swine. J. Vet. Diagn. Investig. 24(6):1137-1144.

Magalhães, E.S., J.F. Zimmerman, P. Thomas, C.A.A. Moura, G. Trevisan, D.J. Holtkamp, C. Wang, C. Rademacher, G.S. Silva, and D.C.L. Linhares. 2022. Whole-herd risk factors associated with wean-to-finish mortality under the conditions of a Midwestern USA swine production system. Prev. Vet. Med. 198:105545.

McCue, M., J.L. Reichert, and T.D. Crenshaw. 2019. Impact of dietary vitamin D_3 supplements in nursery diets on subsequent growth and bone responses of pigs during an immune challenge. J. Anim. Sci. 97(12):4895-4903.

Meuter, A., J.-M. Thoby, J.-Y. Lichou, and B. Renouf. 2016. Effect of dietary 25-Hydroxycholecalciferol on performance of gestating and lactating sows and their litters. J. Rech. Porc. 48:153-154.

Nair, P., B. Venkatesh, and J.R. Center. 2018. Vitamin D deficiency and supplementation in critical illness-the known knowns and known unknowns. Crit. Care. 22:276.

NRC. 2012. Nutrient Requirements of Swine. 11th rev. ed. Natl. Acad. Press, Washington D.C.

NRC. 1998. Nutrient Requirements of Swine. 10th rev. ed. Natl. Acad. Press, Washington D.C.

Oliviero, C., S. Junnikkala, and O. Peltoniemi. 2019. The challenge of large litters on the immune system of the sow and the piglets. Reprod. Domest. Anim. 54(Suppl. 3):12-21.

Oviedo-Rondon, E.O., C. López-Bote, G. Litta, and J.M. Hernandez. 2023. OPTIMUM VITAMIN NUTRITION for More Sustainable Swine Farming. 5m Books Ltd. Great Easton, Essex, UK.

Paiva, R.C., C.A. Moura, P. Thomas, B. Haberl, L. Greiner, C.J. Rademacher, A.P.S.P. Silva, G. Trevisan, D.C.L. Linhares, and G.S. Silva. 2023. Risk factors associated with sow mortality in breeding herds under one production system in the Midwestern United States. Prev. Vet. Med. 213:105883.

Phiri, C.B., C.R. Davis, M. Grahn, B.M. Gannon, B.P. Kokinos, T.D. Crenshaw, and S.A. Tanumihardjo. 2024. Vitamin D Maintains Growth and Bone Mineral Density against a Background of Severe Vitamin A Deficiency and Moderate Toxicity in a Swine Model. Nutrients 16:2037.

Planchenault, D., B. Renouf, A. Meuter, and J.-M. Thoby. 2018. Effect of dietary 25-hydroxycholecalciferol and cholecalciferol on the performance of gestating and lactating sows and their litter. J. Rech. Porc. 50:143-144.

Regassa, A., R. Adhikari, J.-M. Heo, M. Nyachoti, and W.K. Kim. 2013. Effects of 25-hydroxycholecalciferol on growth performance, skeletal integrity, and intestinal transporter gene expression in growing pigs. J. Anim. Sci. 91(Suppl. 2):105.

Roger, L., X. B. de Grave, F. Nell, and B.N. Fisker. 2023. Effect of 25-hydroxycholecalciferol supplementation on vitamin D blood status of the sow and the farrowing process. Anim. - Sci. Proc. 14:713-714.

Rortvedt, L.A., and T.D. Crenshaw. 2012. Expression of kyphosis in young pigs is induced by a reduction of supplemental vitamin D in maternal diets and vitamin D, Ca, and P concentrations in nursery diets. J. Anim. Sci. 90(13):4905-4915.

Rortvedt, L.A., P.M. Cline, C. Ryer, B. Frederick, K.J. Retallick, D.K. Schneider, and T.D. Crenshaw. 2012. An oral dose of vitamin D at birth increased serum 25-OH D at weaning but failed to alter bone mineral density in pigs. J. Anim. Sci. 95(Suppl. 1):115.

Rortvedt, L.A., L.A. Zappitelli, J.L. Reichert, J.R. Booth, and T.D. Crenshaw. 2010. Expression of kyphosis in young pigs is altered by carryover effects of sow vitamin D status. J. Anim. Sci. 88(E-Suppl. 2, late-breaking abstract):665.

Rosenberg, S.J. von, G.M. Weber, A. Erhardt, U. Höller, U.A. Wehr, and W.A. Rambeck. 2016. Tolerance evaluation of overdosed dietary levels of 25-hydroxyvitamin D_3 in growing piglets. J. Anim. Physiol. Anim. Nutr. 100:371-380.

Shepard, R.M., and H.F. DeLuca. 1980. Plasma Concentrations of Vitamin D_3 and Its Metabolites in the Rat as Influenced by Vitamin D_3 or 25-Hydroxyvitamin D_3 Intakes. Arch. Biochem. Biophys. 202(1):43-53.

Sugiyama, T., S. Kusuhara, T.K. Chung, H. Yonekura, E. Azem, and T. Hayakawa. 2013. Effects of 25-hydoxy-cholecalciferol on the development of osteochondrosis in swine. Anim. Sci. J. 84:341-349.

Tang, J.C.Y., S. Jackson, N.P. Walsh, J. Greeves, W.D. Fraser, & Bioanalytical Facility team. 2019. The dynamic relationships between the active and catabolic vitamin D metabolites, their ratios, and associations with PTH. Sci. Rep. 9:6974.

Thayer, M.T., J.L. Nelssen, A.J. Langemeier, J.M. Morton, J.M Gonzalez, S.R. Kruger, Z. Ou, A.J. Makowski, and J.R. Bergstrom. 2019. The effects of maternal dietary supplementation of cholecalciferol (vitamin D_3) and $25(OH)D_3$ on sow and progeny performance. Transl. Anim. Sci. 3:693-708.

Tsai, T.C., B.A. Sam, S. Dritz, C.M. Vier, J.R. Bergstrom, S.D. Hough, U.A. Orlando, K. Blake, N. Davis, H. Grogan, K. Coble, and C.V. Maxwell. 2023. Nutrient Intervention During Gilt Development Leads to Effects on Progeny Growth Responses. J. Anim. Sci. 101(Suppl. 2):307-308.

Tuckey, R.C., C.Y.S. Cheng, and A.T. Slominski. 2019. The serum vitamin D metabolome: What we know and what is still to discover. J. Steroid Biochem. Mol. Biol. 186:4-21.

Uçar, N., and M.F. Holick. 2025. Illuminating the Connection: Cutaneous Vitamin D_3 Synthesis and Its Role in Skin Cancer Prevention. Nutrients 17:386.

Vasquez, A. 2021. Vitamin D3 Pharmacology Infographic: Physiologic Dosing versus Bolus Roulette. Int. J. Hum. Nutr. Funct. Med. 9:3.

Weber, G.M., A.-K. M. Witschi, C. Wenk, and H. Martens. 2014. Effects of dietary 25-hydroxycholecalciferol and cholecalciferol on blood vitamin D and mineral status, bone turnover, milk composition, and reproductive performance of sows. J. Anim. Sci. 92:899-909.

Williams, H.R., M.D. Tokach, J.C. Woodworth, J.M. DeRouchey, R.D. Goodband, J.R. Bergstrom, M.C. Rahe, C.L. Siepker, P. Sitthicharoenchai, S.M. Ensley, S.L. Radke, and J.T. Gebhardt. 2024. Diagnostic survey of analytical methods used to determine bone mineralization in pigs. J. Anim. Sci. 102:skae090.

Wimalawansa, S.J., S.T. Weiss, and B.W. Hollis. 2024. Integrating Endocrine, Genomic, and Extra-Skeletal Benefits of Vitamin D into National and Regional Clinical Guidelines. Nutrients 16:3969.

Wimsatt, J., S.L. Marks, T.W. Campbell, J.D. Johnson, and R.F. Nachreiner. 1998. Dietary Vitamin D Toxicity in a Household of Pot-Bellied Pigs (*Sus scrofa*). J. Vet. Intern. Med. 12:42-44.

Witshci, A.-K.M., A. Liesegang, S. Gebert, G.M. Weber, and C. Wenk. 2011. Effect of source and quantity of dietary vitamin D in maternal and creep diets on bone metabolism and growth in piglets. J. Anim. Sci. 89(6):1844-1852.

Zering, K. 2019 (Spring). Hog Farming, Past, Present, and Future: An Economist's View. J. Land Use & Environ. Law. 34(2):313-324.

Zhou, H., Y. Chen, G. Lv, Y. Zhuo, Y. Lin, B. Feng, Z. Fang, L. Che, J. Li, S. Xu, and D. Wu. 2016. Improving maternal vitamin D status promotes prenatal and postnatal skeletal muscle development of pig offspring. Nutr. 32:1144-1152.

Zhou, H., Y. Chen, Y. Zhuo, G. Lv, Y. Lin, B. Feng, Z. Fang, L. Che, J. Li, S. Xu, and D. Wu. 2017. Effects of 25-hydroxycholecalciferol supplementation in maternal diets on milk quality and serum bone status markers of sows and bone quality of piglets. Anim. Sci. J. 88:476-483.

Utilization of diet and enteric neurochemistry as a new mechanistic-based feed strategy to accelerate the application of the gut microbiome in poultry and livestock production

Utilisation de l'alimentation et de la neurochimie entérique en tant que nouvelle stratégie alimentaire basée sur la mécanistique pour accélérer l'application du microbiome intestinal à la production de volaille et de bétail

J. M. Lyte[1]

[1] *Poultry Production and Product Safety Research Unit, Agricultural Research Service, United States Department of Agriculture, Fayetteville, AR, USA*

joshua.lyte@usda.gov

Abstract

The gut microbiome of food production animals varies greatly between individual animal species. This is due, in part, to nutritional requirements and feed composition of each particular species as well as other physiological factors. The identification of shared mechanistic components between poultry and livestock, as well as ruminants and non-ruminants, would enable the design of a tractable nutritional platform that could be easily adapted to a number of farm production animals under a range of environmental conditions. This talk will posit that the use of the evolutionary, neurochemical-based crosstalk between the microbiome and the host, known as microbial endocrinology, provides such a framework by which to design a novel approach in the design of feed that will accelerate the application of the microbiome in food animal production. Microbial endocrinology, which is the intersection of neurobiology and microbiology, has been applied since its introduction to understanding the ability of stress to influence susceptibility to infectious disease. However, feed can also be understood beyond the traditional nutritional properties needed for optimal animal performance to also include an array of naturally occurring neurochemical molecules. These components of feed, which are typically not measured as part of any nutritional diet, encompass neurochemical precursors or metabolites that are useable by the microbiome to affect animal health, behavior, and food safety status. As neurochemicals represent an evolutionary-based interkingdom 'language' for bi-directional host-microbe communication, feed concentrations of neurochemicals are relevant across all food production species. The value of an interkingdom approach to nutrition and the design of feed is increasingly self-evident considering that common feed components such as corn, wheat, soy, animal by-products, and other plant and animal ingredients are rich sources of neuroactive compounds including neurochemicals, precursors, and/or metabolites. Through the lens of microbial endocrinology, the presence of naturally occurring neurochemicals and related compounds in feed can be explored to create next-generation feed strategies to target causal pathways between microbial taxa and the host relevant in performance, health, and food safety. For example, if poultry or livestock are expected to encounter stressful environmental conditions, feed compositions might be tailored to concentrations of neuroactive molecules, such as histamine or dopamine, that could potentially affect inflammation and animal gut health and susceptibility to foodborne pathogen carriage. Feed-based applications that leverage neurochemical-based host-microbe crosstalk hold promise as a tractable functional platform that can be utilized for all food production animals to improve feed performance and overall health as thereby also benefit food safety. Thus, there exists a need for the development of evidence-based, function-driven hypotheses to accelerate the application of microbiome research in animal nutrition. The use of microbial endocrinology provides for such a functional framework by which the microbiome can be effectively modulated through understanding the potential of neurochemistry to be applied to feed formulation.

Keywords: microbial endocrinology, microbiome, feed, neurochemical

Résumé

Le microbiome intestinal des animaux d'élevage diffère considérablement d'une espèce à l'autre. Cette variation est principalement attribuable aux besoins nutritionnels et à la composition des aliments spécifiques à chaque animal, ainsi qu'à des facteurs physiologiques. L'identification de mécanismes communs à la volaille et au bétail ainsi qu'aux ruminants et aux non-ruminants permettrait de concevoir une plateforme nutritionnelle flexible et susceptible d'être facilement adaptée à un certain nombre d'espèces animales élevées dans diverses conditions environnementales. Dans cet exposé, nous avançons que l'étude des mécanismes de communication évolutionniste de type neurochimique entre le microbiome et l'hôte, soit l'endocrinologie microbienne, fournit un cadre permettant de concevoir une nouvelle approche de l'alimentation animale qui accélérera l'application du microbiome à la production d'animaux destinés à l'alimentation. Depuis ses débuts, l'endocrinologie microbienne, qui se situe à l'intersection de la neurobiologie et de la microbiologie, a été appliquée pour comprendre comment le stress parvient à influencer la sensibilité aux maladies infectieuses. Cependant, les aliments des animaux ne se limitent pas à leurs propriétés nutritionnelles traditionnelles nécessaires pour atteindre une performance zootechnique optimale. Ils contiennent également un ensemble de molécules neurochimiques naturelles. Ces composants, qui ne sont généralement pas mesurés dans le cadre de la formulation d'une ration, comprennent des précurseurs neurochimiques ou des métabolites qui peuvent être utilisés par le microbiome pour influer sur la santé et le comportement des animaux ainsi que sur la salubrité alimentaire. Compte tenu que les substances neurochimiques représentent un « langage » interrègnes issu de l'évolution pour la communication bidirectionnelle entre l'hôte et les microbes, les concentrations de substances neurochimiques dans les aliments sont pertinentes pour toutes les espèces animales destinées à la production alimentaire. La valeur d'une approche interrègnes de la nutrition et de la conception d'aliments est de plus en plus évidente si l'on considère que les ingrédients habituels, tels que le maïs, le blé, le soya, les sous-produits animaux et d'autres ingrédients d'origine végétale et animale, sont des sources riches en composés neuroactifs, y compris en substances neurochimiques, en précurseurs ou en métabolites. Sous l'angle de l'endocrinologie microbienne, la présence naturelle de substances neurochimiques et de composés apparentés dans les aliments peut être étudiée pour créer des stratégies alimentaires novatrices visant à cibler les voies de causalité entre les taxons microbiens et l'hôte en ce qui concerne les performances, la santé et l'innocuité des aliments. Par exemple, si l'on s'attend à ce que les volailles ou le bétail soient confrontés à des conditions environnementales stressantes, la composition des aliments pourrait être adaptée en fonction de concentrations de molécules neuroactives, telles que l'histamine ou la dopamine, susceptibles de modifier la réponse inflammatoire et la santé intestinale de l'animal ainsi que sa sensibilité aux agents pathogènes d'origine alimentaire. Les applications basées sur l'alimentation qui exploitent la communication neurochimique entre l'hôte et les microbes sont prometteuses en tant que plateforme fonctionnelle flexible pouvant être utilisée chez tous les animaux d'élevage, afin d'améliorer la performance des aliments, la santé générale, et, éventuellement, l'innocuité alimentaire. Il est donc nécessaire de développer des hypothèses axées sur la fonction et appuyées par des preuves pour accélérer les applications de la recherche sur le microbiome à la nutrition animale. L'endocrinologie microbienne offre un tel cadre fonctionnel qui pourrait permettre de réguler efficacement le microbiome, si l'on comprend le potentiel de la neurochimie pour l'appliquer à la formulation des aliments pour animaux.

Introduction

Poultry and livestock nutritionists share the common goal of meeting the animal's nutritional requirements to optimize performance. In pursuit of this goal, feed quality evaluation provides the nutritionist a comprehensive, quantitative profile of nutritive and non-nutritive factors that enable effective feed formulation and management. It should not be surprising that the toolbox of tested variables that encompass feed quality testing is ever expanding. Microbial Endocrinology, which is the intersection of neurobiology and microbiology, offers the food animal nutritionist a non-traditional lens through which to leverage the nexus of the microbiome and biogenic amines in feed formulation to advance animal performance, food safety, and other facets of production.

For perspective, the presence of a narrow set of biogenic amines (i.e. histamine, tyramine, putrescine, and others) already has an established utility for the nutritionist in testing feed ingredients for potential toxicologic effects on poultry and livestock health and performance (Scherer et al., 2015; Souza et al., 2022). Biogenic amines occupy several roles outside of potential toxic effects on the animal, however, many of which are neurochemicals or neurochemical-precursors that act as interkingdom signaling molecules (Sudo, 2019). Monoamine neurochemicals, such as dopamine are synthesized and used in the physiologies of many species of feed crops, bacteria, and animals; and it is important to highlight that the dopamine, for example, that is found in plants, animals, and bacteria is structurally identical regardless of its origin (Roshchina, 2010). For the animal nutritionist, it would be highly advantageous to have a quantifiable target that naturally occurs in a feed ingredient that can directly modulate intestinal inflammation, impact foodborne or animal pathogen colonization, or preferentially target specific members of the microbiota to benefit animal behavior and performance. There exists a veritable catalog of biogenic amines that are not currently tested for in feed yet can be naturally present in a variety of feed ingredients that affect local gastrointestinal host-microbiome interactions, local enteric inflammation, as well as broader physiological axes, such as the microbiota-gut-brain axis, the significance of which is increasingly recognized for poultry (Cao et al., 2021) and livestock production (Lyte & Lyte, 2019; Welch et al., 2022). Moreover, while feed safety testing routinely assesses the presence of foodborne pathogens or disease-causing microorganisms, little is known regarding the microbial communities that are present in finished feed formulations. Considering the widespread use of probiotic, prebiotic, and other in-feed additives to improve many aspects of food animal production, understanding the microbial communities that are in feed, both from a compositional (i.e. who's there) and functional (i.e. what are they doing) perspective, is likely in the near future to become part of the animal nutritionist's scientific lexicon. Together, Microbial Endocrinology enables a novel synergistic approach in animal nutrition for understanding the utility of the intersection of biogenic amines and microbiome in order to bridge the gap from foundational science to application in poultry and livestock production.

Neurochemicals in Feed? A Brief Detour to Reflect on What You Ate Today

Before delving into the relevance of in-feed neurochemicals to poultry and livestock production, it is worth asking the reader: What did you eat today? Consider the foods of your own diet as there is a rich scientific literature dating back nearly a century documenting the presence of biogenic amines, including monoamine neurochemicals, in food and impact on human and animal health (Lyte, 2018). Norepinephrine is a 'fight-or-flight' neurochemical that can raise blood pressure. As early as the 1950's bananas were found to contain norepinephrine and serotonin (Waalkes et al., 1958). To illustrate the relevance of 'structurally-identical' neurochemicals in plants on animal health, norepinephrine was isolated from bananas and administered to dogs, resulting in the animal's rise in blood pressure (Waalkes et al., 1958). Separately, if your physician has prescribed you a monoamine oxidase inhibitor (MAO-I), you are likely on a MAO-I modified diet (Van den Eynde et al., 2022). What does this mean? Monoamine oxidases are enzymes that deaminate many biogenic amines. Biogenic amines affect blood pressure, and ingestion of biogenic amine-rich foods in conjunction with inhibition of monoamine oxidase through a course of MAO-I can lead to hypertensive crises in the human patient. Many biogenic amines, such as tyramine, are found in different foods such as a variety of cheeses; which means that slice of pizza you might have eaten could have had medically-consequential biogenic amines in it (Shulman & Walker, 1999). Hence, the point of this brief detour is to illustrate that common foods that you eat can contain neurochemicals that can have medically-important consequences for your health. So, what about what your animals are eating?

Neurochemicals in Feed – What We Know and What We Need to Find Out

Biogenic amines that are tested for in poultry and livestock feeds, including histamine, spermidine, and select other poly- or monoamines, are traditionally thought to arise from microbial utilization of amino acids. Yet, histamine, spermidine, and other biogenic amines are also synthesized and utilized in many feed crop (Bouchereau et al., 2000) and animal physiologies (Beaver & Wostmann, 1962) independent of any microbe. Considering feed is a composite of plant and animal materials, and that feeds do contain bacterial microbiomes, ascribing an origin for any single biogenic amine in a finished feed formulation is a complex task. This should be viewed as a positive aspect, as there are several opportunities for the tailoring of in-feed neurochemical concentrations including feedstuff testing and selection, as well as in-feed microbiome testing and modulation. This leads us to ask the question: What do we know regarding neurochemicals in feed components?

Feed Crops

For decades, neurochemicals have been recognized to be widely distributed in plants. Although foods for human consumption have received the majority of attention, a growing body of literature exists covering crops that are utilized in poultry and livestock feed formulation. It is important to highlight that food-derived neurochemical studies in human clinical literature do hold translational value to poultry and livestock production. Therefore, in this proceedings, human clinical studies are referenced only to fill-in existing gaps in poultry and livestock literature. Soybean is taken as a prototypical example of plant-derived neurochemicals for poultry and livestock producers for two main reasons: 1) Soybean is one of the most frequently used crops in poultry and livestock feeds; and 2) Gamma amino-butyric acid (GABA), a neurochemical, is already used effectively as an exogenous feed additive in poultry, ruminant, and swine production, illustrating the immediate utility, for example, of harnessing naturally occurring GABA concentrations in soybeans across food animal species.

GABA is a neurochemical that is naturally produced in soybean (Wu et al., 2020). The anxiolytic effects of GABA consumption via foods in people are well documented (Mazzoli & Pessione, 2016). Incorporation of exogenous GABA in feeds has been reported in poultry to reduce stocking density stress (Jeong et al., 2020) and heat stress-related (Dai et al., 2011) effects, in dairy cows to improve early lactation (Wang et al., 2013), in ruminants to impact feed intake and other factors (Scherer et al., 2015), as well as in swine to attenuate the impact of transport stress (Bi et al., 2020). Nevertheless, no study to date has examined the impact of naturally occurring GABA in soybeans on poultry or livestock production. Understanding endogenous concentrations of GABA in the batch of soybeans used is an important consideration for the interpretation of different studies' results reporting the impact of the addition of GABA in feed on animal performance or physiological response to stress. This is because both cultivar and harvest year are known to impact soybean concentrations of biogenic amines, including histamine, tyramine, and others (Scherer et al., 2015). Why should harvest year impact biogenic amine concentrations? GABA, among other neurochemicals and precursors, play functional roles in plant physiology including adaptation to environmental stressors (Dehghanian et al., 2024). Hence, harvest years which experience drought, heat waves, or even increased pest insect populations could elicit batch-to-batch variation in crop concentrations of neurochemicals.

In addition to acting as delivery vehicles of neurochemicals, such as GABA, soybeans can also influence the bioavailability of catecholamine (i.e. 'fight-or-flight' neurochemical) precursors. In human clinical literature, it was reported that the consumption of ground soybean meal increased the absorption of L-3,4-dihydroxyphenylalanine (L-dopa) (Nagashima et al., 2016). In addition, soybeans reduced L-dopa metabolism in patients by affecting a key neurochemical degrading enzyme, catechol-O-methyl transferase (COMT). Considering many plants, including some used in feed, naturally contain L-dopa (Soares et al., 2014), the bioavailability of L-dopa in the food animal is immediately relevant to the poultry and livestock producer. L-dopa is converted by the animal to dopamine, and downstream to norepinephrine. In poultry, L-dopa was shown to mediate dopamine signaling in the brain to affect feed intake behavior (Alam et al., 2011). Why would a producer be interested in L-dopa if it is provided in feed if dopamine signaling is occurring in the brain? Because L-dopa is absorbed in the gut and can enter the brain. Hence, in poultry if a L-dopa containing feedstuff is provided as part of a soybean-based feed, there is potential for soybeans to modulate L-dopa stability and bioavailability in the bird to directly mediate a gut-brain axis mechanism that affects feed intake behavior. Indeed, in swine feed-supplementation of L-dopa was shown to directly affect brain concentrations of catecholamines (Erlander et al., 1985).

Dopamine (Liu et al., 2020) and norepinephrine (Płonka & Michalski, 2017), in addition to other neurochemicals such as serotonin (Ramakrishna et al., 2011), are also widely distributed throughout foods that are finding increasing use as components of poultry and livestock feeds. For example, fruit pomaces have been used in poultry (Diarra, 2018; Mnisi et al., 2022), swine (Rebolledo-Leiva et al., 2024), and ruminant (Vinyard et al., 2021) feeds. The catecholamines play several roles in gut health, foodborne pathogen carriage, growth performance, and other factors in poultry (Lyte et al., 2022), swine (Lyte & Lyte, 2019), and ruminants (Lyte et al., 2018). The relevance of in-feed concentrations of catecholamines is immediately apparent to ruminant producers as beta-adrenergic agonists, which are structurally similar to catecholamines, are additives in cattle feed. Beta-adrenergic agonists are well recognized to affect growth performance and feed efficiency, in addition to altering the rumen microbiome (Pfau et al., 2023).

Animal-based Ingredients

Poultry and livestock feeds commonly contain both plant and animal ingredients. Like plant ingredients, animal sources are abundant sources of biogenic amines. Indeed, it should not be surprising that poultry meal, fish meal, feather meal, blood meal, and other animal products contain histamine, tyramine, and other biogenic amines as these neurochemicals are widely used in animal physiologies (Feddern et al., 2019). Importantly, it is the concentration and not simply the presence or absence of these biogenic amines that helps determine if a product is spoiled. Surprisingly, almost nothing is known in these feed ingredients regarding neurochemicals that are not related to spoilage. Also, similar to harvest year variation in biogenic amine content in feed crops, it is important to appreciate that many of the biogenic amines in animal physiology are modulated in response to stress. Hence, it is reasonable to hypothesize that batch-to-batch variation in biogenic amine content of animal-based protein sources such as poultry meals, could, in part, be dependent on the conditions in which animals used to generate the by-products were raised.

Blood consists of thrombocytes or platelets which serve as one of the largest stores of serotonin in avian (Lyte et al., 2025) and mammalian species, respectively. Therefore, when used in feed formulation, blood meal is likely a significant source of serotonin in feed. In rodent studies fluoxetine, a selective serotonin reuptake inhibitor which increases serotonin availability and prevents its degradation, was used to demonstrate (McVey Neufeld et al., 2019) that serotonin signaling in the gut can drive changes in the brain to affect behavior via the vagus nerve, which is a major component of the bi-directional gut-brain axis in mammals (Bonaz et al., 2018) and avians (Orosz & Bradshaw, 2007). Interestingly, in dairy calves fed either fluoxetine or 5-hydroxtryptophan (5-HTP), the latter which is the immediate precursor to serotonin, it was shown that fluoxetine or 5-HTP caused increases in serotonin which correlated with increases in average daily gain and changes in select behaviors (Marrero et al., 2019). Likewise, 5-HTP as a feed additive in poultry both reduced fearfulness (Lundgren & Løvlie, 2023) and modulated foodborne pathogen carriage (Donato et al., 2015). Hence, bloodmeal could potentially serve the animal nutritionist as a dual-purpose protein source and serotonin delivery system to modulate the food animal gut-brain axis.

Poultry offal which can be processed into poultry meal or oil often contain biogenic amines such as histamine and tyramine (Rocker et al., 2021). Like blood meal, little is known regarding non-spoilage related neurochemicals including serotonin and the catecholamines. As poultry offal can refer to a variety of organ tissues, it is important to appreciate that organs such as gizzard, heart, intestinal tract, and other organs are rich sources of monoamine neurochemicals in poultry (Lyte et al., 2022). It would not be surprising then if poultry meal or oil contained measurable amounts of serotonin or catecholamines. Considering the catecholamines play several roles in poultry intestinal function (Dennis, 2016), and host-microbiome interactions—the latter which is discussed later on in this proceedings paper—catecholamine content in animal-based ingredients holds potential to impact poultry and livestock health and performance.

Likewise, a variety of insects, which serve as emerging protein sources in feed for food animal production (Belhadj Slimen et al., 2023), contain several catecholamines including dopamine as well as L-dopa (Barbero et al., 2023; Verlinden, 2018). As with poultry- or blood meal, little is known regarding neurochemical concentrations in insect-based protein sources. Results (submitted for publication) from our laboratory demonstrated that mealworm and cricket feeds for poultry contain substantial quantities of dopamine as compared to a typical corn-soy feed.

The Feed Microbiome – Do You Know 'Who' is in Your Feed?

The utilization of amino acids by bacteria in the production of biogenic amines is well appreciated in feed safety. Hence, it should not be surprising that many bacteria can synthesize monoamine neurochemicals, as well as possess structurally similar transport and uptake pathways (Lyte & Brown, 2018) as well as nearly identical metabolic enzymes as those found in animals, such as MAO (Murooka et al., 1979) for the metabolism of neurochemicals. As feeds do contain microbial communities, it therefore worthwhile to ask the questions: what bacteria are in your feed, and what are they doing?

The monogastric and ruminant microbiomes are impacted by a variety of environmental factors, including type of feed. A little considered component outside of feed safety and the presence of bacterial pathogens in feed, is the contribution of the feed microbiome (Olson et al., 2022) to the food animal gut microbiome and host-microbe interactions through the provision of different bacterial species and metabolites. In an excellent recent study, the microbial communities in a variety of poultry feedstuffs including but not limited to wheat, corn, canola, barley, soybean, sorghum, poultry oil, and meat meal were investigated (Haberecht et al., 2020). The authors found distinct microbial communities between feedstuffs such as meat meal and grain, and that the microbiota in each feed did contribute to the development of the poultry gut microbiota. Interestingly, several bacterial genera identified in feed included *Bifidobacterium*, *Enterococcus*, *Lactobacillus*, and *Clostridium* (Olson et al., 2021). Species of these bacterial genera have been reported to respond to, synthesize, and/or transport neurochemicals including serotonin and several catecholamines (Roshchina, 2010; Scardaci et al., 2022). Furthermore, *Lactobacillus* spp., *Bifidobacterium* spp., and *Enterococcus* spp. are frequently utilized as in-feed probiotics for poultry and livestock (El Jeni et al., 2024; Yaqoob et al., 2022), emphasizing the neuroactive potential of bacteria taxa as a possible key mechanistic pathway that likely contributes to their beneficial impact on the host. Similarly, *Clostridium* spp. includes commensal species that in rodents are known to modulate host intestinal synthesis of serotonin, and recently in poultry were shown to positively correlate with serotonin concentrations in the ceca (Seyoum et al., 2025).

Nevertheless, the studies that have so far characterized the feed microbiota have only examined compositional characteristics, typically through the use of 16S rRNA gene sequencing. Functional studies are needed to assess global microbial metabolic potential through the use of combined metagenomic and culture-based tools. Indeed, the cooperative use of shotgun metagenomic sequencing and culture-based methods will allow for understanding of functional utility of the feed microbiome for poultry and livestock production. For example, pre-,pro-, and post-biotics are commonly used as feed additives that are judged for efficacy on action in the animal intestinal tract. This view should be expanded to include the use of pre-,pro-, and post-biotics as also having utility on effecting changes in the feed microbiome. After all, the feed microbiome affects development of the gut microbiome. Hence, for the animal nutritionist the question of feed formulation becomes not only: 'What' are you feeding your animal, but 'who' are you feeding to your animal?

Leveraging the Intersection of the Microbiome and Feed to Improve Poultry and Livestock Production

Through the lens of Microbial Endocrinology, the neurochemical concentrations in crop and animal ingredients, and the microbiome both in feed and the gastrointestinal tract, can be synergistically applied by the animal nutritionist for promising next-generation strategies in poultry and livestock production. For example, glutamine is a non-essential amino acid in poultry but is used as a feed additive (Selle et al., 2024), and in the poultry gut exists *Bifidobacterium* spp. (Lyte et al., 2024). Recently, *Bifidobacterium* spp. was used (Ikegami et al., 2024) in rodents to increase intestinal concentrations of GABA to cause reductions in animal anxiety. As glutamine is the first molecular building block in the GABA synthesis pathway, the nutritionist could potentially leverage the addition of feed-added glutamine to specifically provide intestinal populations of *Bifidobacterium* spp. the raw material for GABA synthesis (Lyte, 2011; Mazzoli & Pessione, 2016), in order to target a microbiota-gut-brain axis mechanism in poultry. Likewise, *Bifidobacterium* spp. were found in poultry feed, hence there is also potential to modulate in-feed concentrations of GABA.

Catecholamines are known to drive foodborne pathogen growth and virulence in poultry (Cogan et al., 2007) and livestock (Vlisidou et al., 2004), as well as affect the gastrointestinal microbiome (Neuman et al., 2015). From the perspective of animal ingestion of neurochemicals in feed, plant-derived catecholamines have been reported to affect the growth of commensal intestinal bacteria including *Lactobacillus* spp. and *Escherichia coli* (Sandhya, 2020) similarly to how host-derived catecholamines affect a variety of non-pathogenic strains of *E. coli* (Freestone et al., 2002). Relatedly, the modulation of the feed microbiome for enrichment of taxa with neurochemical metabolism functionality, such as bacteria that naturally express MAO (Murooka et al., 1979), could plausibly help reduce host 'fight-or-flight' neurochemicals released in the gastrointestinal tract in response to stress that known to impact the food animal microbiome (Rath et al., 2016).

Microbiome functionality could also hold promise for giving the producer control of in-feed neurochemical concentrations. Norepinephrine concentrations that naturally occur in fruits, and hence potentially in pomaces fed to food animals, might be reduced through targeted enrichment of bacterial taxa in the feed microbiome that express MAO. Hence there is potential to modulate the concentrations of these structurally-identical neurochemicals in the feed to prevent their provision to the animal and downstream impact on the gut microbiome. Serotonin and dopamine have roles in the gastrointestinal tract to control inflammation (Jadhav et al., 2022; Lyte, 2025). Presence of tryptophan or L-dopa in feed could be harnessed for the *in situ* synthesis in the gut by microbial species into serotonin or dopamine, respectively. Almost 30 years ago, addition of tryptophan to feed was shown in broiler chickens to increase brain concentrations of serotonin (Rosebrough, 1996). Recently, it was reported that a probiotic *Enterococcus* spp. that synthesizes dopamine from L-dopa, effectively reduced inflammation driven changes in the poultry gut (Lyte, 2025). As *Enterococcus* spp. are also found in poultry feed, it is reasonable to suggest that potential exists for the feed microbiome to affect food animal gut inflammation through conversion of neurochemical precursors, such as L-dopa.

Today's animal nutritionist wears many proverbial hats ranging from understanding feed composition and safety to animal physiology and performance. It may therefore be proposed that the next-generation animal nutritionist will likely need to add a few more feathers to their hat one of which may be Microbial Endocrinology.

References

Alam, M. R., Yoshizawa, F., & Sugahara, K. (2011). L-DOPA induced extracellular dopamine increases in the ventromedial hypothalamus affects food intake by chickens on a lysine-free diet. *Neurosci Lett*, *495*(2), 126-129. https://doi.org/10.1016/j.neulet.2011.03.053

Barbero, F., Mannino, G., & Casacci, L. P. (2023). The Role of Biogenic Amines in Social Insects: With a Special Focus on Ants. *Insects*, *14*(4). https://doi.org/10.3390/insects14040386

Beaver, M. H., & Wostmann, B. S. (1962). HISTAMINE AND 5-HYDROXYTRYPTAMINE IN THE INTESTINAL TRACT OF GERM-FREE ANIMALS, ANIMALS HARBOURING ONE MICROBIAL SPECIES AND CONVENTIONAL ANIMALS. *British Journal of Pharmacology and Chemotherapy*, *19*(3), 385-393. https://doi.org/https://doi.org/10.1111/j.1476-5381.1962.tb01443.x

Belhadj Slimen, I., Yerou, H., Ben Larbi, M., M'Hamdi, N., & Najar, T. (2023). Insects as an alternative protein source for poultry nutrition: a review [Review]. *Frontiers in Veterinary Science*, *10*. https://doi.org/10.3389/fvets.2023.1200031

Bi, C., Yin, J., Yang, W., Shi, B., & Shan, A. (2020). Effects of dietary γ-aminobutyric acid supplementation on antioxidant status, blood hormones and meat quality in growing-finishing pigs undergoing transport stress. *J Anim Physiol Anim Nutr (Berl)*, *104*(2), 590-596. https://doi.org/10.1111/jpn.13280

Bonaz, B., Bazin, T., & Pellissier, S. (2018). The Vagus Nerve at the Interface of the Microbiota-Gut-Brain Axis [Mini Review]. *Frontiers in Neuroscience*, *12*. https://www.frontiersin.org/journals/neuroscience/articles/10.3389/fnins.2018.00049

Bouchereau, A., Guénot, P., & Larher, F. (2000). Analysis of amines in plant materials. *Journal of Chromatography B: Biomedical Sciences and Applications*, *747*(1), 49-67. https://doi.org/https://doi.org/10.1016/S0378-4347(00)00286-3

Cao, C., Chowdhury, V. S., Cline, M. A., & Gilbert, E. R. (2021). The Microbiota-Gut-Brain Axis During Heat Stress in Chickens: A Review. *Front Physiol*, *12*, 752265. https://doi.org/10.3389/fphys.2021.752265

Cogan, T. A., Thomas, A. O., Rees, L. E., Taylor, A. H., Jepson, M. A., Williams, P. H., Ketley, J., & Humphrey, T. J. (2007). Norepinephrine increases the pathogenic potential of Campylobacter jejuni. *Gut*, *56*(8), 1060-1065. https://doi.org/10.1136/gut.2006.114926

Dai, S. F., Gao, F., Zhang, W. H., Song, S. X., Xu, X. L., & Zhou, G. H. (2011). Effects of dietary glutamine and gamma-aminobutyric acid on performance, carcass characteristics and serum parameters in broilers under circular heat stress. *Animal Feed Science and Technology*, *168*(1), 51-60. https://doi.org/https://doi.org/10.1016/j.anifeedsci.2011.03.005

Dehghanian, Z., Ahmadabadi, M., Asgari Lajayer, B., Bagheri, N., Chamani, M., Gougerdchi, V., Hamedpour-Darabi, M., Shu, W., Price, G. W., & Dell, B. (2024). Role of Neurotransmitters (Biomediators) in Plant Responses to Stress. *Plants*, *13*(22).

Dennis, R. L. (2016). Adrenergic and noradrenergic regulation of poultry behavior and production. *Domestic Animal Endocrinology*, *56*, S94-S100. https://doi.org/https://doi.org/10.1016/j.domaniend.2016.02.007

Diarra, S. S. (2018). Peel meals as feed ingredients in poultry diets: Chemical composition, dietary recommendations and prospects. *Journal of Animal Physiology and Animal Nutrition*, *102*(5), 1284-1295. https://doi.org/https://doi.org/10.1111/jpn.12954

Donato, T. C., Baptista, A. A. S., Garcia, K. C. O. D., Smaniotto, B. D., Okamoto, A. S., Sequeira, J. L., & Andreatti Filho, R. L. (2015). Effects of 5-hydroxytryptophan and m-hydroxybenzylhydrazine associated to Lactobacillus spp. on the humoral response of broilers challenged with Salmonella Enteritidis. *Poultry Science*, *94*(9), 2081-2087. https://doi.org/https://doi.org/10.3382/ps/pev206

El Jeni, R., Villot, C., Koyun, O. Y., Osorio-Doblado, A., Baloyi, J. J., Lourenco, J. M., Steele, M., & Callaway, T. R. (2024). Invited review: “Probiotic” approaches to improving dairy production: Reassessing “magic foo-foo dust”. *Journal of Dairy Science*, *107*(4), 1832-1856. https://doi.org/https://doi.org/10.3168/jds.2023-23831

Erlander, M. G., Parliman, J. A., Draper, D. D., Christian, L. L., Murrin, L. C., Pfeiffer, R. F., & Beitz, D. C. (1985). Effects of L-dopa supplementation on concentrations of brain catechols, dopamine receptor binding and the incidence of pale, soft and exudative meat in stress-susceptible pigs. *J Anim Sci*, *61*(4), 914-923. https://doi.org/10.2527/jas1985.614914x

Feddern, V., Mazzuco, H., Fonseca, F. N., & de Lima, G. J. M. M. (2019). A review on biogenic amines in food and feed: toxicological aspects, impact on health and control measures. *Animal Production Science*, *59*(4), 608-618. https://doi.org/10.1071/AN18076

Freestone, P. P., Williams, P. H., Haigh, R. D., Maggs, A. F., Neal, C. P., & Lyte, M. (2002). Growth stimulation of intestinal commensal Escherichia coli by catecholamines: a possible contributory factor in trauma-induced sepsis. *Shock*, *18*(5), 465-470. https://doi.org/10.1097/00024382-200211000-00014

Haberecht, S., Bajagai, Y. S., Moore, R. J., Van, T. T. H., & Stanley, D. (2020). Poultry feeds carry diverse microbial communities that influence chicken intestinal microbiota colonisation and maturation. *AMB Express*, *10*(1), 143. https://doi.org/10.1186/s13568-020-01077-5

Ikegami, M., Narabayashi, H., Nakata, K., Yamashita, M., Sugi, Y., Fuji, Y., Matsufuji, H., Harata, G., Yoda, K., Miyazawa, K., Nakanishi, Y., & Takahashi, K. (2024). Intervention in gut microbiota increases intestinal γ-aminobutyric acid and alleviates anxiety behavior: a possible mechanism via the action on intestinal epithelial cells. *Front Cell Infect Microbiol*, *14*, 1421791. https://doi.org/10.3389/fcimb.2024.1421791

Jadhav, V. V., Han, J., Fasina, Y., & Harrison, S. H. (2022). Connecting gut microbiomes and short chain fatty acids with the serotonergic system and behavior in Gallus gallus and other avian species. *Front Physiol*, *13*, 1035538. https://doi.org/10.3389/fphys.2022.1035538

Jeong, S.-B., Kim, Y. B., Lee, J.-W., Kim, D.-H., Moon, B.-H., Chang, H.-H., Choi, Y.-H., & Lee, K.-W. (2020). Role of dietary gamma-aminobutyric acid in broiler chickens raised under high stocking density. *Animal Nutrition*, *6*(3), 293-304. https://doi.org/https://doi.org/10.1016/j.aninu.2020.03.008

Liu, Q., Gao, T., Liu, W., Liu, Y., Zhao, Y., Liu, Y., Li, W., Ding, K., Ma, F., & Li, C. (2020). Functions of dopamine in plants: a review. *Plant Signal Behav*, *15*(12), 1827782. https://doi.org/10.1080/15592324.2020.1827782

Lundgren, K. A., & Løvlie, H. (2023). Increased dietary 5-hydroxytryptophan reduces fearfulness in red junglefowl hens (Gallus gallus). *Front Physiol*, *14*, 1213986. https://doi.org/10.3389/fphys.2023.1213986

Lyte, J. M. (2018). Eating for 3.8 × 10(13): Examining the Impact of Diet and Nutrition on the Microbiota-Gut-Brain Axis Through the Lens of Microbial Endocrinology. *Front Endocrinol (Lausanne)*, *9*, 796. https://doi.org/10.3389/fendo.2018.00796

Lyte, J. M., Assumpcao, A. L. F. V., Caputi, V., Ashwell, C. M., Seyoum, M., Honaker, C. F., Daniels, K. M., Lyte, M., Siegel, P. B., & Taylor, R. L. (2025). Co-evolution of the humoral immune and serotonergic systems in chickens selected for high or low blood antibody titer response to sheep red blood cells. *Poultry Science*, *104*(2), 104699. https://doi.org/https://doi.org/10.1016/j.psj.2024.104699

Lyte, J. M., Eckenberger, J., Keane, J., Robinson, K., Bacon, T., Assumpcao, A. L. F. V., Donoghue, A. M., Liyanage, R., Daniels, K. M., Caputi, V., & Lyte, M. (2024). Cold stress initiates catecholaminergic and serotonergic responses in the chicken gut that are associated with functional shifts in the microbiome. *Poultry Science*, *103*(3), 103393. https://doi.org/https://doi.org/10.1016/j.psj.2023.103393

Lyte, J. M., & Lyte, M. (2019). Review: Microbial endocrinology: intersection of microbiology and neurobiology matters to swine health from infection to behavior. *Animal*, *13*(11), 2689-2698. https://doi.org/https://doi.org/10.1017/S1751731119000284

Lyte, J. M., Martinez, D. A., Robinson, K., Donoghue, A. M., Daniels, K. M., & Lyte, M. (2022). A neurochemical biogeography of the broiler chicken intestinal tract. *Poultry Science*, *101*(3), 101671. https://doi.org/https://doi.org/10.1016/j.psj.2021.101671

Lyte, M. (2011). Probiotics function mechanistically as delivery vehicles for neuroactive compounds: Microbial endocrinology in the design and use of probiotics. *Bioessays*, *33*(8), 574-581. https://doi.org/10.1002/bies.201100024

Lyte, M., & Brown, D. R. (2018). Evidence for PMAT- and OCT-like biogenic amine transporters in a probiotic strain of Lactobacillus: Implications for interkingdom communication within the microbiota-gut-brain axis. *PLOS ONE*, *13*(1), e0191037. https://doi.org/10.1371/journal.pone.0191037

Lyte, M., Daniels, K., Byrd, J.A., Genovese, K., Swaggerty, C.L., Kogut, M. (2025). Use of a microbial endocrinology designed dopamine-producing probiotic to control gut neurochemical levels associated with the development of gut inflammation. *Poultry Science*, *In Press*.

Lyte, M., Villageliú, D. N., Crooker, B. A., & Brown, D. R. (2018). Symposium review: Microbial endocrinology—Why the integration of microbes, epithelial cells, and neurochemical signals in the digestive tract matters to ruminant health1. *Journal of Dairy Science*, *101*(6), 5619-5628. https://doi.org/https://doi.org/10.3168/jds.2017-13589

Marrero, M. G., Dado-Senn, B., Field, S. L., da Silva, D. R., Skibiel, A. L., & Laporta, J. (2019). Increasing serotonin bioavailability in preweaned dairy calves impacts hematology, growth, and behavior. *Domestic Animal Endocrinology*, *69*, 42-50. https://doi.org/https://doi.org/10.1016/j.domaniend.2019.04.007

Mazzoli, R., & Pessione, E. (2016). The Neuro-endocrinological Role of Microbial Glutamate and GABA Signaling. *Front Microbiol*, *7*, 1934. https://doi.org/10.3389/fmicb.2016.01934

McVey Neufeld, K. A., Bienenstock, J., Bharwani, A., Champagne-Jorgensen, K., Mao, Y., West, C., Liu, Y., Surette, M. G., Kunze, W., & Forsythe, P. (2019). Oral selective serotonin reuptake inhibitors activate vagus nerve dependent gut-brain signalling. *Sci Rep*, *9*(1), 14290. https://doi.org/10.1038/s41598-019-50807-8

Mnisi, C. M., Mhlongo, G., & Manyeula, F. (2022). Fruit Pomaces as Functional Ingredients in Poultry Nutrition: A Review [Review]. *Frontiers in Animal Science*, *3*. https://doi.org/10.3389/fanim.2022.883988

Murooka, Y., Doi, N., & Harada, T. (1979). Distribution of membrane-bound monoamine oxidase in bacteria. *Appl Environ Microbiol*, *38*(4), 565-569. https://doi.org/10.1128/aem.38.4.565-569.1979

Nagashima, Y., Kondo, T., Sakata, M., Koh, J., & Ito, H. (2016). Effects of soybean ingestion on pharmacokinetics of levodopa and motor symptoms of Parkinson's disease--In relation to the effects of Mucuna pruriens. *J Neurol Sci*, *361*, 229-234. https://doi.org/10.1016/j.jns.2016.01.005

Neuman, H., Debelius, J. W., Knight, R., & Koren, O. (2015). Microbial endocrinology: the interplay between the microbiota and the endocrine system. *FEMS Microbiology Reviews*, *39*(4), 509-521. https://doi.org/10.1093/femsre/fuu010

Olson, E. G., Dittoe, D. K., Jendza, J. A., Stock, D. A., & Ricke, S. C. (2022). Application of microbial analyses to feeds and potential implications for poultry nutrition. *Poult Sci*, *101*(5), 101789. https://doi.org/10.1016/j.psj.2022.101789

Olson, E. G., Dittoe, D. K., Micciche, A. C., & Ricke, S. C. (2021). Identification of bacterial isolates from commercial poultry feed via 16S rDNA. *Journal of Environmental Science and Health, Part B*, *56*(3), 272-281. https://doi.org/10.1080/03601234.2020.1868236

Orosz, S. E., & Bradshaw, G. A. (2007). Avian Neuroanatomy Revisited: From Clinical Principles to Avian Cognition. *Veterinary Clinics of North America: Exotic Animal Practice*, *10*(3), 775-802. https://doi.org/https://doi.org/10.1016/j.cvex.2007.06.001

Pfau, A. P., Shepherd, E. A., Martin, M. G., Ascolese, S., Mason, K. M., Egert-McLean, A. M., Voy, B. H., & Myer, P. R. (2023). Beta-Adrenergic Agonists, Dietary Protein, and Rumen Bacterial Community Interactions in Beef Cattle: A Review. *Vet Sci*, *10*(9). https://doi.org/10.3390/vetsci10090579

Płonka, J., & Michalski, A. (2017). The influence of processing technique on the catecholamine and indolamine contents of fruits. *Journal of Food Composition and Analysis*, *57*, 102-108. https://doi.org/https://doi.org/10.1016/j.jfca.2016.12.023

Ramakrishna, A., Giridhar, P., & Ravishankar, G. A. (2011). Phytoserotonin: a review. *Plant Signal Behav*, *6*(6), 800-809. https://doi.org/10.4161/psb.6.6.15242

Rath, L. L., Samuelson, K. L., Salazar, A. L., Lopez, F. A., Scholljegerdes, E. J., & Loest, C. A. (2016). 1647 WS mammalian hormones associated with stress impact microbial fermentation of rumen fluid in vitro. *Journal of Animal Science*, *94*(suppl_5), 803-803. https://doi.org/10.2527/jam2016-1647

Rebolledo-Leiva, R., Hernández, D., Moreira, M. T., & González-García, S. (2024). Using olive and apple pomaces for fattening pig diets: Environmental impacts under an attributional and consequential perspective. *Environmental Technology & Innovation*, *34*, 103549. https://doi.org/https://doi.org/10.1016/j.eti.2024.103549

Rocker, M. M., Lewis, M. J., Mock, T. S., Francis, D. S., Bellagamba, F., Moretti, V. M., Quinn, G. P., Smullen, R. P., & Turchini, G. M. (2021). Poultry offal meal production conditions impact meal quality and digestibility in Atlantic salmon (Salmo salar). *Aquaculture*, *542*, 736909. https://doi.org/https://doi.org/10.1016/j.aquaculture.2021.736909

Rosebrough, R. W. (1996). Crude protein and supplemental dietary tryptophan effects on growth and tissue neurotransmitter levels in the broiler chicken. *Br J Nutr*, *76*(1), 87-96. https://doi.org/10.1079/bjn19960011

Roshchina, V. V. (2010). Evolutionary Considerations of Neurotransmitters in Microbial, Plant, and Animal Cells. In M. Lyte & P. P. E. Freestone (Eds.), *Microbial Endocrinology: Interkingdom Signaling in Infectious Disease and Health* (pp. 17-52). Springer New York. https://doi.org/10.1007/978-1-4419-5576-0_2

Sandhya, M., Jessy, P. (2020). Effect of plant derived catecholamines on growth of *Escherichia coli* and *Lactobacillus* sp. *International Journal of Botany Studies*, *5*(6), 274-276. https://doi.org/https://www.botanyjournals.com/archives/2020/vol5/issue6/5-6-41

Scardaci, R., Bietto, F., Racine, P. J., Boukerb, A. M., Lesouhaitier, O., Feuilloley, M. G. J., Scutera, S., Musso, T., Connil, N., & Pessione, E. (2022). Norepinephrine and Serotonin Can Modulate the Behavior of the Probiotic Enterococcus faecium NCIMB10415 towards the Host: Is a Putative Surface Sensor Involved? *Microorganisms*, *10*(3). https://doi.org/10.3390/microorganisms10030487

Scherer, R., Gerlach, K., & Südekum, K. H. (2015). Biogenic amines and gamma-amino butyric acid in silages: Formation, occurrence and influence on dry matter intake and ruminant production. *Animal Feed Science and Technology*, *210*, 1-16. https://doi.org/https://doi.org/10.1016/j.anifeedsci.2015.10.001

Selle, P. H., Macelline, S. P., Toghyani, M., & Liu, S. Y. (2024). The potential of glutamine supplementation in reduced-crude protein diets for chicken-meat production. *Animal Nutrition*, *18*, 49-56. https://doi.org/https://doi.org/10.1016/j.aninu.2024.03.017

Seyoum, M. M., Assumpcao, A. L. F. V., Caputi, V., Ashwell, C. M., Honaker, C. F., Daniels, K. M., Lyte, M., Siegel, P. B., Taylor Jr, R. L., & Lyte, J. M. (2025). Multigenerational selection for high or low antibody response to sheep red blood cells modulates the chicken cecal microbiome and its relationship to the immune and serotonergic systems. *Poultry Science*, *104*(4), 104943. https://doi.org/https://doi.org/10.1016/j.psj.2025.104943

Shulman, K. I., & Walker, S. E. (1999). Refining the MAOI diet: tyramine content of pizzas and soy products. *J Clin Psychiatry*, *60*(3), 191-193. https://www.ncbi.nlm.nih.gov/pubmed/10192596

Soares, A. R., Marchiosi, R., Siqueira-Soares Rde, C., Barbosa de Lima, R., Dantas dos Santos, W., & Ferrarese-Filho, O. (2014). The role of L-DOPA in plants. *Plant Signal Behav*, *9*(4), e28275. https://doi.org/10.4161/psb.28275

Souza, R. M. d., Glória, M. B. d. A., Oliveira, A. d. L., Almeida, A. C. d., Colen, F., Souza, C. F. A. e., & Oliveira, N. J. F. d. (2022). Bioactive amines in ingredients and feeds of broilers and storage effects on their levels. *Research, Society and Development*, *11*(5), e36211528347. https://doi.org/10.33448/rsd-v11i5.28347

Sudo, N. (2019). Biogenic Amines: Signals Between Commensal Microbiota and Gut Physiology. *Front Endocrinol (Lausanne)*, *10*, 504. https://doi.org/10.3389/fendo.2019.00504

Van den Eynde, V., Gillman, P. K., & Blackwell, B. B. (2022). The Prescriber's Guide to the MAOI Diet-Thinking Through Tyramine Troubles. *Psychopharmacol Bull*, *52*(2), 73-116. https://www.ncbi.nlm.nih.gov/pubmed/35721816

Verlinden, H. (2018). Dopamine signalling in locusts and other insects. *Insect Biochemistry and Molecular Biology*, *97*, 40-52. https://doi.org/https://doi.org/10.1016/j.ibmb.2018.04.005

Vinyard, J. R., Myers, C. A., Murdoch, G. K., Rezamand, P., & Chibisa, G. E. (2021). Optimum grape pomace proportion in feedlot cattle diets: ruminal fermentation, total tract nutrient digestibility, nitrogen utilization, and blood metabolites. *Journal of Animal Science*, *99*(2), skab044. https://doi.org/10.1093/jas/skab044

Vlisidou, I., Lyte, M., van Diemen, P. M., Hawes, P., Monaghan, P., Wallis, T. S., & Stevens, M. P. (2004). The neuroendocrine stress hormone norepinephrine augments

Escherichia coli O157:H7-induced enteritis and adherence in a bovine ligated ileal loop model of infection. *Infect Immun*, *72*(9), 5446-5451. https://doi.org/10.1128/iai.72.9.5446-5451.2004

Waalkes, T. P., Sjoerdsma, A., Creveling, C. R., Weissbach, H., & Udenfriend, S. (1958). Serotonin, Norepinephrine, and Related Compounds in Bananas. *Science*, *127*(3299), 648-650. https://doi.org/10.1126/science.127.3299.648

Wang, D. M., Wang, C., Liu, H. Y., Liu, J. X., & Ferguson, J. D. (2013). Effects of rumen-protected γ-aminobutyric acid on feed intake, lactation performance, and antioxidative status in early lactating dairy cows. *Journal of Dairy Science*, *96*(5), 3222-3227. https://doi.org/https://doi.org/10.3168/jds.2012-6285

Welch, C. B., Ryman, V. E., Pringle, T. D., & Lourenco, J. M. (2022). Utilizing the Gastrointestinal Microbiota to Modulate Cattle Health through the Microbiome-Gut-Organ Axes. *Microorganisms*, *10*(7). https://doi.org/10.3390/microorganisms10071391

Wu, S. J., Chang, C. Y., Lai, Y. T., & Shyu, Y. T. (2020). Increasing γ-Aminobutyric Acid Content in Vegetable Soybeans via High-Pressure Processing and Efficacy of Their Antidepressant-Like Activity in Mice. *Foods*, *9*(11). https://doi.org/10.3390/foods9111673

Yaqoob, M. U., Wang, G., & Wang, M. (2022). An updated review on probiotics as an alternative of antibiotics in poultry - A review. *Anim Biosci*, *35*(8), 1109-1120. https://doi.org/10.5713/ab.21.0485

There's only one Hy-D® for lifetime productivity

When it comes to purity, performance and immunity, Hy-D® has been helping pigs, birds and producers stand strong for years.

8+ /25+

Over 8 years on the North American market in swine and over 25 in poultry, with demonstrated safety and research for increasing vitamin D status

100+

Research trials demonstrating the safety and benefits of Hy-D in diets for poultry, swine, and ruminants globally

2.6 M/3 B

2.6 Million pigs and over 3 Billion birds are fed Hy-D per year in North America, based on dsm-firmenich actual sales, recommended feeding rates and survey data

Watch the swine video

Learn more about the recent CFIA-approved increased vitamin D levels for swine and poultry nutrition in Canada.

Watch the poultry video

dsm-firmenich.com/hy-d

Follow us on:

dsm-firmenich

ANCC

2025 ANIMAL NUTRITION CONFERENCE OF CANADA

CNAC

2025 COLLOQUE DE NUTRITION ANIMALE DU CANADA

Opening Plenary
Plénière d'ouverture

Regulation of digestion kinetics in swine

Régulation de la cinétique de la digestion chez le porc

W. J. J. Gerrits, S. de Vries
Animal Nutrition Group, Wageningen University & Research, Wageningen, The Netherlands,
walter.gerrits@wur.nl

SPONSORED BY / COMMANDITÉ PAR

LALLEMAND

LALLEMAND ANIMAL NUTRITION

Abstract

Growing importance of upcycling agricultural and food processing byproducts through livestock production, increases the variation in the nutritional quality of feed ingredients. In addition to the extent, the kinetics of digestion can affect the metabolic fate of nutrients after absorption. Together with a growing body of evidence of interactions occurring in the lumen of the digestive tract, this urges the need for insights into the regulation of digestion kinetics. Results of recent studies on the regulation of digestion kinetics in swine will be presented and discussed, with a focus on digestion kinetics of starch and proteins, regulated by dietary fibres and particle size distribution. Whereas starch digestion kinetics are highly influenced by the starch source, the feed matrix, and can be accelerated by thermal treatment, digestion kinetics of proteins follows the sequential emptying of liquids and solids from the stomach. Typically, digesta liquids are emptied about 2 h faster from the stomach than solids, but this difference can be reduced by viscous, soluble fibres or enlarged by coarse particles (>0.3 mm). Gastric mixing affects the pH gradient in the stomach and thereby likely the post-stomach hydrolysis rates. Dietary water binding capacity, but not dietary extract viscosity, appears to be a strong predictor of digesta phase separation in the stomach. Small intestinal digesta transit appears to be highly variable and difficult to influence through dietary properties. These insights can facilitate progress in feed ingredient evaluation and feed formulation in complex diets, provided that more information about ingredient properties is registered.

Keywords: starch, protein, dietary fibre, particle size, stomach

Résumé

La valorisation des sous-produits issus de l'agriculture et de la transformation des aliments dans les fermes d'élevage entraîne une variabilité accrue de la qualité nutritionnelle des ingrédients utilisés dans l'alimentation animale. En outre, la cinétique de digestion peut influencer le devenir métabolique des nutriments après l'absorption. Compte tenu de l'accumulation des preuves d'interactions se produisant dans la lumière du tube digestif, il est crucial d'approfondir notre compréhension de la régulation de la cinétique de digestion. Les résultats d'études récentes sur la régulation de la cinétique de digestion chez les porcs seront présentés et analysés, en portant une attention particulière à la cinétique de digestion de l'amidon et des protéines, qui est influencée par les fibres alimentaires et la distribution de la taille des particules. La cinétique de digestion de l'amidon dépend principalement de la source de l'ingrédient, la matrice alimentaire, et peut être accélérée par un traitement thermique. En revanche, la cinétique de digestion des protéines est déterminée par la vidange séquentielle des liquides et des solides de l'estomac. En règle générale, les liquides sont évacués de l'estomac environ deux heures plus rapidement que les solides, mais cette différence peut être atténuée par la présence de fibres visqueuses et solubles ou augmentée par celle de particules grossières (> 0,3 mm). Le brassage du contenu gastrique modifie le gradient de pH dans l'estomac, influençant vraisemblablement le taux d'hydrolyse en aval. Le pouvoir de rétention d'eau de l'aliment, plutôt que la viscosité de l'extrait, semble être un facteur de prédiction important de la séparation des phases du digesta dans l'estomac. Le transit du digesta dans l'intestin grêle semble être très variable et difficilement à modifier par les propriétés des aliments. Ces connaissances peuvent contribuer à améliorer l'évaluation des ingrédients alimentaires et la formulation des rations complexes, à condition de pouvoir recueillir davantage de données sur les propriétés des ingrédients.

Introduction

Upcycling of agricultural by-products, food waste, and food processing by-products through livestock production is of growing importance. Concomitantly, variation in type and nutritional quality of feed ingredients is increasing. Ingredients are commonly evaluated based on their measured extent of digestion, listed in feeding tables that provide the basis for diet formulation. Awareness increases that in addition to the extent, the kinetics of digestion affects the metabolic fate of nutrients after absorption. This includes its influence on feed intake patterns, synchrony of absorption of nutrients and synchrony between nutrient supply and demand. Together with a growing body of evidence of complex interactions occurring within the lumen of the digestive tract (see e.g. De Vries et al. (2016)), this urges the need of developing feed evaluation approaches that take this variation into account. For this purpose, mathematical modelling provides an excellent platform. To date, however, the application mathematical models in feed evaluation is limited by the use of complete feeds as the starting point in most models. Connection to alterations in feeding value of ingredients, or interactions occurring inside the digestive tract, require the use of feed ingredients rather than complete feeds as the starting points of simulations. A first step in this direction was accomplished by Schop et al. (2023). The limiting step in improving this type of models is an appropriate characterization of physico-chemical properties of feed ingredients and the way these impact on digesta transit and hydrolysis kinetics, particularly so in the proximal parts of the digestive tract. This paper summarizes recent work in our lab on protein and starch digestion kinetics with a special role for dietary fibres.

Starch digestion kinetics

Variation in starch digestion kinetics among sources of different botanical origin has been well-established, with intrinsic starch properties like increased concentration of A-type crystalline structure and short amylopectin side-chains increasing hydrolysis rates across botanic origins. Within botanic sources, variation in hydrolysis kinetics is mainly explained by other properties such as the amylose content and the number of pores on the starch granules (Martens et al., 2018). The matrix in which starch granules are embedded has a dominant and ingredient dependent effect on the rate of starch digestion. For example, Martens et al. (2019) compared the digestion kinetics of isolated starch with that of starch embedded in the raw ingredient. For barley, starch digestion of the isolated starch was 16 % points higher half way through the small intestine, and 4 % at the distal small intestine when compared with that of ground barley. For maize, the difference half-way through the small intestine was not observed, but at the distal small intestine, the starch digestibility of the isolated maize starch was 15% points higher compared with that of ground maize. These differences were assessed both by *in vitro* methods (based on Englyst) and *in vivo* by the use indigestible markers in a dissection study. Remarkable differences occurred between the *in vitro* and *in vivo* methodologies, particularly so in the proximal small intestine. This difference is illustrated for isolated and ground barley in Figure 1.

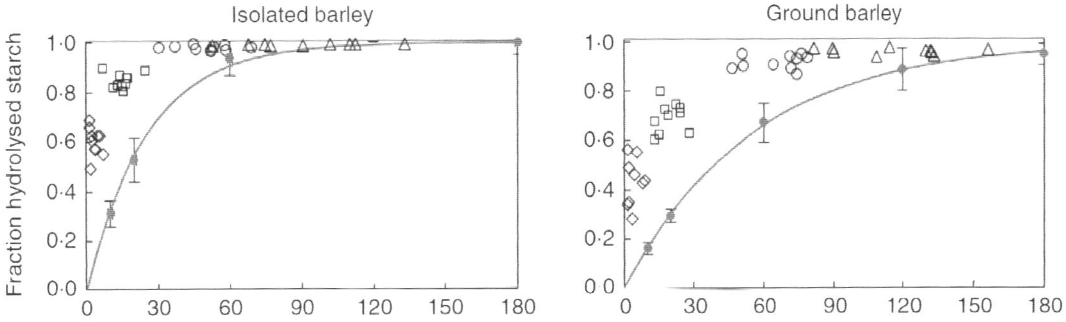

Figure 1. Digestion kinetics of barley starch *in vitro* and *in vivo* in pigs. Digestion coefficients of starch measured in digesta recovered from part 1 (♦), 2 (□), 3 (○) and 4 (Δ) of the small intestine in pigs fed diets containing starch isolated from barley or as ground barley, plotted against the cumulative retention time. Symbols (●) indicate the average of in triplicate measured values of *in vitro* starch hydrolysis against incubation time; lines represent the first-order kinetic model fitted to the data points. Error bars represent the standard deviation of *in vitro* measured starch digestion (source: Martens et al. (2019)).

The comparison between the *in vitro* and *in vivo* estimates of starch hydrolysis kinetics illustrates the limitations of the *in vitro* methodology to adequately represent occurrence of starch disappearance in the very proximal small intestine. This points at an important role for the stomach. Further work with enzymes obtained from stomach extracts by Martens et al. (2020) highlighted the importance of salivary and bacterial amylases in initial solubilization of starch, and appearance of small glucose polymers in the stomach. The importance of feed technologies to steer starch hydrolysis kinetics is obvious but not discussed here.

Protein digestion kinetics

Data on digestion and absorption kinetics measured *in vivo* are scarce. The importance of variation in absorption kinetics of amino acids from the intestinal tract have been underpinned by portal blood appearance studies. While important, these are technically so demanding that between-study variation is large, and variation between protein sources scarce or unreliable. Milk proteins can be regarded as an exception. Modelling work of Le Feunteun et al. (2014) illustrates that variation in amino acid absorption kinetics of dairy proteins in minipigs is dominated by variation in stomach emptying processes. A similar conclusion was derived in the modelling work of Schop et al. (2023) in growing pigs. In a recent meta-analysis, (Dorado-Montenegro et al., 2025) estimated the difference in mean retention time between digesta solids and fluids in the stomach to be on average 1.5 hours, which is nearly 2/3 of the mean retention time of digesta in the entire small intestine. Based on this rather large difference, attention in *in vitro* studies of protein hydrolysis kinetics focuses on solubilization kinetics. Many research efforts concentrate on measurement of protein digestion kinetics *in vitro*, using techniques based on modifications to static protocols by Boisen and Fernández (1997) or INFOGEST (Brodkorb et al., 2019). The *in vitro* work of Chen et al. (2019) illustrates that considerable and relevant variation exists in N solubilization kinetics both during stomach and small intestinal incubation among feed ingredients. This was confirmed by unpublished work of Zhang in our group (Zhang, 2024) showing that the pattern of appearance of low molecular weight nitrogenous compounds largely followed patterns of N solubilization kinetics. She modified the Boisen & Fernandez procedure to include a pH gradient during stomach incubation. The major portion of N was solubilized and hydrolyzed, primarily following a drop in pH from 4 to 2 during incubation in the stomach, with large differences among ingredients. For example, solubilization rates for rapeseed meal, barley, peas and fishmeal were 0.1, 0.3, 0.3 and 0.1%/min during initial incubation at pH=4, respectively. This changed to 0.2, 0.1, 0.1 and 0.3/min during subsequent incubation at pH=2, respectively. In addition, the pH gradient itself contributes to altered N solubility. In the end, 46, 55, 68 and 73% of the N incubated solubilized in the stomach phase for rapeseed meal, barley, peas and fishmeal, respectively. This variation illustrates that measurement of solubilization kinetics in vitro, combined with understanding of pH gradients in the stomach and the regulation of stomach emptying are key to progress in this field.

The role of dietary fibres

Dietary fibres have been shown to decrease the extent of digestion of dietary proteins. In a recent meta-analysis, Zhang et al. (2024) demonstrated that across databases, apparent ileal digestibility of proteins decrease by 0.6 to 0.7%-units per g/100g increase of total dietary fibre for cereal fibre sources and by 0.5 to 0.9%-units for non-cereal fibre sources. Mechanisms include entrapment of proteins in the fibre matrix, effects of physico-chemical properties of the fibre source and stimulation of endogenous protein losses. It is likely that the effects induced by dietary fibres do not only affect the extent, but also kinetics of the digestion process.

As discussed above, stomach emptying processes play a key role, simply by the gastric sieving process, causing a variable but substantial difference in the mean retention time of solids and fluids in the stomach. The meta-analysis by Dorado-Montenegro et al. (2025) using data from 156 growing pigs fed one of 20 diets, revealed that 56% of the variation in digesta phase separation in the stomach was explained by dietary physico-chemical properties. The major portion of this variation was attributed to the water binding capacity of the diets (see Figure 2), and limited variation by other rheological parameters. Notably, extract viscosity of the diets explained no variation in digesta phase separation in the stomach, and none of the physico-chemical properties of the diets explained relevant variation in digesta transit in the small intestine. Determining the mean retention time of dietary fibres requires the use of dedicated tracers. Metals, mordanted to the fibre source of study can be useful in such tracer studies, particularly so in determining the effects of particle size. Using this methodology Sebastian et al. (2023), and (Lannuzel et al., 2024a; Lannuzel et al., 2024b), showed that the effects of particle size are source dependent. While increasing insoluble fibres clearly magnify digesta phase separation in the stomach (Sebastian et al., 2023), large particles of straw (Lannuzel et al., 2024a) and soybean hulls but not oat husks (Sebastian et al., 2023), remained longer in the stomach than their ground counterparts. Increasing gelating properties of the diet by including soluble fibre sources typically decrease digesta phase separation in the stomach (Lannuzel et al., 2024b). Even though fibre sources in animal nutrition do not always contribute to the supply of proteins and starch substantially, by altering digesta passage kinetics, they can act as key-regulators of nutrient absorption kinetics.

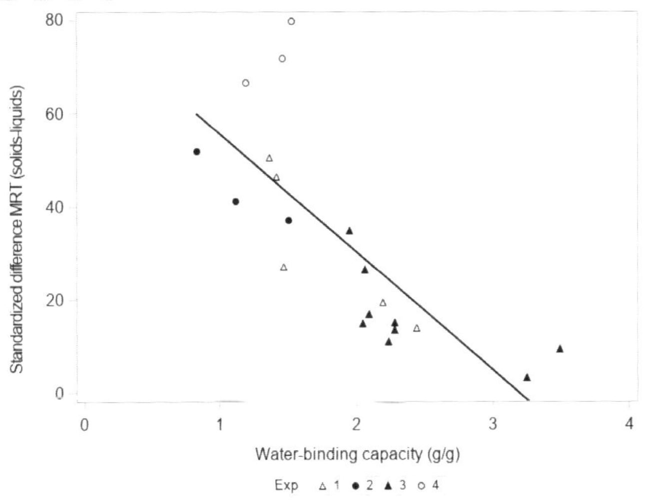

Figure 2. Association between digesta-phase separation in the stomach and the water binding capacity of 20 diets fed in 4 experiments. Difference in mean retention time (MRT) is expressed as (MRTsolids-MRTliquids)/ (MRTsolids+MRTliquids) *100) of gastric digesta; r= -0.76, P < 0.001; Data from Dorado-Montenegro et al. (2025).

References

Boisen, S., and J. A. Fernández. 1997. Prediction of the total tract digestibility of energy in feedstuffs and pig diets by in vitro analyses. Anim. Feed Sci. Technol. 68(3-4):277-286. (Article) doi: 10.1016/S0377-8401(97)00058-8

Brodkorb, A., L. Egger, et al. 2019. INFOGEST static in vitro simulation of gastrointestinal food digestion. Nat. Protoc. 14(4):991-1014. (Article) doi: 10.1038/s41596-018-0119-1

Chen, H., P. A. Wierenga, W. H. Hendriks, and A. J. M. Jansman. 2019. In vitro protein digestion kinetics of protein sources for pigs. Animal 13(6):1154-1164. (Article) doi: 10.1017/S1751731118002811

de Vries, S., W. J. J. Gerrits, M. A. Kabel, T. Vasanthan, and R. T. Zijlstra. 2016. β-Glucans and resistant starch alter the fermentation of recalcitrant fibers in growing pigs. PLoS ONE 11(12)(Article) doi: 10.1371/journal.pone.0167624

Dorado-Montenegro, S., J. A. Dijksman, W. J. J. Gerrits, and S. de Vries. 2025. Predicting digesta retention times from dietary physicochemical properties in pigs. Anim. Feed Sci. Technol. 321(Article) doi: 10.1016/j.anifeedsci.2025.116242

Lannuzel, C., R. J. Veersma, G. M. Bornhorst, G. van Erven, M. A. Kabel, W. J. J. Gerrits, and S. de Vries. 2024a. Particle size of straw and gelation of pectin influence gastric mixing and emptying in pigs. Animal 18(12)(Article) doi: 10.1016/j.animal.2024.101362

Lannuzel, C., R. J. Veersma, N. Wever, G. van Erven, M. A. Kabel, W. J. J. Gerrits, and S. de Vries. 2024b. Particle size of insoluble fibres and gelation of soluble fibres influence digesta passage rate throughout the gastrointestinal tract of finishing pigs. Animal 18(6)(Article) doi: 10.1016/j.animal.2024.101175

Le Feunteun, S., F. Barbé, D. Rémond, O. Ménard, Y. Le Gouar, D. Dupont, and B. Laroche. 2014. Impact of the Dairy Matrix Structure on Milk Protein Digestion Kinetics: Mechanistic Modelling Based on Mini-pig In Vivo Data. Food. Bioprocess Technol. 7(4):1099-1113. (Article) doi: 10.1007/s11947-013-1116-6

Martens, B. M. J., E. M. A. M. Bruininx, W. J. J. Gerrits, and H. A. Schols. 2020. The importance of amylase action in the porcine stomach to starch digestion kinetics. Anim. Feed Sci. Technol. 267(Article) doi: 10.1016/j.anifeedsci.2020.114546

Martens, B. M. J., T. Flécher, S. de Vries, H. A. Schols, E. M. A. M. Bruininx, and W. J. J. Gerrits. 2019. Starch digestion kinetics and mechanisms of hydrolysing enzymes in growing pigs fed processed and native cereal-based diets. Br. J. Nutr. 121(10):1124-1136. (Article) doi: 10.1017/S0007114519000503

Martens, B. M. J., W. J. J. Gerrits, E. M. A. M. Bruininx, and H. A. Schols. 2018. Amylopectin structure and crystallinity explains variation in digestion kinetics of starches across botanic sources in an in vitro pig model. J. Animal Sci. Biotechnol. 9(1)(Article) doi: 10.1186/s40104-018-0303-8

Schop, M., H. Nguyen-Ba, A. J. M. Jansman, S. de Vries, J. L. Ellis, A. Bannink, and W. J. J. Gerrits. 2023. SNAPIG: a model to study nutrient digestion and absorption kinetics in growing pigs based on diet and ingredient properties. Animal 17(Article) doi: 10.1016/j.animal.2023.101025

Sebastian, D. M., L. J. Kim, G. Walter, and S. de Vries. 2023. Insoluble fibers affect digesta transit behavior in the upper gastrointestinal tract of growing pigs, regardless of particle size. J. Anim. Sci. 101(Article) doi: 10.1093/jas/skad299

Zhang, S. 2024. The journey of protein and starch through the gastrointestinal tract of pigs. Studying digestion kinetics by in vitro methodology, Wageningen University, Wageningen.

Zhang, S., S. de Vries, and W. J. J. Gerrits. 2024. Quantifying the effects of dietary fibres on protein digestibility in pigs - A review. Anim. Feed Sci. Technol. 308(Review) doi: 10.1016/j.anifeedsci.2023.115864

From bowl to barn: A shared future for companion and farm animal nutritional management

Du bol à l'étable : un avenir commun pour la gestion nutritionnelle des animaux de compagnie et de ferme

A. K. Shoveller[1], C. Croney[2]
[1] Professor, Professor and Champion Petfoods Chair in Canine and Feline Nutrition, Physiology and Metabolism, and University of Guelph Research Leadership Chair, Department of Animal Biosciences, University of Guelph, Guelph, ON, Canada
[2] Professor, Department of Veterinary Sciences, Purdue University, West Lafayette, IN, USA
ashovell@uoguelph.ca

SPONSORED BY / COMMANDITÉ PAR

PROBIOTECH INTERNATIONAL

Abstract

The bond between humans and their pets has deepened significantly over the past half-century, with dogs and cats increasingly considered integral members of the family. This evolving relationship has heightened scrutiny regarding the use of companion animals in research, even as the demand for species-specific data grows to improve their health, welfare, and environmental management. Despite these needs, research on the health and well-being of growing cats lags far behind studies on humans and agricultural animals. Conducting such research requires not only adherence to evolving laws and ethical policies governing the use of companion animals but also an acute awareness of societal beliefs, which continue to shift. This paper will examine successful research plans involving feline studies, emphasizing species-specific environmental requirements to enhance the welfare of research cats. As our understanding of animals' needs and behavioral complexity advances, societal perceptions and ethical obligations surrounding animal welfare evolve, often sparking public debate. This discussion highlights the importance of balancing scientific progress with ethical considerations to ensure the well-being of animals in research settings. Having a bond with companion animals may prime people to worry about welfare in other species and feel a greater sense of moral obligation to animals and increased sensitivity to well-being issues.

Résumé

Les liens entre les humains et leurs animaux de compagnie se sont considérablement renforcés au cours des cinquante dernières années, les chiens et les chats étant de plus en plus considérés comme des membres à part entière de la famille. Cette évolution de la relation a accru l'attention portée à l'utilisation des animaux de compagnie en recherche, alors même qu'augmentent les besoins de données visant à améliorer la santé, le bien-être et l'environnement propres à chaque espèce. Malgré cela, la recherche sur la santé et le bien-être des chats en croissance est très en retard par rapport aux études portant sur les humains et les animaux de ferme. Pour mener ces recherches, il est non seulement nécessaire de respecter les lois et normes éthiques en vigueur sur l'utilisation des animaux de compagnie, mais aussi d'être bien au fait des valeurs sociétales qui ne cessent d'évoluer. Cet article examine les plans de recherche efficaces qui comportent des études sur les félins mettant l'accent sur les besoins environnementaux spécifiques à l'espèce et qui visent à améliorer leur bien-être en recherche. À mesure que notre compréhension des besoins et de la complexité comportementale des animaux s'approfondit, les perceptions sociétales et les obligations éthiques relatives au bien-être animal évoluent, générant souvent des débats publics. Cette analyse met en avant la nécessité de concilier les avancées scientifiques avec les impératifs éthiques, afin d'assurer le bien-être des animaux utilisés dans les recherches. Le lien établi avec les animaux de compagnie peut amener les individus à s'intéresser davantage au bien-être des autres espèces, à éprouver un sentiment accru d'obligation morale envers les animaux et à être plus attentifs aux enjeux de leur bien-être.

Introduction

The relationship between humans and their companion animals has evolved significantly over the past half-century. Historically, cats and dogs were primarily working animals, serving roles in hunting, herding, and pest control. However, their status has changed dramatically, with many now regarded as integral family members. This deepening bond has led to increased public scrutiny of the use of companion animals in scientific research, and unsurprisingly, has led to calls for greater oversight of studies utilizing them by animal care and use committees. Consequently, maintaining kennels of dogs and cats for research using current guidance for best practices is extremely resource intensive. Nonetheless, given the growing demand for species-specific data to improve understanding of cats and their needs, including the effects genetics, health management, and environmental aspects of their care necessitates more targeted research efforts and careful consideration of the interests of the animals and procedures researchers plan to use in their investigations.

Current Gaps in Feline Research

Despite their popularity as companion animals, the body of research on domestic cats and their health and welfare needs lags behind what exists for humans, dogs, and agricultural animals. Much of the available data on companion cats is extrapolated from other species, which may be inappropriate in some instances, as cats have idiosyncrasies of metabolism and behaviour due to their carnivorous and crepuscular nature. For example, understanding dietary needs for growth and behavioral development in growing cats requires dedicated studies that account for species-specific variables. However, conducting this kind of research requires consideration of their welfare needs and public expectations relative to the acceptability of the conditions and procedures to which cats might be subjected. One way in which we have attempted to address both while facilitating the research conduct is to carefully incorporate accommodations of cats' physical health, behavioral and psychological needs into the design of the housing, test facilities, and data collection practices. For example, in our recent redesign of our facilities that house cats we ensured that their housing not only complied with best-practice guidance for requirements such as sanitization, cleaning and decontamination, but also offered cats opportunities to hide, elevate themselves as needed for safety using dedicated vertical spaces, and to access water from multiple different types of waterers. In addition, we provided multiple scratching posts, different types of litter boxes, and a variety of different textures to explore. Because our laboratory largely focuses on nutrition and metabolism, we also needed the ability to separate cats for individual feeding and we also acclimate them to having floor restriction and associate that restriction with positive experiences. We also chose to install 24-7 video surveillance that reduces the unnecessary disturbance of or distress in cats that can result when unfamiliar people must enter the colony for animal care checks. Further, cats receive social enrichment by familiar caretakers, who are also members of the research team, and low-stress handling to facilitate test procedures. These interventions help create conditions for cats, and the researchers, to experience positive affective states, and they afford additional opportunities to monitor the cats' physical and behavioral states. These up-front considerations will help us maintain this colony for the planned research over the next 5-10 years. They also position us to help reduce distress associated with transitioning cats to homes when adopted after the research is completed and all these steps facilitate successful rehoming outcomes.

Regulations and Ethical Considerations

Research involving companion animals is governed by an evolving set of laws and ethical guidelines. In many countries, regulations exist to ensure humane treatment, including guidelines for housing, medical care, and minimizing distress. However, the ethical concerns surrounding research with domestic cats extend beyond legal frameworks to societal expectations of ethical care and use. Public perception continues to shape notions and practices relating to the acceptability of research practices, making it imperative that studies prioritize animal welfare if they hope to retain a social license to be permitted. Researchers must therefore be aware of this and integrate ethical considerations into their methodologies to maintain public trust and adhere to evolving legal standards. Thus, throughout the cats' lives while they are in the research colony, we consider their value to society, the potential harms incurred to them by the research relative to the value of the information we are generating, and the quality of their lives while in our colony. Correspondingly, we use only the number minimally required to detect statistically significant differences among means, and we proactively and continuously assess the cats' behaviour and health. For any experiencing undue distress, we make decisions about removing individuals from the study and the colony and in our case adopt them out. But pointedly, we consider them as individuals, not just as groups of animals, as we start each study.

Designing Welfare-Oriented Feline Research Plans

While members of society may be apprehensive about research that uses cats, controlled research colonies are needed in basic studies where management is standardized, and the effect of the treatment is more easily isolated. To achieve this controlled population, we use purpose-bred cats whose entire life history is known. This allows us to eliminate differences in environmental variables that exist in other types of studies, such as epidemiological studies or those conducted with client-owned animals. To address both ethical concerns and scientific needs, research plans must be designed with feline welfare at their core. Ethical research practices include ensuring that animals have as much consistency, predictability and control as possible. To that end we recommend minimizing invasive procedures, providing enriched living environments (enrichment is a powerful tool that helps give animals a sense of control or agency), utilizing low-stress handling, and keeping their environments, schedules, caretakers and other personnel interacting with them consistent and predictable, while prioritizing non-harmful observational studies whenever possible. Designing studies that accommodate the unique physical and psychological requirements of cats will contribute to more humane and effective research outcomes that have greater real-world applicability.

Evolving Animal Welfare Obligations

Advancements in animal behavior research have highlighted the complexity of domestic feline needs, leading to a shift in societal expectations regarding their care. Increased awareness of feline cognition, emotional capacity and individuality dictates greater attention to using moral deliberation and ethical research practices, so at each step of our experimental design process, we seek to minimize procedures that may cause stress in cats, while working to increase opportunities for them to experience positive states using the interventions previously described. Regarding minimizing distress, the individual cat is discussed prior to blood collection procedures to decide on the most appropriate method of restraint and a plan is made that can be tailored to that individual. For example, we determine what methods will provide an optimal level of restraint, avoiding aversive approaches restraint such as aggressive scuffing, and consider alternatives that do not interfere with or confound the data being collected (as sedation might do), instead, considering when it is useful to simply begin by securely wrapping the cat in a towel. It bears reiterating that while research is essential for advancing veterinary medicine and improving feline health, it must be conducted with a commitment to minimizing harm. Balancing these priorities ensures that research findings are both scientifically valuable and ethically sound.

Conclusion and Future Directions

The study of feline health and welfare is a critical but underdeveloped area of scientific research. Addressing current knowledge gaps requires a collaborative approach that includes scientists, policymakers, veterinarians, and the public. Ethical research methodologies should be continuously refined to align with evolving societal expectations. Future research should prioritize species-specific studies that respect the welfare needs and interests of research animals and consider how these might be compromised and/or supported while in pursuit of studies that have the potential to contribute valuable insights to feline physiology, metabolism, behaviour, and overall health. By integrating ethical considerations with scientific progress, researchers can foster a more humane and informed approach to feline research, ensuring that both our scientific and moral obligations are met.

Maximizing the utilization of wheat straw with canola or flax screenings supplementation in finishing beef cattle diets: Growth performance, carcass characteristics and economic analysis

Maximiser l'utilisation de la paille de blé avec une supplémentation en criblures de canola ou de lin dans les rations des bovins de finition : performances de croissance, caractéristiques de carcasse et analyse économique

B.J. Montenegro[1], G. B. Penner[2], H.A. Lardner[2], K. Larson[3], J. J. McKinnon[4], D. Gibb[5], T. A. McAllister[6], G. O. Ribeiro[2]

[1] MSc. Student, Department of Animal and Poultry Science, University of Saskatchewan, Saskatoon, SK

[2] Professor, Department of Animal and Poultry Science, University of Saskatchewan, Saskatoon, SK

[3] Professional Research Associate, Department of Agricultural and Resource Economics, University of Saskatchewan, Saskatoon, SK

[4] Professor Emeritus, Department of Animal and Poultry Science, University of Saskatchewan, Saskatoon, SK

[5] Telus Agriculture & Consumer Goods, Okotoks, AB

[6] Lethbridge Research and Development Center, Agriculture and Agrifood Canada, Lethbridge, AB

SPONSORED BY / COMMANDITÉ PAR

Animal Nutrition Association of Canada

ANAC

Association de nutrition animale du Canada

Abstract

Incorporation of crop residues and byproducts in beef cattle diets has become prevalent due to feed shortages, increased feed prices, and disrupted supply chains. This study evaluated growth performance, carcass characteristics, and economic outcomes of finishing feedlot cattle fed wheat straw diets supplemented with canola or flax screenings compared to a standard western Canadian finishing diet. In a completely randomized design, 300 steers (366 ± 32 kg) were stratified by weight and randomly allocated to 20 pens (15 steers/pen). Each pen was randomly assigned to one of five treatments (n = 4/treatment): control (CTL); low straw canola (LSC); low straw flax (LSF); high straw canola (HSC); and high straw flax (HSF). The CTL diet consisted of conventional finishing ingredients (10% barley silage:90% concentrate, dry matter [DM] basis). Low and high straw diets included 5% and 10% wheat straw respectively, with screenings at 12.50% inclusion (DM basis). Steers were fed for 148 d. Dry matter intake was not different among treatments. Final BW, average daily gain (ADG), and hot carcass weight (HCW) declined ($P<0.05$) with increasing wheat straw inclusion. Gain:feed tended ($P<0.10$) to decrease as wheat straw inclusion increased. No significant differences were observed for dressing percentage, backfat thickness, lean meat yield, marbling score, empty body fat, liver scores, or cost:gain. The proportion of AAA carcasses was highest for LSF and decreased with increasing wheat straw inclusion ($P<0.02$). Despite some negative effects at higher straw inclusion rates, the overall performance of steers fed diets with lower wheat straw inclusion (5%) was comparable to that of controls, suggesting that low inclusion levels of wheat straw with oilseed screenings can be viable during feed shortages.

Résumé

L'incorporation de résidus et de sous-produits de culture dans l'alimentation des bovins de boucherie est devenue courante en raison des pénuries d'aliments, de l'augmentation des prix des aliments et de la perturbation des chaînes d'approvisionnement. Cette étude avait pour objectif d’évaluer les performances de croissance, les caractéristiques de carcasse et les résultats économiques associés à des bovins en parc d'engraissement nourris avec de la paille de blé additionnée de criblures de canola ou de lin, comparativement à un régime de finition standard utilisé dans l'Ouest canadien. Dans le cadre d’un plan expérimental complètement aléatoire, 300 bouvillons (366 ± 32 kg) ont été groupés par poids et répartis au hasard dans 20 enclos (15 bouvillons/enclos). Chaque enclos a été assigné au hasard à un des cinq traitements (n = 4/traitement) : témoin (CTL); faible teneur en paille + canola (LSC); faible teneur en paille + lin (LSF); forte teneur en paille + canola (HSC); et forte teneur en paille + lin (HSF). Le régime témoin était composé d'ingrédients de finition classiques (10 % d'ensilage d'orge : 90 % de concentré, sur une base de la matière sèche [MS]). Les régimes à faible et à forte teneur en paille comprenaient respectivement 5 % et 10 % de paille de blé et 12,50 % de criblures (base MS). Les bouvillons ont été nourris pendant 148 jours. Aucune différence de consommation volontaire de matière sèche n’a été observée entre les traitements. Le poids corporel final, le gain moyen quotidien (GMQ) et le poids de carcasse chaude (PCC) ont diminué ($P<0,05$) avec l'augmentation de la teneur en paille de blé. Le rapport gain/aliment a eu tendance à diminuer ($P<0,10$) avec l'augmentation de la teneur en paille de blé. Aucune différence significative n'a été observée pour le rendement de carcasse, l'épaisseur de gras dorsal, le rendement en viande maigre, le persillage, le gras total de la carcasse, l’évaluation hépatique ni le rapport coût/gain. La proportion de carcasses AAA a été plus élevée avec la LSF et a diminué avec l'augmentation de la teneur en paille de blé ($P<0,02$). Malgré certains effets négatifs observés avec les teneurs en paille plus élevées, la performance globale des bouvillons nourris avec les rations les plus pauvres en paille de blé (5 %) a été comparable à celle obtenue chez les témoins, ce qui suggère qu'une ration à faible teneur en paille de blé conjuguée à des criblures d'oléagineux peut être viable en cas de pénurie d'aliments.

Introduction

Climate change is increasingly affecting weather patterns resulting in erratic weather events and extended droughts. In 2023, most of western Canada had been classified as abnormally dry or undergoing a moderate or exceptional drought (Agriculture and Agrifood Canada, 2023). Resulting feed shortages have challenged producers to find alternative feeds to conventional ingredients that are locally available, economical, nutritious and that can be implemented in backgrounding and finishing feedlot operations. Utilization of low-quality crop residues or crop byproducts, such as wheat straw and oilseed screenings, in feedlot diets provide an opportunity to reduce agricultural waste through the upcycling of feedstuffs that would have otherwise been landfilled to produce high quality animal protein (Ominski et al., 2021a). However, their inclusion in finishing feedlot diets remains a challenge due to their low nutrient content, in the case of wheat straw (McCartney et al., 2006), and inconsistent nutrient composition, in the case of oilseed screenings (Ominski et al., 2021b). Despite this, increasing utilization of low-quality roughage in feedlot diets may be achieved with oilseed screening supplementation due to their high energy and protein content thereby improving wheat straw utilization. Such alternative feeding strategies can provide Canadian beef producers with options that can be implemented during times of low conventional feed ingredient availability.

Materials and Methods

This study was conducted at the University of Saskatchewan Livestock and Forage Centre of Excellence in Clavet, Saskatchewan. Three hundred steers (initial body weight [BW]: 366 ± 32 kg) were stratified by BW and randomly allocated to 20 pens (15 steers/pen). Each pen was randomly assigned to a treatment (n = 4/treatment; Table 1). The diet treatments contained (DM basis): 1) 85.64% DRB, 10.00% barley silage (CTL); 2) 78.14% DRB, 12.50% canola screenings, 5.00% WS, (LSC); 3) 78.14% DRB, 12.50% flax screenings, 5.00% WS (LSF); 4) 73.14% DRB, 12.50% canola screenings, 10.00% WS (HSC); and 5) 73.14% DRB, 12.50% flax screenings, 10.00% WS (HSF). Supplement (4.36% of dietary DM) was included in all diets. Steers were fed for a total of 148 d divided into 5 periods with live weights, feed refusals and fecal samples obtained at the end of every period. Amount of feed offered was tracked daily and weekly as-fed total mixed ration (TMR) and diet ingredient samples were obtained and used to determine dry matter intake (DMI). Steers were humanely harvested at the end of the trial and carcass parameters were obtained from the abattoir. Liver scores were recorded at the time of harvesting using the Elanco liver score system. Statistical analysis was performed using SAS 9.4 using a completely randomized model with pen as the experimental unit and diets as the fixed effect. Quantitative live performance and carcass data was analysed using the MIXED procedure and qualitative carcass data was analyzed using the GLIMMIX procedure. The sums of squares were further partitioned by orthogonal contrasts to analyze differences between specific diet treatments with significance was at $P<0.05$ and trends established at $0.05 < P < 0.10$.

Results and Discussion

Despite the higher levels of NDF and $peNDF_4$ in the wheat straw diets compared to CTL, there were no differences ($P \geq 0.25$) in DMI expressed in kg/d or as a % of BW among the treatment diets (Table 2). High wheat straw diets (10% of diet DM, HS) promoted lower ($P<0.01$) average daily gain (ADG), gain:feed (G:F), and shrunk final body weight (FBW) when compared to the CTL. Diets with low wheat straw (5% of diet DM, LS) inclusion also promoted lower ($P \leq 0.05$) ADG, and G:F than CTL but shrunk final BW was not different ($P=0.27$). The screenings source (canola vs. flax) did not affect ($P \geq 0.27$) any of the growth performance parameters measured.

Similar trends were observed for carcass characteristics with hot carcass weight (HCW) and AAA quality grade proportions being significantly lower ($P<0.05$) for HS diets compared to CTL (Table 3). There were tendencies ($P \leq 0.10$) for marbling score, yield grade 1 (Y1), yield grade 2 (Y2) and AA quality grade proportions to be lower for HS diets compared to the CTL. There were no significant differences ($P \geq 0.34$) between LS diets and CTL in any carcass parameters measured. Lastly, only quality grades were affected by the source of screenings in the diet, with flax inclusion decreasing ($P=0.04$) the proportions AA carcasses and promoting a tendency ($P=0.07$) to increase the proportion of AAA quality grades when compared to canola.

The feed cost per head per day was affected by the price of wheat straw and the level of inclusion, with CTL diet having the highest feed cost per head per day and the cost decreasing as level of what straw inclusion increased ($P<0.05$). Despite these differences, the cost of gain (C:G) across diets did not significantly differ with even HS inclusion diets resulting in similar C:G to the CTL diet. Although these results are encouraging, it's important to note that the economic performance of each diet is obviously dependent on ingredient price and did not take into consideration yardage or transportation costs for some of these ingredients as these costs are highly variable across operations and areas.

Conclusion

Although challenges remain with the inclusion of low-quality forages in finishing feedlot diets, the present study presents strategies to increase their inclusion through supplementation with high-protein crop byproducts, such as oilseed screenings. Although some negative effects were observed at HS inclusion levels, the inclusion of wheat straw at 5% of DM (LS) in the feedlot diets resulted in comparable live and carcass performance to that of the CTL diet. Moreover, the C:G of LS inclusion diets did not significantly differ from that of the CTL diet thereby providing a viable strategy that producers can utilize during times of conventional feed shortages, drought, or high conventional feed prices. Although this study specifically assessed only non-processed whole oilseed screenings to reduce diet costs, more research is needed to determine if oilseed processing, and therefore improved digestibility, would result in improved performance which could justify the additional cost of processing.

Table 1. Ingredients, chemical composition, and physical characteristics of experimental diets (mean ± SD).

Item	Control (CTL)	5% Wheat Straw		10% Wheat Straw	
		Canola (LSC)	Flax (LSF)	Canola (HSC)	Flax (HSF)
Diet Ingredient, % of DM [1]					
Barley grain	85.64	78.14	78.14	73.14	73.14
Barley silage	10.00	0.00	0.00	0.00	0.00
Wheat straw	0.00	5.00	5.00	10.00	10.00
Canola screenings	0.00	12.50	0.00	12.50	0.00
Flax screenings	0.00	0.00	12.50	0.00	12.50
Canola meal	2.50	2.50	2.50	2.50	2.50
Urea	0.25	0.25	0.25	0.25	0.25
Salt, white	0.14	0.14	0.14	0.14	0.14
Limestone	1.43	1.43	1.43	1.43	1.43
Supplement[2]	0.04	0.04	0.04	0.04	0.04
Nutritional composition [3]					
DM, %	73.8 ± 1.79	78.6 ± 3.29	78.3 ± 3.54	78.0 ± 3.00	78.1 ± 3.41
ADF [1], % of DM	8.3 ± 0.38	11.1 ± 0.40	10.1 ± 0.59	13.9 ± 0.41	12.5 ± 1.08
aNDFom [4], % of DM	16.9 ± 0.55	20.7 ± 0.39	20.5 ± 0.67	24.0 ± 0.68	22.3 ± 1.45
CP [1], % of DM	13.4 ± 0.49	15.1 ± 0.16	15.0 ± 0.25	14.7 ± 0.16	14.5 ± 0.78
Starch, % of DM	53.3 ± 0.65	44.3 ± 0.29	46.6 ± 0.40	41.2 ± 0.54	44.6 ± 2.52
Calcium, % of DM	0.88 ± 0.202	0.93 ± 0.053	0.65 ± 0.192	0.94 ± 0.062	1.03 ± 0.230
Phosphorous, % of DM	0.31 ± 0.019	0.36 ± 0.012	0.30 ± 0.073	0.36 ± 0.005	0.35 ± 0.041
NE_m [1], Mcal/kg	2.16 ± 0.035	2.11 ± 0.010	2.18 ± 0.021	2.05 ± 0.018	2.13 ± 0.026
NE_g [1], Mcal/kg	1.48 ± 0.035	1.44 ± 0.010	1.49 ± 0.021	1.39 ± 0.018	1.46 ± 0.018
Physical characteristics [3]					
$peNDF_4$ [5]	7.22 ± 0.263	7.71 ± 0.266	8.19 ± 0.108	9.98 ± 0.448	10.64 ± 0.694
$peNDF_8$ [6]	0.21 ± 0.027	0.98 ± 0.054	1.00 ± 0.051	1.90 ± 0.212	2.43 ± 0.322
Physical effectiveness factor [3,7], %					
>19mm	0.13 ± 0.074	1.33 ± 0.151	1.26 ± 0.116	2.20 ± 0.416	3.64 ± 0.844
< 19mm > 8mm	1.12 ± 0.118	3.39 ± 0.110	3.63 ± 0.132	5.70 ± 0.519	7.27 ± 0.603
<8 mm > 4.0 mm	41.49 ± 1.427	32.52 ± 1.511	35.07 ± 0.516	33.69 ± 1.090	36.81 ± 1.717
< 4.0 mm	57.25 ± 1.553	62.76 ± 1.285	60.04 ± 0.525	58.41 ± 1.867	52.27 ± 3.110

[1] ADF, acid detergent fibre; CP, crude protein, DM, dry matter; NE_g, net energy for gain; NE_m, net energy for maintenance.

[2] Supplement contained Monensin: 33 mg/kg and Lubabegron: 3.5mg/kg (Elanco Animal Health, Greenfield, IN).

[3] Diet nutrient composition, physical characteristics and physical effectiveness factor expressed as means ± SD.

[4] Neutral detergent fibre (NDF) corrected for starch, using α-amylase, and ash content.

[5] Physically effective NDF determined by multiplying aNDFom content by proportions retained on 19, 8 and 4mm screens of the Penn State Particle Separator.

[6] Physically effective NDF determined by multiplying aNDFom content by proportions retained on the 19 and 8mm screen of the Penn State Particle Separator.

[7] Determined using a Penn State Particle Separator using 19, 8, and 4mm screens.

Table 2. Summary of live performance parameters for each treatment diet[1].

Item	Treatments[1]						*P*-value[4]			
	CTL	LSC	LSF	HSC	HSF	SEM[2]	CTL vs. LS	CTL vs. HS	LS vs. HS	C vs. F
	366	366	366	367	366	1.58	0.69	0.53	0.79	0.79
Final BW, kg	663	651	663	648	644	4.29	0.27	<0.01	0.02	0.39
DMI[3], kg/d	11.59	11.59	11.91	11.68	11.52	0.158	0.42	0.98	0.34	0.63
DMI, % BW	2.30	2.31	2.36	2.35	2.33	0.031	0.37	0.25	0.75	0.63
ADG[3], kg	2.02	1.91	1.97	1.91	1.90	0.025	0.02	<0.01	0.20	0.27
Gain:Feed	0.173	0.164	0.168	0.162	0.162	0.0027	0.05	<0.01	0.16	0.48

[1] **CTL**, control diet; **HSC**, 10% straw with canola screenings; **HSF**, 10% straw with flax screenings; **LSC**, 5% straw with canola screenings; **LSF**, 5% straw with flax screenings.

[2] Treatment means were compared using the LSMEANS procedure adjusted for the Tukey-Kramer method. SEM, standard error of means reported is for least square means analysis.

[3] **ADG**, average daily gain; **BW**, body weight; **DMI**, dry matter intake.

[4] Orthogonal contrasts of treatment diets. **CTL**, control; **LS**, low straw inclusion diets (LSC and LSF); **HS**, high straw inclusion diets (HSC and HSF); **C**, diets containing canola screenings (LSC and HSC); **F**, diets containing flax screenings (LSF and HSF).

Table 3. Summary of carcass performance for each treatment diet[1].

Item	Treatments[1]						*P*-value[5]			
	CTL	LSC	LSF	HSC	HSF	SEM[2]	CTL vs. LS	CTL vs. HS	LS vs. HS	C vs. F
	399.71	396.11	400.63	388.78	387.12	3.32	0.75	0.01	<0.01	0.67
Dressing %	60.29	60.81	60.44	60.01	60.08	0.28	0.34	0.47	0.05	0.59
REA[3], cm²	100.53	102.14	101.83	98.99	95.78	1.48	0.44	0.10	<0.01	0.25
Marbling score	385.51	380.02	394.18	366.66	367.39	7.91	0.87	0.08	0.03	0.36
Yield Score[4]										
Y1	55.17	40.00	57.63	43.10	39.29	-	0.36	0.07	0.30	0.36
Y2	37.93	55.00	37.29	53.45	57.14	-	0.37	0.05	0.13	0.23
Y3	6.90	5.00	5.08	3.45	3.57	-	0.62	0.33	0.57	0.97
Quality Grade[4]										
AAA	51.72	36.67	57.63	32.76	37.5	-	0.47	0.03	0.10	0.07
AA	48.28	63.33	42.37	67.24	62.5	-	0.67	0.05	0.05	0.04

[1] **CTL**, control diet; **HSC**, 10% straw with canola screenings; **HSF**, 10% straw with flax screenings; **LSC**, 5% straw with canola screenings; **LSF**, 5% straw with flax screenings.

[2] Treatment means were compared using LSMEANS procedure adjusted for the Tukey-Kramer method. SEM, standard error of means reported is for least square means analysis.

[3] **HCW**, hot carcass weight; **REA**, rib eye area.

[4] Yield grade scores and quality grades were assigned according to the current Canadian Beef Grading Agency standards.

[5] Orthogonal contrasts of treatment diets. **CTL**, control; **LS**, low straw inclusion diets (LSC and LSF); **HS**, high straw inclusion diets (HSC and HSF); **C**, diets containing canola screenings (LSC and HSC); **F**, diets containing flax screenings (LSF and HSF).

Table 4. Summary of economic performance for each treatment diet[1].

Item	Treatments[1]						*P*-value[4]			
	CTL	LSC	LSF	HSC	HSF	SEM[2]	CTL vs. LS	CTL vs. HS	LS vs. HS	C vs. F
Feed Cost/Hd/d[3], CAD	4.29	4.16	4.26	4.05	3.99	0.059	0.26	<0.01	<0.01	0.71
Cost:Gain, CAD/kg	1.71	1.77	1.72	1.74	1.73	0.028	0.35	0.42	0.87	0.33

[1] **CTL**, control diet; **HSC**, 10% straw with canola screenings; **HSF**, 10% straw with flax screenings; **LSC**, 5% straw with canola screenings; **LSF**, 5% straw with flax screenings.

[2] SEM, standard error of means reported is for least square means analysis.

[3] Feed cost per head per day calculated based on measured dry matter intake.

[4] Orthogonal contrasts of treatment diets. **CTL**, control; **LS**, low straw inclusion diets (LSC and LSF); **HS**, high straw inclusion diets (HSC and HSF); **C**, diets containing canola screenings (LSC and HSC); **F**, diets containing flax screenings (LSF and HSF).

References

Agriculture and Agri-Food Canada. (2023, October 12). *Current drought conditions.*

Ominski, K., Gunte, K., Wittenberg, K., Legesse, G., Mengistu, G., and McAllister, T. (2021a). The role of livestock in sustainable food production systems in Canada. *Canadian Journal of Animal Science, 101*(4), 591–601. https://doi.org/10.1139/cjas-2021-0005

Ominski, K., McAllister, T., Stanford, K., Mengistu, G., Kebebe, E. G., Omonijo, F., Cordeiro, M., Legesse, G., and Wittenberg, K. (2021b). Utilization of by-products and food waste in livestock production systems: a Canadian perspective. *Animal Frontiers, 11*(2), 55-63. https://doi.org/10.1093/af/vfab004

McCartney, D. H., Block, H. C., Dubeski, P. L., and Ohama, A. J. (2006). Review: The composition and availability of straw and chaff from small grain cereals for beef cattle in western Canada. *Canadian Journal of Animal Science, 86*(4), 443–455. https://doi.org/10.4141/A05-092

Use of machine learning in the modern commercial feed mill

Utilisation de l'apprentissage automatique dans les meuneries commerciales modernes

J.L. Ellis[1], D. Tulpan[1], C. Krziyzek[2], M. Capozzalo[3], M. Malpass[4], M. Martin[5], and J. You[1]
[1] Department of Animal Biosciences, University of Guelph, Guelph, ON
[2] Quality Assurance, Trouw Nutrition Canada Inc., Puslinch, ON
[3] Department of Nutrition and Production Research, Molesworth Farm Supply Ltd, Molesworth, ON
[4] Nutrition & Innovation, Masterfeeds, an Alltech Company, London, ON
[5] Operations, Jones Feed Mills Ltd., Linwood, ON
Corresponding Author: jellis@uoguelph.ca

SPONSORED BY / COMMANDITÉ PAR

Diamond V

Abstract

With the arrival of the 4th Agriculture Revolution (Agriculture 4.0) and its envisioned fusion of farming with emerging digital technologies and automation, new opportunities have arisen to increase the efficiency and precision with which we feed animals. The modern feed mill is a 'big data generator' which until recently, has been largely untapped in terms of systems considered for process control and product optimization. The barriers of the past have been a lack of automation and efficient means of data capture given the multitude of variables in a commercial mill, as well as appropriate analytical tools with which to drive decision support and system optimization in such a complex environment. However, feed mills may benefit from data driven assistance with process and efficiency optimization, minimization of CO_2 emissions given global concerns about greenhouse gas emissions, as well as support for product optimization, such as ensuring pellet quality, which has substantial downstream impacts on animal efficiency and performance. Implementation of automated data capture and machine learning to analyze the data produced at the commercial feed mill therefore has the potential to improve all aspects of feed production from ingredient profiling to efficiency and speed optimization to ensuring adequate product quality.

This research group has implemented, for the first time, machine learning based predictions of pellet quality and mill efficiency based on the automated capture of manufacturing conditions, environmental conditions and formulation data across multiple mills, and provides an example of what is now possible.

Résumé

Avec l'arrivée de la quatrième révolution agricole (l'agriculture 4.0) et de l'intégration proposée des technologies numériques émergentes et de l'automatisation dans ce secteur d'activité, de nouvelles possibilités sont apparues pour améliorer l'efficacité et la précision de l'alimentation des animaux. La meunerie moderne est devenue une source importante de mégadonnées qui, jusqu'à récemment, étaient largement inexploitées pour le contrôle des processus et l'optimisation des produits. Les difficultés précédemment rencontrées incluaient l'absence de systèmes automatisés et de méthodes efficaces pour la collecte de données, en raison des nombreuses variables présentes dans une meunerie commerciale. De plus, le manque d'outils analytiques adéquats limitait la prise de décision et compromettait l'optimisation du système dans un environnement complexe. Pourtant, les données des usines de fabrication d'aliments pour animaux peuvent servir à améliorer les processus et l'efficacité, réduire les émissions de CO2 en réponse aux préoccupations mondiales concernant les gaz à effet de serre et optimiser les produits pour, par exemple, assurer la qualité des granulés, un élément qui influence directement l'efficacité et la performance des animaux. L'intégration de la saisie automatisée des données et de l'apprentissage automatique pour analyser les informations générées dans les meuneries commerciales présente un potentiel significatif pour améliorer tous les aspects de la production d'aliments pour animaux. Cela inclut le profilage des ingrédients, l'optimisation de l'efficacité et de la rapidité des processus ainsi que l'assurance d'une qualité de produit conforme aux standards requis. Pour la première fois, ce groupe de recherche a utilisé des prédictions basées sur l'apprentissage automatique pour évaluer la qualité des granulés et l'efficacité de la meunerie en se basant sur la saisie automatisée des conditions de fabrication, des conditions environnementales et des données de formulation dans plusieurs meuneries. Cela montre les nouvelles possibilités offertes par cette approche.

Introduction

The commercial feed mill industry is undergoing a transformation driven by advancements in data science, automation, and machine learning (ML). Historically, feed mills have relied on traditional process control techniques and empirical knowledge of operators to optimize production efficiency and product quality. However, with the increasing complexity of feed formulations, environmental variability, and manufacturing processes, conventional approaches of knowledge apprehension (Abdollahi et al., 2013; Briggs et al., 1999) are often inadequate for real-time decision-making and process optimization.

The emergence of Agriculture 4.0 - a paradigm shift that integrates digital technologies, automation, and artificial intelligence into agricultural processes - presents new opportunities for feed mills to harness data-driven solutions (Wolfert et al., 2017). Modern feed mills generate vast amounts of data related to ingredient composition, environmental conditions, and manufacturing parameters. Until recently, the lack of automated data capture and advanced analytical tools limited the industry's ability to leverage this information effectively. However, recent advancements in ML offer promising solutions for improving operational efficiency, reducing waste, and enhancing feed quality (You et al., 2024).

Machine learning enables predictive modeling of critical feed mill parameters, such as pellet durability index (PDI) and production efficiency, by identifying complex patterns and relationships within large datasets. Unlike traditional regression-based models, ML algorithms can adapt to non-linear interactions and optimize predictions with minimal human intervention (Mjolsness and DeCoste, 2001). This capability is particularly valuable in commercial feed mills, where numerous factors - including ingredient variability, processing conditions, and environmental influences - interact in intricate ways to affect product quality and throughput.

In this paper and associated talk, we explore the application of models in the modern commercial feed mill, focusing on the potential for optimizing pellet quality and mill efficiency. We present novel findings from our research group, which implemented varying strategies to predict pellet quality using automated data captured across multiple feed mills. This work exemplifies the potential of data-driven approaches (in particular, ML) to enhance feed production processes, reduce variability, and improve overall sustainability in the industry.

Predicting Pellet Quality using equations developed from a Meta-Analysis of Published Studies

A classical approach to tackling a problem like predicting pellet quality might be the development of empirical prediction equations based off a number of driving variables identified through previous scientific knowledge as impactful. To explore this approach, a meta-analysis conducted by (You et al., 2025a) in 2024 identified 29 publications that met inclusion criteria (measured pellet quality, as pellet durability index (PDI), corn-based feeds formulated for monogastric species (excluding fish), reported impacts of nutrition or manufacturing condition changes, and used the ASAE method for PDI testing), and using a mixed model approach, treating study as a random effect, developed a number of univariable and multivariable PDI prediction equations. Variables identified that yielded the best model predictions (as determined by a 3-fold cross-validation) across studies included steam temperature as a manufacturing parameter, while ingredients of importance included the inclusion level of corn and soybean meal, and nutrients of importance included the percent lipid and protein in the formulation. The overall best performing equation for the prediction of PDI across studies included the variables protein (positive impact) and lipid (negative impact), with a CCC (concordance correlation coefficient) of 0.29 (-1 to +1 scale) and rMSPE (root mean square prediction error, % of observed mean) of 10.33% on data not used for model development. Such studies help to distill patterns from bodies of scientific evidence, and this equation may be useful to apply across mills, as it identifies the impact of major nutritional composition shifts on PDI and is simple to apply in practice in low-automation environments. However, the low CCC value (0.29) when tested on independent data indicates that there is much variation left unexplained by this simple model.

Predicting Pellet Quality using Statistical Approaches on Commercial Mill Data

While the meta-analysis approach is useful to understanding the major drivers of differences in PDI across studies, due to the limited number of variables able to be considered within each equation, it certainly lacks ability to pick up on the numerous additional drivers of variation in PDI in the mill. To investigate more advanced statistical modelling approaches on a commercial feed mill (Trouw Nutrition, St Mary's Plant #2), a database containing 2691 observations and 55 variables spanning nutrition/formulation, manufacturing as well as environmental conditions was collected by You et al. (2025c), and explored the use of multiple linear regression approaches with variable selection and dimensionality reduction to develop PDI prediction equations. Forward Selection, Principal Component Analysis and Partial Least Squares were investigated and models built/evaluated using a 5-fold cross-validation approach. The best model was the model developed via Forward Selection, and included 9 statistically significant variables including variables such as expanding temperature (°C), ADF content of the diet (%), the inclusion of bakery meal (%) in the diet and the humidity (%) inside the mill close to the pelletizer which all had positive impacts on predicted PDI, and fat content (%), ambient (outdoor) humidity (%) and the level of processing aid water (%) inclusion which all had negative impacts on predicted PDI. This model had a CCC of 0.55 and rMSPE of 2.66% on data not used for model development. Such approaches may be useful for moderately sized commercial datasets where greater predictive performance is desired, but the simplicity of a single multivariable equation is also desired. Indeed, although this study is not directly comparable to the meta-analysis study described above, the CCC is notably higher. There have been limited attempts in the literature to develop such models. Cavalcanti and Behnke (2005a, 2005b) developed empirical equations reflecting the relationship between PDI and a few nutrient factors, including starch, fat, and protein under experimental settings, and Schroeder et al., (2020), who reported statistical equations using variable selection based on a larger dataset available from a commercial mill.

Predicting Pellet Quality using Machine Learning on Commercial Mill Data

However, with complex systems such as the feed mill, with a high number of variables changing between runs (nutrition/formulation, manufacturing conditions, environmental conditions), other approaches, such as ML, may be a more appropriate fit. You et al., (2022) explored for the first time the use of ML for the prediction of PDI using a commercially available dataset. The hand-collected data included 2471 observations, and 16 features collected over an 8-month period spanning manufacturing parameters, ingredient inclusion levels and temperature/humidity from a regional weather station close to our research partner's commercial facility (Trouw Nutrition, St Mary's Plant #2). The research explored 12 different ML algorithms for the prediction of PDI and employed a 5-fold cross-validation approach on 80% of the data and an independent evaluation on the remaining 20% unseen data. You et al. (2022) identified the SVR algorithm as yielding the best PDI predictions, with a CCC of 0.64 and rMSPE of 4.80%, and feature importance analysis was used to identify influential driving features (variables). Influential features included the average outdoor temperature (°C), bakery byproduct and wheat inclusion levels (%), as well as production line, while expected features like fat added to the mixer were less important, presumably due to controls already in place in the mill to limit this variable. An important outcome of this research was the identification that hand-collection of data was not sustainable within the commercial setting, nor did it facilitate the capture of the real breadth of information captured by the mill. A moderate revision to the existing mill software (Engie) allowed for autonomous capture of higher amounts of data thereafter, allowing for exploration of a higher number of features in subsequent work. In addition, this research interestingly identified the impact of environmental outdoor temperature as important, which was curious. Although this impact was intuitively known by mill workers, supposedly through impacting ingredient temperature and air temperature pulled into the mill for cooling, it has yet to be reported in the scientific literature. To allow further exploration of this impact, temperature/humidity probes were placed throughout the mill in strategic locations, automatically capturing data every minute.

This automation of data capture allowed the database used by You et al., (2025c) for expanded statistical modelling, as well as additional exploration of PDI prediction using a higher number of features for ML. In You et al., (2025b), feature engineering was employed, and 74 new features were generated (such as the pairwise interaction between nutrient composition features, and the delta between temperature/humidity probes throughout the mill) on top of the 55 features already available, and machine learning models were built. Again, a 5-fold cross-validation (using 80% of the data) and independent evaluation (using 20% unseen data) was employed for model construction and evaluation. The SVR model was identified as the top performer for the prediction of PDI and independent evaluation resulted in a CCC of 0.61 and rMSPE of 2.56%. Based on SHAP values, features ranked as important for the prediction of PDI included bakery meal inclusion (%), ambient temperature (°C), diet crude protein*diet starch content (%), and conditioning temperature (°C)*diet ADF content (%).

Interestingly however, the expanded list of features did not improve PDI prediction over the previous ML study (You et al., 2022) (though the database was not the same), though it was a moderate improvement over the statistical models developed in You et al., (2025c). This suggests perhaps that the features available were not exhaustive in accounting for influential factors on PDI, and indeed, one important feature was missing from all of these studies, particle size, as it was not routinely measured, and variation in ingredient composition is also difficult to capture.

The Next Steps Forward

The application of models, and in particular ML, in commercial feed mills represents a significant step forward in optimizing feed production processes. By leveraging automated data capture and advanced predictive models, feed mills can achieve greater efficiency, minimize resource waste, and improve overall product consistency. These technologies enable real-time monitoring and adjustments, reducing reliance on traditional trial-and-error approaches and empowering decision-makers with actionable insights.

Despite the advantages, challenges remain in the widespread adoption of ML in feed mills. Data quality and consistency are critical factors that influence model performance. Ensuring accurate and comprehensive data collection through automated sensors and integration with existing mill management systems is essential for reliable predictions. Additionally, the interpretability of machine learning models must be considered, as stakeholders in the industry typically desire transparency in decision-making processes.

Future research should focus on further refining ML algorithms to enhance prediction accuracy and expand their applications beyond pellet quality and mill efficiency. Integration with Internet of Things (IoT) technologies and cloud computing could provide even greater data accessibility and processing power, allowing for real-time adaptive control of feed manufacturing processes. The further automation and digitization of the mill (for example, in-line NIRS) may also assist with improving predictions and automated decision making.

Current work by this group involves expanding PDI prediction models from a single mill to multiple mills from multiple feed companies and investigating whether the above identified relationships hold true across environments, with the overall goal of developing a generalized ML model for PDI prediction across mills. The research conducted by this group successfully 'opens up the feed mill' for further application of ML, as a prime environment for data-driven decision-making assistance on aspects of manufacturing such as formulation, mill efficiency, and CO2 emissions.

Conclusions

In conclusion, the adoption of machine learning in commercial feed mills holds great promise for transforming the industry. As technological advancements continue, feed mills that embrace data-driven decision-making will be better positioned to improve efficiency, reduce environmental impact, and enhance overall product quality. Continued collaboration between industry experts, data scientists, and technology providers will be crucial in realizing the full potential of these innovations.

References

Abdollahi, M.R., Ravindran, V., Svihus, B. 2013. Pelleting of broiler diets: An overview with emphasis on pellet quality and nutritional value. Anim. Feed Sci. Technol. 179, 1–23. https://doi.org/10.1016/j.anifeedsci.2012.10.011

Briggs, J.L., Maier, D.E., Watkins, B.A., Behnke, K.C. 1999. Effect of ingredients and processing parameters on pellet quality. Poult. Sci. 78, 1464–1471. https://doi.org/10.1093/ps/78.10.1464

Cavalcanti, W.B., Behnke, K.C. 2005a. Effect of composition of feed model systems on pellet quality: A mixture experimental approach. I. Cereal Chem. 82, 455–461. https://doi.org/10.1094/CC-82-0455

Cavalcanti, W.B., Behnke, K.C. 2005b. Effect of composition of feed model systems on pellet quality: A mixture experimental approach. II. Cereal Chem. 82, 462–467. https://doi.org/10.1094/CC-82-0462

Mjolsness, E., DeCoste, D. 2001. Machine learning for science: State of the art and future prospects. Science 293, 2051–2055. https://doi.org/10.1126/science.293.5537.2051

Schroeder, B., Andretta, I., Kipper, M., Franceschi, C.H., Remus, A. 2020. Empirical modelling the quality of pelleted feed for broilers and pigs. Anim. Feed Sci. Technol. 265, 114522. https://doi.org/10.1016/j.anifeedsci.2020.114522

Wolfert, S., Ge, L., Verdouw, C., Bogaardt, M.-J. 2017. Big Data in Smart Farming – A review. Agric. Syst. 153, 69–80. https://doi.org/10.1016/j.agsy.2017.01.023

You, J., Ellis, J.L., Tulpan, D., Malpass, M.C. 2024. Review: recent advances and future technologies in poultry feed manufacturing.

You, J., Hall, K., Civiero, J., Malpass, M.C., Tulpan, D., Ellis, J.L. 2025a. Pellet Durability Index (PDI) of pelleted feeds for corn-based monogastric feeds: A meta-analysis.

You, J., Tulpan, D., Kriziyzek, C., Ellis, J.L., 2025b. Pellet quality prediction using machine learning regression models and feature engineering.

You, J., Tulpan, D., Krziyzek, C., Ellis, J.L. 2025. Prediction of Pellet Durability Index (PDI) in a Commercial Feed Mill Using Multiple Linear Regression with Variable Selection and Dimensionality Reduction. J. Anim. Sci. skaf021. https://doi.org/10.1093/jas/skaf021

You, J., Tulpan, D., Malpass, M.C., Ellis, J.L. 2022. Using machine learning regression models to predict the pellet quality of pelleted feeds. Anim. Feed Sci. Technol. 293, 115443. https://doi.org/10.1016/j.anifeedsci.2022.115443

Sustained impacts of sire nutrition on offspring via paternal programming

Effets durables de l'alimentation du père sur la progéniture par le biais de la programmation paternelle

C. R. Dahlen[1]

[1] *Professor of Reproductive Physiology, Center for Nutrition and Pregnancy and Department of Animal Sciences, North Dakota State University, FARGO, ND 58108,* Carl.Dahlen@ndsu.edu

Abstract

Developmental programming is the concept wherein the experiences of parents before and during pregnancy have impacts on the offspring that can persist throughout their lives. The body of research demonstrating altered offspring outcomes as a result of maternal nutrition is relatively mature and ever-expanding. The concept of paternal programming, however, is a more recent development. Evidence from human epidemiology and research in rodent species has revealed that paternal nutrition, environment, stress, and substance abuse during the time of spermatogenesis can have effects on the offspring subsequently conceived. Epigenetic alterations (DNA methylation, histone modification, and non-coding RNA), oxidative stress, immune response, and microbial transfer have been indicated as potential mechanisms of transferring sire-related signals to the oocyte at fertilization and/or to the uterine environment during early pregnancy. In rodents, paternal nutrition and environment have been demonstrated to impact offspring body weight, fat and muscle deposition, reproductive traits, and metabolic responses. If translated to livestock species these impacts could have major economic importance. Though the information is very limited, early work in livestock species has demonstrated paternal programming effects resulting from specific feed ingredients and different rates of gain. Nutrition during the pre-pubertal period and in the interval immediately before breeding in mature sires have both resulted in offspring characteristics. Our research has demonstrated that divergent planes of nutrition in mature rams can impacts lamb feed intake, growth, and metabolism. However, much more research is needed to determine the extent to which sire nutrition can impact offspring outcomes.

Résumé

La programmation du développement est un concept selon lequel les expériences vécues par les parents avant et pendant la gestation ont sur la progéniture des répercussions qui peuvent persister durant toute leur vie. Les travaux de recherche démontrant les répercussions de l'alimentation maternelle chez la progéniture sont relativement avancés et ne cessent de se multiplier. Le concept de programmation paternelle est toutefois plus récent. Des données issues de l'épidémiologie humaine et de recherches menées sur des rongeurs révèlent que la nutrition paternelle, l'environnement, le stress et l'abus de substance pendant la période de spermatogenèse peuvent avoir des effets sur la progéniture à venir. Les altérations épigénétiques (méthylation de l'ADN, modification des histones et ARN non codant), le stress oxydatif, la réponse immunitaire et le transfert microbien ont été identifiés comme mécanismes potentiels de transfert des signaux liés au père vers l'ovocyte au moment de la fécondation ou vers l'environnement utérin au début de la gestation. Chez les rongeurs, il a été démontré que l'alimentation et l'environnement paternels ont une incidence sur le poids corporel, l'accumulation de tissus adipeux et musculaires, les caractères reproducteurs et les réponses métaboliques chez la progéniture. Si ces effets se traduisent chez les animaux d'élevage, ils pourraient avoir une importance économique majeure. Bien que les connaissances soient encore très limitées, les travaux initiaux sur les animaux d'élevage ont démontré les effets de la programmation paternelle obtenus à partir d'ingrédients alimentaires spécifiques et associés à différents taux de gain. L'alimentation pendant la période prépubère et dans l'intervalle précédant immédiatement la saillie chez les mâles matures a eu des répercussions sur les caractéristiques de la progéniture. Nos recherches ont démontré que différents régimes alimentaires chez les béliers matures peuvent avoir des répercussions sur la consommation alimentaire, la croissance et le métabolisme des agneaux. Cependant, des recherches beaucoup plus approfondies sont nécessaires pour déterminer dans quelle mesure l'alimentation des reproducteurs peut influer sur la progéniture.

Introduction

Developmental programming is genererally though to be the concept whereby *in utero* environment influences offspring development and physiology. David Barker and colleagues were foundational in bringing this concept to light by reporting that extreme nutrient deficit as a result of war-time famine had major health impacts on children of women who were pregnant during this period (Barker, 1995; Hales and Barker, 2012). Subsequently, a body of reserch in livestock species has demonstrated that maternal nutrition and environment during pregnancy can impact offspring growth, mucle and fat development, metabolism, reproduction, and brain development (Greenwood and Cafe, 2007; Gionbelli et al., 2018; Hammer et al., 2023).

More recently, evidence from human epidemiology and research in rodent species has revealed that sire nutrition, environment, stress, and other exposures during the time of spermatogenesis can also impact offspring (Watkins et al., 2020). Early evidence of this phenomenon, called paternal programming, indicated that pre-breeding sire nutrition and has the potential to impact economically important traits in offspring of livestock species (for recent review see Dahlen et al., 2025).

Mechanisms of Action

One of the reported mechanisms of paternal programming is that sire environment influences the signals carried with the sperm/semen to the site of fertilization. Specifically, the field of epigenetics is implicated where no changes are made to the genetic code, but alterations in sperm DNA methylation, histone modification, and non-coding RNA are brought with the sperm during the fertilization process and subsequently alter offspring gene expression and phenotype (Kretschmer and Gapp, 2022). Sire nutrition could also induce damage to sperm DNA and oxidative stress, both which could impact fertility and resultant offspring (Billah et al., 2022). In many livestock systems natural service breeding is used, which results in a greater volume of seminal plasma being present in semen compared with artificial insemination. Seminal plasma is known to elicit an immune response and semen also carries a distinct microbial population, which could serve as mechanisms of paternal programming (Kilama et al., 2025).

Evidence from Rodent Models

The body of available literature related to paternal programming originates from rodent models. Several authors have provided excellent reviews on the topic (Lismer and Kimmins, 2023) but a summary of key models and findings is presented below. The following are a ***non-exhaustive*** list of rodent models that have demonstrated paternal programming effects:

- Over and undernutrition
- High fat or low protein diets
- Targeted supplementation of folic acid, vitamin B-12, methionine, etc.
- Acute or chronic stress and depression
- Exercise
- Drug/alcohol, endocrine disruptor exposure
- Microbiome manipulation
- Maternal nutrition

Specific characteristics of offspring that have been affected by paternal programming include:

- Body weight
- Muscle mass
- Adiposity
- Bone density
- Metabolic response (insulin, glucose, IGF-1, etc.)
- Reproductive outcomes
- Temperament

If similar impacts are present in livestock models there is potential for paternal programming to impact operation performance and profitability.

Early Research in Livestock

Research in the field of sire nutrition seldom extends to the point where offspring performance outcomes are evaluated. Efforts to characterize the impacts of sire nutrition on hormones, metabolites, sperm characteristics, epigenetic signatures in sperm, and/or fertility are most common, but all fall short of answering questions related to paternal programming.

We recently reviewed the body of scientific literature (Dahlen et al., 2024; Dahlen et al., 2025) and found several reports where paternal diet had impacts on embryo development and their epigenetic signatures. More demonstrative of the paternal programming phenomenon, however, are the reports where postnatal offspring characteristics were impacted.

The most complete experiments have been conducted using sheep. Hasan Khatib and his group from the University of Wisconsin have been evaluating the impacts of feeding rumen-protected methionine to young ram lambs during the interval from weaning until puberty. Interestingly, they observed that feeding protected methionine pre-pubertally had an influence on body weight and scrotal circumference of their offspring (Gross et al., 2020). In addition, consuming the protected methionine product by a single generation of developing rams had effects on scrotal circumference and DNA methylation in the F1, F2, and F3 generations (Braz et al., 2022).
Our research group compared feeding strategies to increase (gaining 12%), maintain, or decrease (losing 12%) body weight in the period leading up to natural service breeding in mature Rambouillet rams. After diets were in place for an 84-d period rams were placed into individual pens to each breed 10 ewes. Lambing data revealed that sire nutrition influenced lamb birth weight and body size characteristics (Bochantin-Winders et al., 2024a). Further evaluation of F1 ewe lambs in their own pre-breeding period showed that daily gain, circulating concentrations of IGF-1 and metabolic response to a glucose tolerance challenge were all impacted by sire treatment (Bochantin-Winders et al, 2024b).

Though early results are intriguing, much more work needs to be done to characterize potential paternal programming responses to other common sire paradigms. Based on results from research in rodent species there are many situations under direct control of livestock managers that could result in paternal programming responses in offspring. However, a rat is not a bull; livestock-specific research needs to be conducted.

References

Barker, D. J. 1995. Fetal origins of coronary heart disease. Bmj 311(6998):171-174. doi: 10.1136/bmj.311.6998.171

Billah, M. M., S. Khatiwada, M. J. Morris, and C. A. Maloney. 2022. Effects of paternal overnutrition and interventions on future generations. Int J Obesity 46(5):901-917. doi: 10.1038/s41366-021-01042-7

Bochantin, K.A., K.R.M. Slavick, J.L. Hurlbert, A.C.B. Menezes, J.D. Kirsch, P.P. Borowicz, C.S. Schauer, and C.R. Dahlen. 2024b. Influence of sire plane of nutrition and targeted body weight gain on ewe lamb growth, glucose metabolism, and ovarian reserve. J. Anim. Sci. skae301: https://doi.org/10.1093/jas/skae301

Bochantin-Winders, K. A., F. Baumgaertner, J. L. Hurlbert, A. C. B. Menezes, J. D. Kirsch, S. T. Dorsam, C. S. Schauer, and C. R. Dahlen. 2024a. Divergent planes of nutrition in mature rams influences body composition, hormone and metabolite concentrations, and offspring birth measurements, but not semen characteristics or offspring growth. Journal of Animal Science 102doi: 10.1093/jas/skae207

Braz, C. U., T. Taylor, H. Namous, J. Townsend, T. Crenshaw, and H. Khatib. 2022. Paternal diet induces transgenerational epigenetic inheritance of DNA methylation signatures and phenotypes in sheep model. PNAS Nexus 1(2)doi: 10.1093/pnasnexus/pgac040

Dahlen, C. R., S. Amat, J. S. Caton, M. S. Crouse, W. Diniz, and L. P. Reynolds. 2023. Paternal effects on fetal programming. Anim Reprod 20(2):e20230076. doi: 10.1590/1984-3143-ar2023-0076

Dahlen, C.R., G.D. Ramírez-Zamudio, K.A. Bochantin-Winders, J.L. Hurlbert, M.S. Crouse, K.J. McLean, W.J.S. Diniz, S. Amat, A.P. Snider, J.S. Caton, and L.P. Reynolds. 2024. International Symposium on Ruminant Physiology: Paternal Nutrient Supply: Impacts on Physiological and Whole Animal Outcomes in Offspring. J. Dairy Sci. https://doi.org/10.3168/jds.2024-25800

Gionbelli, T. R. S., C. M. Veloso, P. P. Rotta, S. C. Valadares Filho, B. C. Carvalho, M. I. Marcondes, C. S. Cunha, M. A. S. Novaes, L. D. Prezotto, M. S. Duarte, and M. P. Gionbelli. 2018. Foetal development of skeletal muscle in bovines as a function of maternal nutrition, foetal sex and gestational age. Journal of Animal Physiology and Animal Nutrition 102(2):545-556. doi: https://doi.org/10.1111/jpn.12786

Greenwood, P. L., and L. M. Cafe. 2007. Prenatal and pre-weaning growth and nutrition of cattle: long-term consequences for beef production. animal 1(9):1283-1296. doi: 10.1017/S175173110700050X

Gross, N., T. Taylor, T. Crenshaw, and H. Khatib. 2020. The Intergenerational Impacts of Paternal Diet on DNA Methylation and Offspring Phenotypes in Sheep. Front Genet 11:597943. doi: 10.3389/fgene.2020.597943

Hales, C., and D. Barker. 2012. Type 2 (non-insulin-dependent) diabetes mellitus: the thrifty phenotype hypothesis. International Journal of Epidemiology 42(5):1215-1222. doi: 10.1093/ije/dyt133

Hammer, C. J., J. S. Caton, C. R. Dahlen, A. K. Ward, P. P. Borowicz, and L. P. Reynolds. 2023. DOHaD: A MENAGERIE OF ADAPTATIONS AND PERSPECTIVES: Large animal models of developmental programming: sustenance, stress, and sex matter. Reproduction 165(6):F1-f13. doi: 10.1530/rep-22-0453

Kilama, J., C. R. Dahlen, L. P. Reynolds, and S. Amat. 2024. Contribution of the seminal microbiome to paternal programming. Biol Reprod 111(2):242-268. doi: 10.1093/biolre/ioae068

Kretschmer, M., and K. Gapp. 2022. Deciphering the RNA universe in sperm in its role as a vertical information carrier. Environ Epigenet 8(1):dvac011. doi: 10.1093/eep/dvac011

Lismer, A., and S. Kimmins. 2023. Emerging evidence that the mammalian sperm epigenome serves as a template for embryo development. Nature Communications 14(1):2142. doi: 10.1038/s41467-023-37820-2

Watkins, A.J., E. Rubini, E.D. Hosier, and H.L. Morgan. 2020. Paternal programming of offspring health. Early Human Dev. 150 105185. https://doi.org/10.1016/j.earlhumdev.2020.105185

ANCC

2025 ANIMAL NUTRITION CONFERENCE OF CANADA

CNAC

2025 COLLOQUE DE NUTRITION ANIMALE DU CANADA

Monogastric Session
Séance sur les monogastriques

Practical on-farm and regional biosecurity: Where are we heading?

Biosécurité pratique à la ferme et régionale: dans quelle direction allons-nous?

J.-P. Vaillancourt[1]

[1] *Professor of Population Medicine, Department of Clinical Sciences, Faculty of Veterinary Medicine, Université de Montréal, St-Hyacinthe, QC*

jean-pierre.vaillancourt@umontreal.ca

SPONSORED BY / COMMANDITÉ PAR

Jefo

Abstract

The last few decades have seen the emergence and re-emergence of several infectious pathogens of importance in swine and poultry. In response, producers have been asked to put in place biosecurity measures with two key objectives: reducing sources of contamination and maintaining a separation between these sources and healthy animals. These two principles have been known for centuries. Yet, we are still debating what works best and how to convince all stakeholders to implement these measures.

Several studies have highlighted the lack of consistent application of biosecurity plans. Currently, many investigations center around ways to improve biosecurity compliance via technology and interventions focusing on human behaviour.

There is also a need to develop a regional perspective to biosecurity. This is why research priorities include technologies, mainly sensors, to generate and share regionally the necessary information to reduce the probability of farm-to-farm transmission.

Relatively recently, feed and ingredients have been identified as substantial sources of infection in pigs. In response, three approaches are considered: a) prevention of entry to the feed system; b) mitigation by thermal processing, decontamination by chemical additives, or storage of high-risk ingredients for weeks at a given temperature; c) control over people sanitation and flow at feed mills. Another vector of transmission known in swine, aerosols, has been of interest in poultry due to highly pathogenic avian influenza. This has triggered research into viral aerosol detection, air pretreatment and various filtration processes. This presentation will highlight what works, what doesn't, and which new initiatives may improve biosecurity.

Keywords: beef cattle, wheat, mycotoxins, performance

Résumé

Les dernières décennies ont vu l'émergence et la réémergence de plusieurs agents pathogènes infectieux importants pour les porcs et les volailles. En réponse, les producteurs ont été invités à mettre en place des mesures de biosécurité avec deux objectifs clés : réduire les sources de contamination et maintenir une séparation entre ces sources et les animaux sains. Ces deux principes sont connus depuis des siècles. Pourtant, nous débattons encore de ce qui fonctionne le mieux et de la manière de convaincre tous les acteurs de ces industries de mettre en œuvre ces mesures.

Plusieurs études ont mis en évidence le manque de cohérence dans l'application des plans de biosécurité. Actuellement, de nombreuses recherches portent sur les moyens d'améliorer le respect des règles de biosécurité par le biais de technologies et d'interventions axées sur le comportement humain.

Il est également nécessaire de développer une perspective régionale de la biosécurité. C'est pourquoi les priorités en recherche incluent des technologies, principalement l'usage de capteurs, qui permettent de générer et de partager au niveau régional les informations nécessaires pour réduire la probabilité de transmission d'une exploitation à l'autre.

Relativement récemment, les aliments et les ingrédients ont été identifiés comme des sources importantes d'infection chez les porcs. En réponse, trois approches sont envisagées : a) la prévention de l'entrée dans le système d'alimentation ; b) l'atténuation par le traitement thermique, la décontamination par des additifs chimiques ou le stockage d'ingrédients à haut risque pendant des semaines à une température donnée ; c) le contrôle de l'hygiène des personnes et des flux dans les meuneries. Un autre vecteur de transmission connu chez le porc, les aérosols, suscite de l'intérêt en production avicole à cause de l'influenza aviaire hautement pathogène. Cela a déclenché des recherches sur la détection des aérosols viraux, le prétraitement de l'air et divers procédés de filtration. Cette présentation mettra en évidence ce qui fonctionne, ce qui ne fonctionne pas et les nouvelles initiatives susceptibles d'améliorer la biosécurité.

Introduction

Biosecurity is any actions or health plans designed to protect a population against infectious and transmissible agents (Toma et al, 1999). They are based on principles that have been known for centuries. Sanitation measures are detailed in the Old Testament (Ojewole, 2011); most biosecurity measures found in modern swine and poultry farms are like recommendations given to lepers during the mass of separation in the Middle Ages (Brody, 1974). Yet, we are still struggling today to get people to perform simple tasks that we know work for preventing contagious diseases on farms (Racicot et al., 2011).

In most cases, critical elements of biosecurity programs have to do with people. Many human factors widely impact biosecurity compliance. Yet, too often, these are neglected or not considered a priority in the field. So, this paper will present the risks factors associated with infectious diseases, the well-established measures to mitigate them, and what is being developed to improve the efficacy of on-farm and regional biosecurity.

Risk factors

Remarkably, swine and poultry share essentially the same types of risk factors (Table 1). They are nuances, but, for example, very large production sites in high density production areas are more at risk of having flocks or herds infected. Only the diseases are different. The risks listed in Table 1 cover all main sources and determinant factors associated with important infectious diseases. For example, a large farrowing unit in a region where porcine reproduction and respiratory syndrome (PRRS) is prevalent and that is located near a few large finishing units with herds made of pigs from multiple origins would face a high likelihood of infection to this virus. The consequences of such an infection would also be severe. This assessment could justify installing air filtration (Desrosiers and Cousin, 2023). Although this is by no means a simple or easy measure to apply, the challenge is largely technical. The problem is when risk mitigation requires long term and sustained compliance by people in the application of a series of measures.

Table 1: Risk factors associated with poultry and swine infectious diseases and categories of biosecurity measures associated with these risks.

Category	Risk	POULTRY	SWINE	Reduce[2]	Separate[2]	Regional[2]	Human[3]
Location	Multi-species site	+++[1]	+++		X		X
	Size of farm	++++	++++		X		X
	Access outdoor	++++	++		X	X	X
	Proximity to neighbor/road	Laryngotracheitis Avian influenza	PRRS			X	X
	Size of neighbor's farm	Poultry farms	Swine farms			X	

	Type of neighbor's production	Poultry farms	Swine farms			X	
	Regional density	++++	++++			X	X
	Aerosols	Limited evidence	PRRS	X	X	X	
	Insects/vectors	Beetles/flies	Flies/mosquitoes	X	X		
	Wildlife (e.g., rodents)	Waterfowl	Wild boars	X	X		
Animals	Type of production	Birds > 12 wks	Farrowing		X		X
	Origin of flock/herd	Imported eggs/chicks	Semen	X	X		X
	New arrivals	Spiking males	Gilts	X	X		X
Management	People: Owner, employee, visitor	++++	++++	X	X	X	X
	Equipment, vehicles	++++	++++	X	X	X	X
	Feed and Water	+++	++++	X	Feed		X
	Litter/ manure storage & pread	+++	+++	X	X	X	X
	Dead animal disposal	Proximity to barns		X	X	X	X
	Low downtime	< 14 days between flocks	< 3 days between farrowing batches	X	X	X	X

[1]Degree of significance (++: significant; +++: very significant; ++++: highly significant)
[2]Basic principles: Reduce = diminish sources of infection; Separate = prevent contact between sources of infection and healthy animals; Regional = regional measures of biosecurity
[3]Human = factors that largely involve human actions

Basic principles & application

"The biosecurity program for a sow farm is only as strong as its weakest link and filtering incoming air will not be a wise investment unless all other routes of viral introduction are eliminated." (Morrison, 2011)

Typically, all on-farm biosecurity measures aim at either reducing or eliminating sources of infection or maintaining a distance (separation) between any potential source of infection and the herd or flock we want to protect. Both reduction and separation are often part of biosecurity measures designed to mitigate management related risks (Table 1).

Reducing sources of contamination

Cleaning and disinfection of barns and equipment: The decontamination of barns and equipment is a multi-step process. Several publications detail how to best achieve consistent results (Payne et al., 2005; Watkins and Venne, 2015; Blondel et al., 2018). The points deserving attention are: 1) the dry removal of visible organic material; 2) the application of a detergent followed by stripping and drying; it makes it possible to reduce the bacterial concentration of a contaminated surface by a factor of 10 to 1000 compared to a wash with only clean water (Course et al., 2021); 3) the adequate contact time of the detergent and disinfectant is paramount; 4) water that is too cold slows down chemical reactions; water at 40 °C is best (Böhm, 1998); 5) disinfection can be carried out by vaporization or by fumigation / misting. On average, 150 to 300 ml / m² are applied (Blondel et al., 2018); 6) drying after cleaning and after disinfection is critical for best results (Böhm, 1998).

Regarding tires of vehicles, wheel baths are assumed to be useful (Pinto and Urcelay, 2003), but there is virtually no evidence that they work. The contact time is too limited for disinfectants to reduce wheel contamination (Chaber and Seagerman, 2017). Ramsey et al. (2021) conducted an excellent study demonstrating how difficult it is to decontaminate the undercarriage of a truck, even when using a high-pressure automated truck wash with a two-stage (washing & disinfection) decontamination system. Drying swine transportation truck has been shown to be a very effective decontamination procedure (Dee et al., 2004). Wheel baths do the opposite by wetting surfaces, and often by adding contaminants to the vehicles. This is why, when possible, vehicles and equipment should be exposed to the sun, which helps the drying process as well as the decontamination via the ultraviolet radiation from sunlight (Kuney et al, 2002).

Water hygiene: Water quality and sanitation would require its own paper. Briefly, when using chlorine, it is best to assess the oxidation reduction potential (ORP) of water. A strong oxidizer will effectively destroy microbes. An ORP value of 650 millivolts or greater is required. The goal should be to maintain 2-5 ppm of free chlorine in the water supply. For removing scale and biofilm, it is best to drop the water pH below 5 but not below 4 to prevent equipment damage. Several disinfectants are available to clean water lines, but some of the most effective and safe products for drinker systems are concentrated, stabilized hydrogen peroxides (Watkins and Venne, 2015).

Pest control: Because of space limitations, the reader is directed to the following publications for valuable information:_Villa and Velasco, 1994; Allen et al., 2005; Axtell, 1999; Corrigan, 2002).

Disposal of dead animals: In many countries, rendering is the most prevalent approach to dead animal disposal. However, it has been associated with a greater risk of disease transmission in different species, including swine (Lambert et al., 2012) and poultry (McQuiston et al, 2005). Proximity of the rendering container to poultry barns may partly explain the greater risk of contamination when renderers visit the site. Another important factor is the absence of an adequate sanitation protocol for personnel visiting the site where the rendering container is located as they return to the flock or herd. So, if possible, disposing of dead animals on site is a better alternative. Incineration is subject to technology-specific regulations in Canada. Burying is often not an option because of environmental regulations. So, composting for the routine disposal of on-farm poultry and swine carcasses has increased in prevalence over the past 30 years (Tablante et al, 2006; Kalbasi et al., 2005).

Separating healthy birds from sources of contamination

Zoning production sites

Basically, it is about preventing the contact between susceptible birds and microbes. The barns where the animals are located must be the most restricted zone (restricted access zone or RAZ); the farm where we find the barns is a controlled access zone (CAZ). Everything else around these two zones must be considered an environment that is source of contamination (Anon., 2018). Only essential vehicles, personnel and equipment should get within the CAZ. The inadequate management of vehicles, including where they are parked on a production site (too close to barn entrances), has been associated with a greater risk of disease transmission between farms (Guinat et al, 2020). Rose et al. (2000) demonstrated an association between the surface covered by vehicles on a broiler chicken production site and the probability of isolating *Salmonella*. Snow et al. (2010) also identified a link between on-farm transportation and *Salmonella* in egg-layer farms. More recently, Yoon et al. (2022) demonstrated a higher highly pathogenic avian influenza infection attack rate for vehicles allowed to enter the CAZ compared to those kept outside.

Managing barn entrances: The separation between the RAZ and outside must be clearly delineated. When a shower is present, the shower itself is the line separating the two. In the absence of a shower, it is important to delineate the zones with at least a line. Anterooms separated in only two zones (external or contaminated and internal or clean) can be problematic because cross-contamination between the two zones is frequent. Often known as Danish entrance, the three-zone design offers an area between the contaminated and clean zones where visitors and personnel can wash their hands. This step, while transitioning between the contaminated and clean zones, is important because it ensures that hands are decontaminated before handling boots, clothing, or equipment on the clean side of the anteroom (Anon., 2018).

Footbaths: The effectiveness of a footbath is influenced by several factors. The two main factors are the presence of organic matter (on the boots or in the footbath) and the insufficient contact time (Amass, 1999). Other factors also influence the effectiveness of footbaths, but one key problem often seen on farms is the fact that they are not positioned in such a way as to easily go from one zone to the other without recontaminating the footwear. Several studies have confirmed the inadequacy of traditional footbaths. Hauck et al. (2017) reported that quaternary ammonium and quaternary ammonium-glutaraldehyde footbaths were not able to eliminate avian influenza viruses. Owen and Lawlor (2012) compared phenol and quaternary ammonium filled footbaths to dry footbaths filled with a dry bleach powder. The average residual life of the liquid footbaths was less than two hours and did not impact significantly the bacterial count on the boots. By contrast, the dry bleach footbath reduced the bacterial count by around 98% and had a residual life of 14 days.

Downtime: This is the period when a barn or an entire production site (all-in / all-out) is without animals (Anon., 2018). This approach has a protective effect regarding *Salmonella* contamination (Snow et al. 2010) and avian influenza (Kim et al, 2018). Poultry veterinarians participating in a Delphi study indicated that a 14-day downtime between meat bird flocks should be the norm, with 21 days between breeder flocks (Vaillancourt and Martinez, 2001). In swine, a downtime of at least three days is recommended between farrowing batches.

Feed Biosecurity

Feed as long been known as a source of *Salmonella* contamination in swine and poultry. A systematic review has shown that feed accounts for 4.8% of the occurrence of *Salmonella* in poultry production systems (Wang et al, 2023). Relatively recently, feed and ingredients have been identified as substantial sources of viral infections in pigs (Dee et al 2014). In its October 2021version of Fundamentals of the Commercial Feed Industry in Canada, the Animal Nutrition Association of Canada mentions African swine fever virus (ASFv) and porcine epidemic diarrhea virus (PEDv) as important pathogens known to be transmitted over long distances via feed or ingredients (Dee et al 2018). The consequences are not negligible. For example, PEDv spread from China to the United States in 2013, and within one year, the virus had impacted about half of breeding herds, resulting in the deaths of several million piglets (Goede and Morrison, 2016). Dee et al (2024) also provided evidence that feed could be a source of the porcine reproductive and respiratory syndrome virus (PRRSv).

Ingredient sanitation

Since feedmills have often been built without biosecurity in mind, the priority should be to prevent entry of contaminated ingredients to the mill. It is important to select trustworthy suppliers and to be aware if a disease such as ASF has been confirmed in the region of origin of the ingredients.

To reduce the likelihood of virus introduction, high-risk ingredients are often stored for extended periods, generally about 30 days, to inactivate viruses prior to use in diets. Dee et al. (2022) demonstrated that storage temperature is an important consideration. Keeping soybean meals contaminated with PRRSv and Senecavirus A (SVA) for 30 days at 23.9◦C, 15.5◦C or 10◦C, they concluded that a temperature of 23.9◦C was more likely to reduce the risk of viral transmission via plant-based feed ingredients.

Additional mitigations may include thermal processing and decontamination by chemical additives. Heat treatment at a temperature of about 85°C has been shown to be effective against *Salmonella* (Vukmirović et al., 2017). But the temperature must exceed 130°C to reduce PEDv survival (Trudeau et al., 2016). Although thermal processing can significantly reduce the presence of many biological hazards, it is a point-in-time mitigation step that does not prevent post-processing cross-contamination.

In a controlled study mimicking real-world conditions, Dee et al. (2021) tested 15 additives for mitigating the risk of feed contaminated with PRRSv, SVA and PEDv. Fourteen of the 15 additives tested had a significant impact, offering better performance and lower infection levels compared to non-supplemented controls.

Mill sanitation

Keeping a feed processing facility dry and free of any standing water is important to minimize contamination of people and vehicles and to deter birds, rodents and any other animals from congregating and contributing to the contamination of the site and the spreading of pathogens such as *Salmonella* (Obermeyer, 2021). These animals are all potential carriers or vectors of bacteria and viruses. Once these pests infest a facility, it is very difficult to eliminate them. So, buildings should always be closed and secured with bird and rodent proofing procedures. People and vehicles are also potential vectors. It is paramount to effectively control the traffic flow and segregate the different sections of the mill with clear zoning instructions to personnel, as well as having physical barriers such as gates, fencing and locked doors.

The receiving pit must be dry and clean. The pit should be covered when not in use. A funnel should be used to minimize spillage during unloading and it should be cleaned up (avoiding dust and debris from entering) before any further vehicle movement in the receiving pit area.

Dust is a substrate for some pathogens, and it can be an effective means of transportation between different sections of the mill. Dry cleaning removing dust and debris from walls and equipment must be done by hand, using brushes, scrapers, shovels, brooms, and a vacuum. For this step, water and compressed air should be avoided because it may facilitate the growth and dissemination of pathogens in the mill. All surfaces of the feed mill, including walls, bins, and floors, must be regularly cleaned.

Given the complexity of equipment and surfaces, it is best to disinfect by fumigation, thermal fog, or ozone generators. All disinfectants require a defined contact time to be effective. Frequent causes of non-compliance are applying the product at the wrong concentration or in a wet environment, effectively diluting the disinfectant, and using an inadequate quantity reducing contact time.

Feed delivery

Ingredients and finished feeds should be delivered by clean trucks and in a sequence minimizing the risk of contamination. This requires farm managers to report sites with suspected or confirmed disease outbreaks so that the sequence of truck delivery and decontamination can be adjusted to reduce the risk of biological-hazard transfer between farms and back to the feed mill. Clean footwear, plastic boots, or covers within feed mills or during delivery can reduce pathogen transfer (Jones, 2011). Feed truck drivers should not be allowed to enter barns and should avoid passing near exhaust fans, dead-stock disposal, or cross-traffic with manure disposal equipment. After all runs are completed, the driver goes directly back to the feed truck area. Trucks must be washed completely and disinfected weekly, unless the last site visited has birds infected with a disease of significance to the industry. In this case, truck decontamination must be performed after this last delivery. Breeder farms are normally visited first before meat bird operations. Delivery to diseased or quarantined farms should be done as the last load of the day (Anon., 2018). Feed previously in a bin on a farm must not be allowed back to the mill.

Finally, in-house auditing and third-party audits every three to six months should be conducted. Results can then be used to adjust the biosecurity plan and may serve as material for training purposes.

Air filtration

Robert Desrosiers, a swine veterinarian from Canada, was among the first to point out that the PPRSv could be airborne and transmitted over long distances. There are now several epidemiological studies that have confirmed the airborne transmission of swine pathogens during periods of cold weather when insects should not realistically be involved (Desrosiers, 2008).

During a two-year period, Dee et al. (2024) assessed 56 breeding herds were biosecurity measures included air filtration and 20 herds with no air filtration. The PRRS incidence risk in the air filtered herds was 8.9% as compared to 40.0% in the other group. Such results have caught the attention of poultry growers dealing with the emergence of highly pathogenic avian influenza viruses (HPAIs). So far, there is no strong evidence that HPAIs can travel in the air and remain infective for distances exceeding 100 m (Torremorell et al., 2016). Yet, the last few years have seen the conduct of several studies applied to poultry farm conditions (Guo et al., 2022). However, so far, there is little evidence that any of the air filtration or decontamination devices that have been tested could reduce significantly the risk of HPAI infection.

On-farm compliance issues

Human factors

In many cases, mainly for management related measures, the most critical elements are linked to human factors (Table 2). These are human characteristics that widely impact biosecurity compliance. Too often, these issues are neglected. To address them effectively, a mix of technical and behavioral methods may be needed.

Personality traits and emotional intelligence are two of the main determinants of compliance. Emotional intelligence refers to the level of self-awareness, self-management, social awareness, and relationship management in the context of job performance. High emotional intelligence has a positive impact on employee effectiveness and efficiency, better task coordination, and improved goal achievement. Emotional intelligence is also important for leadership, motivation, communication, decision making, interpersonal relationships, and change management. It is a learned characteristic that can be developed over time. Given its malleability, it is possible to modulate training programs, supervision and reinforcement of employees in order to have an impact on employee performance and compliance. To increase motivation towards desired behavior such as compliance to biosecurity measures, the ability to achieve goals and a sense of self-efficacy are important (Bandura and Cervone, 1983). It is by manipulating the feeling of self-efficacy (e.g. providing a feedback allowing one to compare his/her performance versus others), rather than by giving feedback regarding a risky behavior, that a change in behavior is likely" (Pervin and John, 2005). Table 3 offers other suggestions to improve compliance by considering the link between determinants of poor compliance and universal human desires.

Technical considerations

A study by Racicot (2011, 2012a, b) described the application of biosecurity measures in anterooms of poultry barns, using hidden cameras. Results showed that when the barn entrance design makes it difficult to perform a given measure, individuals are 13 times less likely to comply. The type of zone separation has also a significant impact on compliance. A bench is preferable to a simple line on the floor.

In addition to a set-up favoring the application of required biosecurity measures, real-time feedback contributes to compliance. A radio frequency identification (RFID) continuous automated monitoring system has been developed to monitor compliance with sanitation measures and provide regular feedback. The study aimed at monitoring and evaluating the frequency of two biosecurity measures: hand washing and changing boots according to the different biosecurity zones (clean/contaminated) during a visit to a barn and providing real-time feedback. Employees were asked to participate on a voluntary basis. Each employee was assigned a pair of farm boots with a chip inserted in the sole of each boot for the duration of the project. The data were collected using the RFID system equipped with an antenna installed on the floor of the clean area and connected to the alcohol-based sanitizer, detecting hand sanitization and chipped farm boots. Detection of chipped farm boots by the antenna located in the clean area indicated farm boot compliance. Detection of chipped farm boots by the antenna (entering or leaving the barn) without pressure detected on the hand sanitizer device indicated a lack of compliance with hand sanitation. In the event the sanitizing device was not used, immediate feedback via a sound alarm was triggered, reminding people to comply. The system was also designed to monitor if the biosecurity breach was corrected after hearing the alarm.

For the egg layer farm, 254 entries and exits (i.e., 127 visits) by four different employees were recorded over a 17-day period. Boots were put on in the clean area 122 times out of 131 opportunities (93%). Hands were sanitized 173 times out of 254 opportunities (68%). Three out of four employees fully complied with boot change as their personal shoes were not detected by the RFID system during the entire trial. For the 9 non-compliant shoe events, farms boots were changed twice (2/9; 22%) after hearing the alarm. Out of the 81 non-compliant hand sanitizing occurrences, 26 (32%) were corrected after the alarm rang.

For the broiler farm, 56 entries and exits (i.e. 28 visits) by three different employees were recorded over 13 days. Boot compliance was 100% for the entire period and hand sanitizing compliance was 73.2% (41 out of 56 opportunities). Out of the 15 non-compliant hand sanitizing occurrences, one was corrected after the alarm rang. So, providing real-time feedback seems to have a variable impact on the participants. Some employees were highly influenced by the alarm and the feedback mechanism helped correcting their behavior, while others were not influenced at all. Psychological characteristics, such as personality traits (Racicot et al., 2012b), are closely linked to biosecurity compliance. It may also be the case for the receptivity to real-time audio feedback. Still, for both farms, these compliance levels were much higher than recorded previously (Racicot et al., 2011).

Table 2: Human factors having an impact on the determinants of on-farm biosecurity compliance.

Determinants of biosecurity compliance	Personality	Emotional Intelligence	Attitude	Normative belief	Perceived control	Intention
	Responsibility; complexity; action orientation	Employee effectiveness; leadership; motivation; higher goal achievement; decision making; interpersonal relations	How favorable an individual perceives a certain behavior	How important it is to: coworkers; supervisor; management; owner	Perception of the ability to perform: knowledge; capacity/skills; material/support available	Best predictor of behavior
Economic constraints			X	X	X	X
Lack of time	X	X	X	X	X	X
Difficulty to apply measure	X	X	X	X	X	X
Absence of necessary material				X	X	X
Belief, attitude, perception		X	X	X	X	X
Education, experience	X	X	X			
Personality traits	X					
Weak leadership	X	X	X	X	X	X
Lack of knowledge	X	X	X	X	X	X
Inadequate training	X	X	X	X	X	X
Incoherent communication	X	X	X	X	X	X
No valid audits				X	X	X
Lack of incentives	X	X	X	X	X	X

Table 3: Connection between determinants of poor on-farm biosecurity compliance and universal human desires, with examples of interventions.

Determinant of poor on-farm biosecurity compliance	Examples of Interventions	Universal human desires[i]							
		Activity	Ownership	Power	Affiliation	Competence	Achievement	Meaning	Recognition
Weak leadership	Promote biosecurity through leadership (select and train your best leaders not your best workers)	X		X					
Incoherent communication	Involve your employees in decision making for the biosecurity program		X						
No valid audits	Measure performance individually and compare		X	X	X	X	X		
Inadequate training	Influence attitude through Training/Communication / Incentives					X			X
Lack of incentives	Set realistic goals						X		X
Lack of knowledge	Train workers for: Knowledge Ability (practical how to)					X	X	X	

[i]Activity: meaningful position; Ownership: decision process; Power: responsibilities; Affiliation: team spirit; Competence: knowledge, know how; Achievement: reaching targets/reinforcement; Meaning: sense of direction; Recognition: Care; privilege; time off; meals; award

Regional biosecurity

Regional biosecurity is the coordination of activities in a given region in order to minimize risks of cross-contamination between vehicles and personnel, which may act as mechanical vectors.

The relationship between proximity to a poultry production site and the probability of contamination has been well established (Vandekerchove et al., 2004; Sanchez et al., 2005). Commercial poultry farms infected with the infectious laryngotracheitis virus where 36 times more likely to be within 1.6 km from a backyard flock than disease free farms (Johnson et al., 2005).

The vast majority of infectious diseases are horizontally transmitted. Cross-traffic between farms and the joint use of equipment are certainly significant modes of disease transmission (Guinat et al, 2020). Disease control efforts in regions of high farm concentration cannot rely solely on farm level measures. A regional approach to biosecurity includes establishing separate geographical working zones for service personnel. When an important disease is suspected on a farm, a self-imposed quarantine matched with communication to key personnel will go a long way to prevent a major epidemic (Vaillancourt et al., 2018). In 2005, during an outbreak of infectious laryngotracheitis in California, two integrations coordinated to perform an extended downtime in a region and to set up biosecurity audits. Cooperation and communication between companies has helped eradicate the disease. The importance of communication is also highlighted in a 2001 survey of 72 North American veterinarians on the importance given to certain elements of biosecurity (Vaillancourt and Martinez, 2001).

Geographical information systems (GIS) are routine tools out of the necessity to quickly determine the location of infected or diseased flocks/herds. This technology serves several purposes: 1) it is used primarily by industry to determine at-risk areas and to establish the best routes to avoid infected flocks or to avoid at-risk farms with contaminated material or infected animals; 2) it is also useful to assess the geographical distribution of epidemics; 3) it identifies clustering and permits the statistical evaluation of flock/herd health status associated with proximity to specific sites; 4) and it is a powerful investigative tool when coupled with molecular techniques providing critical information on spread of disease agents.

Over the past decade, sensors and GPS technology have allowed the development of real-time data harvesting and geopositioning of farms, vehicles and people. For example, Farm Health Guardian, LLC (FHG; https://farmhealthguardian.com/) have created a geofence on all properties associated with a commercial system, as well as installing devices on vehicles and a phone application for personnel permitting real-time detection when they are crossing the virtual fences set around each farm. Data include the time of entry on and exit off these properties. Personnel associated with the system and visitors not registered in the system are logged in using a separate device tablet called a kiosk located in the anteroom of the barns. All data are securely encrypted so that the companies involved maintain total control. Using this technology over about one year on 455 properties, Prezioso et al. (2025) could map out an analyze 320 001 trips between properties. They could show that the network of farms, vehicles and people was very well connected, more so than expected based on animal movement alone. It was easy to identify hubs of activities that may increase disease risks and specific potential 'super-spreader' properties. The high connectivity shows that having the ability to monitor human, as well as animal movements, provides a more complete and accurate understanding of swine biosecurity risks at the farm and the regional levels.

Conclusion

On-farm biosecurity measures in swine and poultry are based on principles that have been known for ages. The consistent long-term application of measures requiring human involvement remains a challenge. Human factors are important, but for the most part the swine and poultry industries have not capitalized on existing knowledge regarding interventions based on these factors.

Using sensors, it should be possible to develop cost-effective techniques to improve compliance by providing real-time feedback to farm personnel and by providing data to inform training programs. The same basic technology with the usage of geographical information systems should also assist in the development of a regional perspective to biosecurity.

Over the past few years, feed has emerged as a significant source of viral diseases in swine. Mitigation strategies are efficacious. The same can be said about air filtration on swine premises where risk factors make the investment cost-effective. This is less likely for poultry operations, largely because of the added challenge of dust and feathers, and because the cost-effectiveness has not been shown to date.

References

Allen, V. and Newell, D. 2005. *Evidence for the effectiveness of biosecurity to exclude Campylobacter from poultry flocks*. Commissioned project MS0004. Food Standards Agency Report. Available at: /paper/Evidence-for-the-effectiveness-of-biosecurity-to-Allen-Newell/42361dd1fb80d51742c54629ee46cb315ef936a (Accessed: 22 March 2021).

Amass, S. F. 1999. 'Biosecurity considerations for pork production units', *Swine Health and Production*, 7(5), 12.

Anon. **2018**. National On-Farm Avian Biosecurity Standards. https://inspection.canada.ca/animal-health/terrestrial-animals/biosecurity/standards-and-principles/national-avian-on-farm-biosecurity-standard/eng/1528732756921/1528732872665?chap=3

Axtell R.C. 1999. Poultry integrated pest management: Status and future, Integrated Pest Management Reviews 4: 53–73.

Bandura, A. and Cervone, D. 1983. 'Self-evaluative and self-efficacy mechanisms governing the motivational effects of goal systems', *Journal of Personality and Social Psychology*, 45(5), pp. 1017–1028. doi: 10.1037/0022-3514.45.5.1017.

Blondel, V., Huard, G., Vaillancourt, J.-. P., et Racicot, M. 2018. Base du nettoyage et de la désinfection dans les exploitations agricoles. https://www.agrireseau.net/documents/98011/bases-du-nettoyage-et-de-la-desinfection-dans-les-exploitations-agricoles

Böhm, R. 1998. Disinfection and hygiene in the veterinary field and disinfection of animal houses and transport vehicles. *International biodeterioration & biodegradation*, *41*(3-4), 217-224.

Brody, S. N. 1974. *The disease of the soul: Leprosy in medieval literature*. F First Edition. Ithaca N.Y.: Cornell University Press.

Chaber, A. L. and Saegerman, C. 2017. Biosecurity Measures Applied in the United Arab Emirates – a Comparative Study Between Livestock and Wildlife Sectors, Transboundary and Emerging Diseases, 64(4), pp. 1184–1190. doi: 10.1111/tbed.12488.

Cochrane, R. A., Dritz, S. S., Woodworth, J. C., Stark, C. R., Huss, A. R., Cano, J. P., ... and Jones, C. K. 2016. Feed mill biosecurity plans: A systematic approach to prevent biological pathogens in swine feed. *Journal of Swine Health and Production*, *24*(3), 154-164.

Corrigan, R. M. 2002. Overview of rodent control for commercial pork operation. Swine health and epidemiology report. Special biosecurity issue. 8.

Course, C. E., Boerlin, P., Slavic, D., Vaillancourt, J. P., and Guerin, M. T. 2021. Factors associated with Salmonella enterica and Escherichia coli during downtime in commercial broiler chicken barns in Ontario. *Poultry Science*, 101065.

Dee, S., Clement, T., Schelkopf, A., Nerem, J., Knudsen, D., Hennings, J., and Nelson, E. 2014. An evaluation of contaminated complete feed as a vehicle for porcine epidemic diarrhea virus infection of naïve pigs following consumption via natural feeding behavior: Proof of concept. *BMC Vet Res*,10:176.

Dee, S.A., Bauermann, F., Niederwerder, M.C., Singrey, A., Clement, T., …, and Diel, D.G. 2018. Survival of viral pathogens in animal feed ingredients under transboundary shipping models. *PLOS ONE*, 13(3): e0194509.

Dee, S., Deen, J. and Pijoan, C. 2004. 'Evaluation of 4 intervention strategies to prevent the mechanical transmission of porcine reproductive and respiratory syndrome virus', *Canadian Journal of Veterinary Research*, 68(1), 19–26.

Dee, S., Brands, L., Nerem, J., Schelkopf, A., Spronk, G., Kikuti, M., ... and Havas, K. 2024. Improvements in swine herd biosecurity reduce the incidence risk of porcine reproductive and respiratory syndrome virus in breeding herds in the Midwestern United States. *Journal of the American Veterinary Medical Association*, *262*(4), 520-525.

Dee, S. A., Niederwerder, M. C., Edler, R., Hanson, D., Singrey, A., Cochrane, R., ... and Nelson, E. 2021. An evaluation of additives for mitigating the risk of virus-contaminated feed using an ice-block challenge model. *Transboundary and Emerging Diseases*, *68*(2), 833-845.

Dee, N., Havas, K., Shah, A., Singrey, A., Spronk, G., Niederwerder, M., ... and Dee, S. 2022. Evaluating the effect of temperature on viral survival in plant-based feed during storage. *Transboundary and Emerging Diseases*, *69*(5), e2105-e2110.

Dee, S., Clement, T., and Nelson, E. 2024. Transmission of porcine reproductive and respiratory syndrome virus in domestic pigs via oral ingestion of feed material. *Journal of the American Veterinary Medical Association*, *262*(1), 1-4.

Delpont, M., Racicot, M., Durivage, A., Fornili, L., Guerin, J. L., Vaillancourt, J. P., and Paul, M. C. 2021. Determinants of biosecurity practices in French duck farms after a H5N8 Highly Pathogenic Avian Influenza epidemic: The effect of farmer knowledge, attitudes and personality traits. *Transboundary and emerging diseases*, *68*(1), 51-61.

Desrosiers, R. 2008. Epidemiology, diagnosis and control of enteric diseases in growing-finishing pigs. In *Memorias del IX Congreso Nacional de Producción Porcina* (95-109).

Desrosiers, R., and Cousin, V. 2023. Air filtration to prevent porcine reproductive and respiratory syndrome virus infection. *Journal of Swine Health and Production*, *31*(2), 77-81.

Goede D, and Morrison RB 2016. Production impact & time to stability in sow herds infected with porcine epidemic diarrhea virus (PEDV). Prev Vet Med 123:202–207.

Guo, L., Zhao, B., Jia, Y., He, F., and Chen, W. 2022. Mitigation strategies of air pollutants for mechanical ventilated livestock and poultry housing—A review. *Atmosphere*, *13*(3), 452.

Guinat, C. *et al.* 2020. 'Biosecurity risk factors for highly pathogenic avian influenza (H5N8) virus infection in duck farms, France', *Transboundary and Emerging Diseases*, 67(6), 2961–2970. doi: https://doi.org/10.1111/tbed.13672.

Hauck, R., Crossley, B., Rejmanek, D., Zhou, H., and Gallardo, R.A. 2017. Persistence of highly pathogenic and low pathogenic avian influenza viruses in footbaths and poultry manure. Avian Dis. 61, 64–69. https://doi.org/10.1637/11495-091916-Reg.

Johnson, Y. J. *et al.* 2005. 'Wind-borne transmission of infectious laryngotracheitis between commercial poultry operations', *International Journal of Poultry Science*, 4(5), 263–267. doi: 10.3923/ijps.2005.263.267.

Julio Pinto, C. and Santiago Urcelay, V. 2003. Biosecurity practices on intensive pig production systems in Chile, *Preventive Veterinary Medicine*, 59(3), 139–145. doi: 10.1016/S0167-5877(03)00074-6.

Kalbasi, A., Mukhtar, S., Hawkins, S. E., and Auvermann, B. W. 2005. Carcass composting for management of farm mortalities: A review. *Compost science & utilization*, *13*(3), 180-193.

Kim, W. H., An, J. U., Kim, J., Moon, O. K., Bae, S. H., Bender, J. B., and Cho, S. 2018. Risk factors associated with highly pathogenic avian influenza subtype H5N8 outbreaks on broiler duck farms in South Korea. *Transboundary and emerging diseases*, *65*(5), 1329-1338.

Kuney, D. R., and Jeffrey, J. S. 2002. Cleaning and disinfecting poultry facilities. In *Commercial Chicken Meat and Egg Production* (557-564). Springer, Boston, MA.

Lambert, M. È., Poljak, Z., Arsenault, J., and D'Allaire, S. 2012. Epidemiological investigations in regard to porcine reproductive and respiratory syndrome (PRRS) in Quebec, Canada. Part 1: Biosecurity practices and their geographical distribution in two areas of different swine density. *Preventive Veterinary Medicine*, *104*(1-2), 74-83.

McQuiston, J. H. *et al.* 2005. 'Evaluation of risk factors for the spread of low pathogenicity H7N2 avian influenza virus among commercial poultry farms', *Journal of the American Veterinary Medical Association*, 226(5), 767–772.

Morrison, R. B. 2011. Control and elimination of PRRS virus. Advances in Pork Production, 22.

Obermeyer R. 2012. 4 ways to optimise poultry feed mill biosecurity. In Poultry World. https://www.poultryworld.net/health-nutrition/4-ways-to-optimise-poultry-feed-mill-biosecurity/

Ojewole, A. O. 2011. 'Biblical mandates for sustainable sanitation', *Millenium development goals (MDGs) as instrument for development in Africa*, 598–613.

Owen, R.L., Lawlor, J. 2012. A novel approach to foot dipping. https://fr.slideserve.com/bernad/a-novel-approach-to-foot-dipping

Payne, J. B., Kroger, E. C., and Watkins, S. E. 2005. Evaluation of disinfectant efficacy when applied to the floor of poultry grow-out facilities. *Journal of applied poultry research*, *14*(2), 322-329.

Pervin, L. A. and John, O. P. 2001. *Personality: Theory and Research*. 8th Edition. New York: John Wiley & Sons Inc.

Prezioso, T., Boakes, A., Wrathall, J., Reger, W. J., Bhowmick, S., and Smith, R. L. 2025. A network evaluation of human and animal movement data across multiple swine farm systems in North America. *Preventive Veterinary Medicine*, *234*, 106370.

Racicot, M. *et al.* 2011. 'Description of 44 biosecurity errors while entering and exiting poultry barns based on video surveillance in Quebec, Canada', *Preventive Veterinary Medicine*, 100(3–4), pp. 193–199. doi: 10.1016/j.prevetmed.2011.04.011.

Racicot, M. *et al.* 2012a. 'Evaluation of strategies to enhance biosecurity compliance on poultry farms in Québec: Effect of audits and cameras', *Preventive Veterinary Medicine*, 103(2–3), 208–218. doi: 10.1016/j.prevetmed.2011.08.004.

Racicot, M. *et al.* 2012b. 'Evaluation of the relationship between personality traits, experience, education and biosecurity compliance on poultry farms in Québec, Canada', *Preventive Veterinary Medicine*, 103(2–3), 201–207. doi: 10.1016/j.prevetmed.2011.08.011.

Ramsey, C.L., Serre, S., Rosenberg, J., Daniell, N., Busher, A., and Battaglia, M. 2021. Evaluation of an Automated Truck Wash Modified with a Two-Stage Decontamination System for Sanitizing Transport Trucks At Large Farms or Animal Contaminant Facilities. Glob J Agric Innov Res Dev. 8: 95-103. DOI: https://doi.org/10.15377/2409-9813.2021.08.7

Rose, N. *et al.* 2000. 'Risk factors for Salmonella persistence after cleansing and disinfection in French broiler-chicken houses', *Preventive Veterinary Medicine*, 44(1–2), 9–20. doi: 10.1016/S0167-5877(00)00100-8.

Sanchez, J. *et al.* 2005. 'Temporal and spatial analysis of the 1999 outbreak of acute clinical infectious bursal disease in broiler flocks in Denmark', *Preventive Veterinary Medicine*, 71(3), 209–223. doi: 10.1016/j.prevetmed.2005.07.006.

Snow, L. C. *et al.* 2010. 'Investigation of risk factors for Salmonella on commercial egg-laying farms in Great Britain, 2004-2005', *The Veterinary Record*, 166(19), 579–586. doi: 10.1136/vr.b4801.

Tablante, N. L., and Malone, G. W. 2006. Controlling avian influenza through in-house composting of depopulated flocks: Sharing Delmarva's experience. In *Proceedings of 2006 National Symposium on Carcass Disposal*. https://www.animalmortmgmt.org/wp-content/uploads/2014/04/Controlling-AI-through-in-House-Composting-of-De-Populated-F.pdf

Toma, B. *et al.* 1999. *Dictionary of Veterinary Epidemiology*. Wiley.

Torremorell, M., Alonso, C., Davies, P. R., Raynor, P. C., Patnayak, D., Torchetti, M., and McCluskey, B. 2016. Investigation into the airborne dissemination of H5N2 highly pathogenic avian influenza virus during the 2015 spring outbreaks in the Midwestern United States. *Avian Diseases*, *60*(3), 637-643.

Trudeau, M. P., Verma, H., Sampedro, F., Urriola, P. E., Shurson, G. C., McKelvey, J., and Goyal, S. M. 2016. Comparison of thermal and non-thermal processing of swine feed and the use of selected feed additives on inactivation of porcine epidemic diarrhea virus (PEDV). *PLoS One*, *11*(6), e0158128.

Vaillancourt, J. and Martinez, A. 2001. 'Relative importance of biosecurity measures: A Delphi Study', in. *Annual Meeting of the American Association of Avian Pathologists, 138th Annual Convention AVMA*, Boston.

Vaillancourt, J.-P., Delpont, M., Racicot, M., Paul, M., and Guérin, J.-L. 2018. Une perspective régionale de la biosécurité. Le nouveau praticien vétérinaire; 10/n°40156.

Vandekerchove, D. *et al.* 2004. 'Colibacillosis in caged layer hens: characteristics of the disease and the aetiological agent', *Avian Pathology: Journal of the W.V.P.A*, 33(2), 117–125. doi: 10.1080/03079450310001642149.

Villa, B. and Velasco, A. 1994. 'Integrated pest management of the rat Rattus norvegicus in poultry farms', *Veterinaria México*, 25(3), 247–249.

Vukmirović, Đ. M., Rakita, S. M., Spasevski, N. J., Kokić, B. M., Banjac, V. V., and Čabarkapa, I. S. 2017. A review of possibilities for control of Salmonella and other pathogenic bacteria in pig feed. *Food and Feed research*, *2017*(2017-2), 151-162.

Wang, J., Vaddu, S., Bhumanapalli, S., Mishra, A., Applegate, T., Singh, M., and Thippareddi, H. 2023. A systematic review and meta-analysis of the sources of Salmonella in poultry production (pre-harvest) and their relative contributions to the microbial risk of poultry meat. *Poultry Science*, *102*(5), 102566.

Watkins, S. and Venne, D. 2015. Water quality. In Brugère-Picoux, J., Vaillancourt, J.-P., Bouzouaia, M., Shivaprasad, H. L. et Venne, D. (2015). Manuel de pathologie aviaire. Paris: AFAS.

Yoon, H., Lee, I., Kang, H., Kim, K. S., and Lee, E. 2022. Big data-based risk assessment of poultry farms during the 2020/2021 highly pathogenic avian influenza epidemic in Korea. *Plos one*, *17*(6), e0269311.

Keeping track of fibres: How fibres shape digesta transit and digestion in chickens

Garder les fibres à l'œil : comment les fibres influencent le transit gastro-intestinal et la digestion chez les poulets

S. de Vries[1], W.J.J. Gerrits[1]
[1]Animal Nutrition Group, Wageningen University & Research, Wageningen, The Netherlands
sonja.devries@wur.nl

Abstract

By altering the gastrointestinal environment, digesta passage behaviour, and the site of nutrient digestion, physicochemical properties of fibrous ingredients, such as particle size or gelling behaviour, are likely to influence the absorption kinetics of nutrients and their metabolic use. Results of recent studies on the regulation of digesta passage behaviour and nutrient digestion by dietary fibres in chickens will be presented and discussed. Through their presumed effects on gizzard grinding activity, insoluble dietary fibres at moderate inclusion levels typically improve the digestibility of proteins and, to a lesser extent, starch and fat, whereas soluble fibres hamper nutrient digestibility already at inclusion levels beyond 20 g/kg DM. Beyond their impact on gizzard functioning and ileal digestibility, fibres – particularly soluble fibres – may impact caecal filling and retention time. When studying digesta passage behaviour in chickens, selective retention of various fractions throughout the gastrointestinal tract should be anticipated. In the proximal gastrointestinal tract, coarse fibre particles may be retained longer in the crop or proventriculus and gizzard compared with fine particles and liquids. In the distal gastrointestinal tract, reverse peristalsis and selective entrance of liquids and fine particles into the caeca elicit considerably longer total tract retention times for these fractions (~14-31h) than typically observed for solid digesta (~3-8h). However, our provisional estimates indicate that only ~30% of the liquid digesta enters the ceca. These insights enhance understanding of the digestion process and thereby aid in optimizing feeding strategies, particularly when feeding ingredients with more diverse properties, to support gastrointestinal function and health.

Keywords: non-starch polysaccharides, broilers, laying hens, particle size

Résumé

En modifiant le milieu gastro-intestinal, le transit du digesta et le site de digestion des nutriments, les propriétés physico-chimiques des ingrédients fibreux, telles que la taille des particules ou leur comportement de gélification, sont susceptibles d'influencer la cinétique d'absorption des nutriments et leur utilisation métabolique. Les résultats d'études récentes sur la régulation du transit gastro-intestinal et de la digestion des nutriments par les fibres alimentaires chez les poulets seront présentés et discutés. Les fibres alimentaires insolubles, en raison de leurs effets présumés sur l'activité de broyage du gésier, améliorent généralement la digestibilité des protéines lorsqu'elles sont présentes en concentration modérée. Dans une moindre mesure, ces fibres favorisent également la digestion de l'amidon et des graisses. En revanche, les fibres solubles réduisent la digestibilité des nutriments dès qu'elles dépassent une concentration de 20 g/kg de MS. Au-delà de leur impact sur le fonctionnement du gésier et la digestibilité iléale, les fibres - en particulier les fibres solubles - peuvent intervenir sur le remplissage du cæcum et le temps de rétention des aliments. Dans le cadre de l'analyse du transit gastro-intestinal chez les poulets, il faut s'attendre à une rétention sélective des différentes fractions le long du tube digestif. Dans la partie proximale du tube digestif, les particules de fibre grossières peuvent être retenues plus longtemps dans le jabot ou le proventricule et le gésier que les particules fines et les liquides. Dans la portion distale, le péristaltisme inverse et l'entrée sélective des liquides et des particules fines dans le cæcum entraînent des temps de rétention totale considérablement plus longs pour ces fractions (~14 à 31h) que ce qui est généralement observé pour le digesta solide (~3 à 8h). Cependant, nos estimations provisoires indiquent que seulement 30 % du digesta liquide pénètre dans le cæcum. Ces connaissances permettent de mieux comprendre le processus de digestion et donc d'optimiser les stratégies d'alimentation, en particulier lorsqu'il s'agit d'ingrédients aux propriétés plus diverses, afin de favoriser la fonction et la santé gastro-intestinales.

Introduction

By altering the gastrointestinal environment, digesta passage behaviour, and the site of nutrient digestion, physicochemical properties of fibres are likely to influence the absorption kinetics of nutrients and their subsequent metabolic use. In chickens, the role of fibres, particularly insoluble fibres, on gizzard development and function has been extensively studied (reviewed by Hetland, et al., 2004; Svihus, 2011) but the influence of specific fibre physicochemical properties - such as particle size and -stiffness, gelling- and hydration properties, and fermentability - remains insufficiently understood. Moreover, fibres may impact nutrient (re)absorption beyond the small intestine by influencing the composition- and flow behaviour of digesta in the hindgut, yet these processes are considerably less studied and remain poorly understood. Advancing our understanding of these mechanisms is crucial to effectively integrate fibre aspects into feed formulation strategies.

Fibres as modulator of nutrient digestion and absorption kinetics?

Despite the limited degradation of fibres in chickens (reviewed by de Vries, et al., 2012; Lannuzel, et al., 2022), the effects of fibres on the digestion of other nutrients may be considerable. Well-known are the detrimental effects of various (soluble) fibre sources on the (apparent) digestion of protein, starch, and fat in the upper gastrointestinal (a.o. Choct and Annison, 1992; Smits and Annison, 1996; Smits, et al., 1997) by amongst others influencing digesta passage behaviour, accessibility of nutrients by enzymes and solute diffusion, microbial colonization, bile salt functionality, and endogenous losses (Grala, et al., 1998; Smits, et al., 1998; Lentle and Janssen, 2010). These adverse effects are mainly ascribed to the gelling properties of specific polysaccharides, such as soluble β-glucans and arabinoxylans commonly found in cereal grains, that modulate the digestive environment for example by increasing digesta viscosity. To limit these effects, fibre-degrading enzymes (such as β-glucanases and xylanases) are commonly added to poultry diets. Nevertheless, potential beneficial effects of fibres on intestinal health but also nutrient digestion, are increasingly addressed, and fibre-rich ingredients are nowadays considered as functional nutrients with the capacity to modulate the physical characteristics of the feed and the digesta, also in a positive manner (Mateos, et al., 2012; Kheravii, et al., 2018; Jha and Mishra, 2021; Lannuzel, et al., 2022). Through their presumed effects on development of the muscular gizzard, gizzard grinding activity, and secretion of pepsin and pancreatic enzymes, insoluble dietary fibres - at moderate inclusion levels - typically improve the digestibility of proteins and, to a lesser extent, starch and fat (reviewed by Vivares, et al., 2025, Figure 1). Inclusion of insoluble fibre sources, coarse particles, or inclusion of whole cereal grains, are, therefore, commonly used strategies to improve gastrointestinal functioning and nutrient use (Hetland, et al., 2002; Svihus, 2011; Akbari Moghaddam Kakhki, et al., 2024).

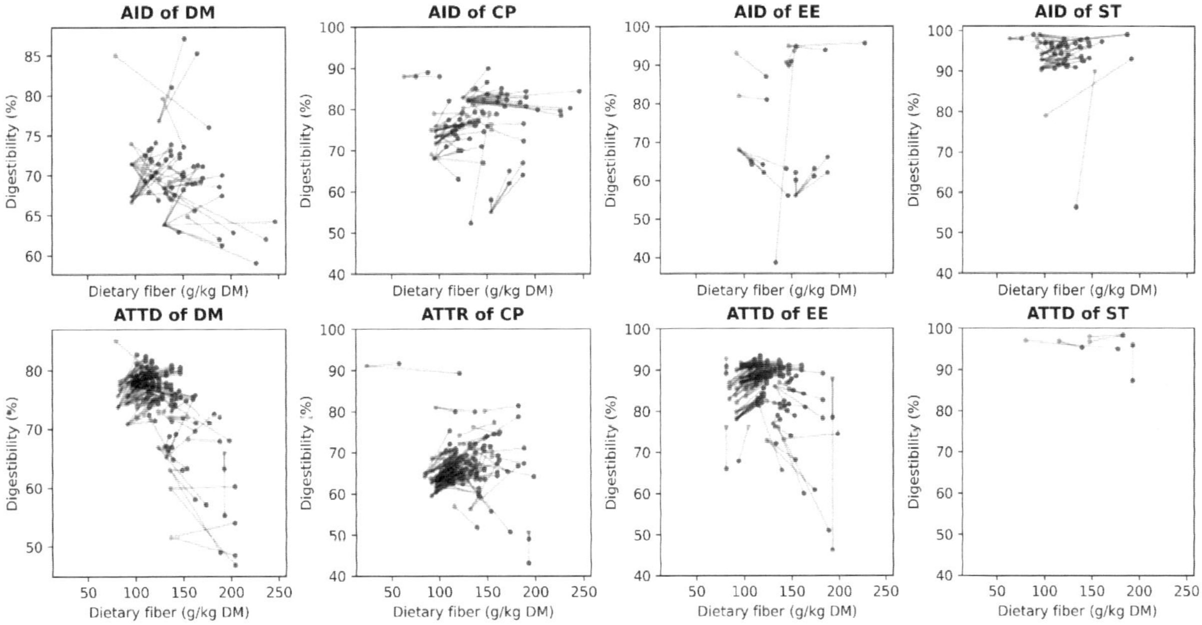

Figure 1. Relations between dietary fiber (DF) concentration (g/kg DM) and nutrient digestibility (%), as obtained from meta-analysis of a total of 45 studies and 198 experimental treatments. Orange dots indicate control diets and blue dots indicate treatment diets where DF was included, control and treatment diets within studies are connected by solid grey lines. Abbreviations: AID, apparent ileal digestibility; ATTD, apparent total tract digestibility; DM, dry matter; CP, crude protein; EE, ether extract; St, starch. Data from: Vivares et al., (2025).

Both soluble and insoluble dietary fibres have the ability to modulate the digestive environment, by affecting morphology and functionality of the stomach-complex and intestines, physicochemical properties of the digesta, and the gastrointestinal microbiome. Understanding the effects of fibres on digesta properties and digesta flow behaviour allows for better understanding of the final consequences for nutrient absorption as well as animal performance and gut health. For example, the selective retention of coarse particles and more rapid emptying of liquid digesta in the upper gastrointestinal tract unequivocally affects the timing of nutrients entering the small intestine, where nutrients that flow predominantly with the liquid phase, e.g. a large part of the dietary proteins (de Vries, et al., 2023), arrive earlier in the small intestine than those flowing with the solid phase. Such phenomena are often neglected when total extent of nutrient absorption is considered but are relevant to predict nutrient absorption kinetics and subsequent metabolic fate of nutrients after absorption (Schop, et al., 2023). By shaping the digestive environment and modifying digesta flow, nutrient hydrolysis and -absorption, fibres are important regulators of nutrient absorption kinetics, and accounting for this variation is crucial to accurately predict the nutritional value of feed ingredients, particularly when increasingly relying on fibre-rich feed ingredients in diets fed to chickens.

Modulation of digesta flow behaviour depending on fibre physicochemical properties

Predicting modulation of digesta properties and -flow by the diverse mix of fibres present in feed ingredients is, however, not straightforward. After all, digesta is a composite of particulate matter - with particles varying in e.g. size, shape, and buoyancy - suspended in liquids, where fibres contribute to the heterogeneity of the solid phase and the viscosity of the liquid phase. It follows that a one-dimensional parameter as water-solubility will not fully capture the diverse nature of effects observed. For commonly used feed ingredients, however separating between insoluble dietary fibres (iDF) and soluble dietary fibres (sDF), allows to distinguish between (1) those ingredients that predominantly provide structure/bulk and are not, or only limitedly, fermented and (2) those that exert strong gelling properties and thus increase digesta viscosity, and serve as substrate for the microbiome. Coarse insoluble fibre sources clearly increase gizzard weight (González-Alvarado, et al., 2010; Jiménez-Moreno, et al., 2010; Jiménez-Moreno, et al., 2019) – considered a proxy for its muscularity – and may prolong retention of solid and liquid digesta in the foregut (Mens, et al., 2022), whereas viscous soluble fibres reduced gizzard weight and prolonged retention time of solid and liquid digesta throughout the gastrointestinal tract (van der Klis and van Voorst, 1993; Dorado-Montenegro, et al., 2024). That a sole distinguishment between iDF and sDF is not sufficient for all situations is illustrated by the lack of effect of purified cellulose - a fine DF source - on gizzard weight (Jiménez-Moreno, et al., 2010) and the source-dependent effects of iDF on proventriculus + gizzard weight, total tract digesta retention time, and nutrient digestibility (Akbari Moghaddam Kakhki, et al., 2024). These observations highlight that it is the combination between chemical and physical properties of fibres, such as hydration properties (including solubility and swelling), particle size and -shape distributions, gelling properties and stiffness of particles – originating from the polymers constituting the fibre fraction and their crosslinks within the plant cell wall - that finally determine the multifaceted effects of fibres in the animal.

An often understated aspect is the selective entrance of liquids and fine particles into the ceca, whereas the ceca are largely bypassed by coarse particles (Vergara, et al., 1989; de Vries, et al., 2014; Vanderghinste, et al., 2025). After passing the ileocecal junction, digesta passes through the colon to the cloaca, where coarse particles are assumed to be directly voided, resulting in typical total tract mean retention times of ~3-8 h when estimated based on tracers for solid digesta (reviewed by Svihus and Itani, 2019). Reflux from the cloaca through the colon selectively delivers fluids, including urine, and small particles to the ceca. In this way, solubilized materials and fine particles can enter the ceca and be subjected to fermentation by the microbiota residing in the ceca. Estimated mean retention times in the ceca for liquid digesta range from 500-1500 min, implying that total tract mean retention times of digesta that enter the ceca may reach ~14-31 h (Garçon, et al., 2023). However, our provisional estimates indicate that only ~30% of the liquid digesta fraction entered the ceca (Garçon, et al., 2023). Following the rapid excretion of coarse solid digesta, iDF sources may be expected to have typically little impact on digesta retention in the hindgut. Soluble DF sources may in contrast be to affect digesta flow behaviour beyond the small intestine. For example, sugar beet pulp – having a relatively high sDF content (~ 40% of total non-starch polysaccharides) but also high water binding capacity – increased gizzard weight and tended to accelerate prececal, and - more importantly - cecal retention time of digesta (Garçon, et al., 2023; Akbari Moghaddam Kakhki, et al., 2024; Dorado-Montenegro, et al., 2024). Again highlighting the complexity of predicting the multifarious effect of the heterogenous fibre fractions found in feed ingredients.

Formulating with fibres: quantifying fibre parameters for feed evaluation purposes

It is evident that in order to account for the diverse effects that fibres can have in the animal, one should look beyond crude fibre or neutral detergent fibre characterizations, when formulating diets. Although these commonly used analytical procedures are robust with respect to analytical variation and might correlate well with the fate of specific fibre fractions in the animal (Mertens, 2003), they recover only a part of the fibre fraction, largely neglecting soluble fibres. A recent meta-analysis showed that considering both insoluble and soluble dietary fibres (DF), instead of total DF, allowed to significantly better explain variation in nutrient digestibility found (Vivares, et al., 2025). When added to maize-soybean meal diets with ~100 g/kg DF, moderate levels of insoluble DF (up to 50 g/kg DM) increased nutrient digestibility, particularly protein digestibility, up to 10%-units, whereas soluble DF negatively affected nutrient digestibility already at levels beyond 20 g/kg DM. These insights can be used to further refine feed evaluation models and account for interactions between fibre-rich and other feed ingredients. Potentially, concentrations of digestible nutrients in ingredients that are currently reported in feed databases, can be corrected for the presence of fibres at a diet level, like proposed for ileal digestible protein and amino acid values for pigs (Zhang, et al., 2024), possibly extended by including key-physicochemical properties. Contrary to our findings in pigs, where physicochemical properties of the feed explained variation in gastric emptying but were found to be limitedly related to digesta retention in more distal segments of the gastrointestinal tract (Dorado-Montenegro, et al., 2025), preliminary meta-analyses in chickens indicate that physicochemical properties of the feed modulate digesta passage behaviour throughout the gastrointestinal tract. Using data from 224 laying hens (32 replicate pens) and 456 broiler chickens (52 replicate pens) fed one of 10 diets, we found that among all properties evaluated, water binding capacity (WBC) and extract viscosity of the feed and stiffness of coarse particles were most consistently associated with digesta flow behaviour. Most prominent relations between feed physicochemical properties and digesta mean time (MRT) were found in the small intestine where ~80% of the variation in MRT could be explained by WBC, extract viscosity, particle size and -morphometry, and stiffness of coarse particles (Figure 1). In the hindgut extract viscosity largely explained variation in MRT of liquid digesta, whereas WBC and other particle properties (size, morphometry, and stiffness of coarse particles) explained segregation of solid and liquids digesta phases. Speculatively, these point towards fibre properties that modulate cecal access, not only of liquids, but also of (coarse) particles (van der Klis and van Voorst, 1993)

A point of consideration is the appropriate digestibility proxy to be used to predict the nutritional value of feed ingredients of fibre-rich diets. Apparent digestibility estimates include the aggregate effects of fibres on (1) true digestibility of dietary nutrients and (2) fibre-induced endogenous losses, but do not reveal any insights in the mode of action. In line with above described modes of action, preliminary results from a study with diets varying in soluble fibre sources, revealed that addition of soluble arabinoxylans reduced apparent ileal digestibility of proteins by reducing true digestibility, whereas sugar beet pulp reduced apparent digestibility but increased true digestibility (Habibi, et al., Unpublished-a). Simultaneous, converse effects of fibres on true digestibility versus endogenous losses, may therefore mask potential positive effects of fibres on true digestibility. To accurately account for these multifarious effects of fibres in feed formulation, insights in the specific effects of various fibre sources on different aspects of digestive processes, are crucial. Although convenient from a methodological viewpoint, the use of standardized (SID) values does not turn useful here, as SID only accounts for basal endogenous losses related to the quantity of feed ingested but neglects diet-specific endogenous losses. New methodologies to quantify diet-specific endogenous losses under semi-practical conditions (Noorman, et al., 2024; Habibi, et al., Unpublished-b) may prove useful to map nutrient losses from the digestion process for diets varying in fibre level and composition, and separate between undigested dietary-, endogenous-, and microbial components.

These insights can facilitate progress in refining predictions of nutritional values of feed ingredients and diet formulation particularly when relying increasingly on fibre-rich feed ingredients diets with exerting diverse properties physicochemical properties.

References

Akbari Moghaddam Kakhki, R., A. Navarro-Villa, J. de los Mozos, S. de Vries, and A. I. García-Ruiz. 2024. Evaluation of fibrous feed ingredients alternatives to oat hulls as a source of feed structure in broiler diets. Poult Sci 103 104297. doi https://doi.org/10.1016/j.psj.2024.104297

Choct, M., and G. Annison. 1992. The inhibition of nutrient digestion by wheat pentosans. Br J Nutr 67 123-132.

de Vries, S., H. Beukers-van laar, A. I. Garcia Ruiz, and W. J. J. Gerrits. 2023. Short communication: Estimation of ileal digestibility in chickens using single- and dual tracer methods. Animal submitted.

de Vries, S., R. P. Kwakkel, A. M. Pustjens, M. A. Kabel, W. H. Hendriks, and W. J. J. Gerrits. 2014. Separation of digesta fractions complicates estimation of ileal digestibility using marker methods with cr_2o_3 and co-edta in broiler chickens. Poult Sci 93 2010 - 2017.

de Vries, S., A. M. Pustjens, H. A. Schols, W. H. Hendriks, and W. J. J. Gerrits. 2012. Improving digestive utilization of fiber-rich feedstuffs in pigs and poultry by processing and enzyme technologies: A review. Anim Feed Sci Technol 178 123-138.

Dorado-Montenegro, S., J. A. Dijksman, W. J. J. Gerrits, and S. de Vries. 2025. Predicting digesta retention times from dietary physicochemical properties in pigs. Anim Feed Sci Technol 321 116242. doi https://doi.org/10.1016/j.anifeedsci.2025.116242

Dorado-Montenegro, S., M. F. Habibi, W. J. J. Gerrits, and S. de Vries. 2024. Effect of adding soluble viscous fibers to diets containing coarse and finely ground insoluble fibers on digesta transit behavior and nutrient digestibility in broiler chickens. Poult Sci 103 103487. doi https://doi.org/10.1016/j.psj.2024.103487

Garçon, C. J. J., J. L. Ellis, C. D. Powell, A. Navarro Villa, A. I. Garcia Ruiz, J. France, and S. de Vries. 2023. A dynamic model to measure retention of solid and liquid digesta fractions in chickens fed diets with differing fibre sources. animal 17 100867. doi https://doi.org/10.1016/j.animal.2023.100867

González-Alvarado, J. M., E. Jiménez-Moreno, D. González-Sánchez, R. Lázaro, and G. G. Mateos. 2010. Effect of inclusion of oat hulls and sugar beet pulp in the diet on productive performance and digestive traits of broilers from 1 to 42 days of age. Anim Feed Sci Technol 162 37-46. doi https://doi.org/10.1016/j.anifeedsci.2010.08.010

Grala, W., M. W. Verstegen, A. J. Jansman, J. Huisman, and P. van Leeusen. 1998. Ileal apparent protein and amino acid digestibilities and endogenous nitrogen losses in pigs fed soybean and rapeseed products. J Anim Sci 76 557-568.

Habibi, M. F., S. Dorado-Montenegro, A. I. Gracia Ruiz, W. J. J. Gerrits, D. Georgaki, R. P. Kwakkel, and S. de Vries. Unpublished. Viscous soluble dietary fibers increase estimated total endogenous nitrogen losses and decrease true digestibility of nitrogen in broilers. Wageningen University & Research, Wageningen, The Netherlands.

Habibi, M. F., S. Dorado-Montenegro, R. K. Kwakkel, W. J. J. Gerritsa, A. I. Gracia Ruiz, and S. de Vries. Unpublished. Quantification of total endogenous nitrogen losses and true nitrogen digestibility in broilers using a minimally-invasive 15n-isotope dilution method. Wageningen University & Research, Wageningen, The Netherlands.

Hetland, H., M. Choct, and B. Svihus. 2004. Role of insoluble non-starch polysaccharides in poultry nutrition. World Poult Sci J 60 415-422.

Hetland, H., B. Svihus, and V. Olaisen. 2002. Effect of feeding whole cereals on performance, starch digestibility and duodenal particle size distribution in broiler chickens. Br Poult Sci 43 416-423.

Jha, R., and P. Mishra. 2021. Dietary fiber in poultry nutrition and their effects on nutrient utilization, performance, gut health, and on the environment: A review. Journal of Animal Science and Biotechnology 12 51. doi 10.1186/s40104-021-00576-0

Jiménez-Moreno, E., J. M. González-Alvarado, A. de Coca-Sinova, R. P. Lázaro, L. Cámara, and G. G. Mateos. 2019. Insoluble fiber sources in mash or pellets diets for young broilers. 2. Effects on gastrointestinal tract development and nutrient digestibility1. Poult Sci 98 2531-2547. doi https://doi.org/10.3382/ps/pey599

Jiménez-Moreno, E., J. M. González-Alvarado, D. González-Sánchez, R. Lázaro, and G. G. Mateos. 2010. Effects of type and particle size of dietary fiber on growth performance and digestive traits of broilers from 1 to 21 days of age1. Poult Sci 89 2197-2212. doi https://doi.org/10.3382/ps.2010-00771

Kheravii, S. K., M. N. K., S. R. A., C. M., and S. B. and Wu. 2018. Roles of dietary fibre and ingredient particle size in broiler nutrition. World Poult Sci J 74 301-316. doi 10.1017/S0043933918000259

Lannuzel, C., A. Smith, A. L. Mary, E. A. Della Pia, M. A. Kabel, and S. de Vries. 2022. Improving fiber utilization from rapeseed and sunflower seed meals to substitute soybean meal in pig and chicken diets: A review. Anim Feed Sci Technol 285 115213. doi https://doi.org/10.1016/j.anifeedsci.2022.115213

Lentle, R., and P. Janssen. 2010. Manipulating digestion with foods designed to change the physical characteristics of digesta. Crit Rev Food Sci Nutr 50 130-145.

Mateos, G. G., E. Jiménez-Moreno, M. P. Serrano, and R. P. Lázaro. 2012. Poultry response to high levels of dietary fiber sources varying in physical and chemical characteristics. J Appl Poult Res 21 156-174.

Mens, A. J. W., A. J. M. Jansman, E. Willems, and M. M. van Krimpen. 2022. Improvement of protein digestibility by changing digesta passage kinetics in broilers in Feed For FoodureWageningen University & Research, Wageningen, The Netherlands.

Mertens, D. R. 2003. Challenges in insoluble dietary fiber. J Anim Sci 81 3233-3249.

Noorman, L., B. van der Hee, W. J. Gerrits, K. C. Lammers-Jannink, A. K. Kies, N. V. der Wielen, M. Tretola, G. J. Hooiveld, and S. de Vries. 2024. Potential new methods to analyze basal and total endogenous protein losses of host and bacterial origin in pigs. J Nutr 154 3832-3846. doi 10.1016/j.tjnut.2024.10.029

Schop, M., H. Nguyen-Ba, A. J. M. Jansman, S. de Vries, J. L. Ellis, A. Bannink, and W. J. J. Gerrits. 2023. Snapig: A model to study nutrient digestion and absorption kinetics in growing pigs based on diet and ingredient properties. animal 17 101025. doi https://doi.org/10.1016/j.animal.2023.101025

Smits, C. H. M., and G. Annison. 1996. Non-starch plant polysaccharides in broiler nutrition–towards a physiologically valid approach to their determination. World Poult Sci J 52 203-221.

Smits, C. H. M., A. Veldman, H. J. Verkade, and A. C. Beynen. 1998. The inhibitory effect of carboxymethylcellulose with high viscosity on lipid absorption in broiler chickens coincides with reduced bile salt concentration and raised microbial numbers in the small intestine. Poult Sci 77 1534-1539.

Smits, C. H. M., A. Veldman, M. W. A. Verstegen, and A. C. Beynen. 1997. Dietary carboxymethylcellulose with high instead of low viscosity reduces macronutrient digestion in broiler chickens. J Nutr 127 483-487.

Svihus, B. 2011. The gizzard: Function, influence of diet structure and effects on nutrient availability. World Poult Sci J 67 207-223.

Svihus, B., and K. Itani. 2019. Intestinal passage and its relation to digestive processes. Journal of Applied Poultry Research 28 546-555. doi https://doi.org/10.3382/japr/pfy027

van der Klis, J. D., and A. van Voorst. 1993. The effect of carboxy methyl cellulose (a soluble polysaccharide) on the rate of marker excretion from the gastrointestinal tract of broilers. Poult Sci 72 503-512. doi 10.3382/ps.0720503

Vanderghinste, P., A. Bautil, M. R. Bedford, G. González-Ortiz, C. Lamberigts, I. Aslam, M. Roeffaers, and C. M. Courtin. 2025. Revealing the physical restrictions of caecal influx in broilers through the use of solid and soluble markers. Animal Nutrition 21 37-48. doi https://doi.org/10.1016/j.aninu.2024.12.004

Vergara, P., C. Ferrando, M. Jiménez, E. Fernández, and E. Goñalons. 1989. Factors determining gastrointestinal transit time of several markers in the domestic fowl. Quarterly Journal of Experimental Physiology 74 867-874.

Vivares, G., M. P. de Nanclares, A. Conde-Aguilera, and S. de Vries. 2025. Quantifying the interactions between dietary fibers and macronutrient digestibility in broiler chickens: The importance of considering fiber solubility. Anim Feed Sci Technol 321 116241. doi https://doi.org/10.1016/j.anifeedsci.2025.116241

Zhang, S., S. de Vries, and W. J. J. Gerrits. 2024. Quantifying the effects of dietary fibres on protein digestibility in pigs - a review. Anim Feed Sci Technol 308 115864. doi https://doi.org/10.1016/j.anifeedsci.2023.115864

Effects of canola meal processing on the choline bioavailability in canola meal-based broiler diets

Effets de la transformation du tourteau de canola sur la biodisponibilité de la choline dans les rations pour poulets à griller à base de tourteau de canola

S. Karki[1], *R. W. Newkirk*[2]
[1]*Graduate Student, Department of Animal and Poultry Science, University of Saskatchewan, Saskatoon, SK*
qnl457@usask.ca
[2]*Associate Professor, Department of Animal and Poultry Science, University of Saskatchewan, Saskatoon, SK*
rex.newkirk@usask.ca

SPONSORED BY / COMMANDITÉ PAR

trouw nutrition
a Nutreco company

Abstract

Canola meal, a vital protein source in Canadian poultry diets, has significant amounts of choline, primarily as sinapine. However, the bioavailability of choline in a canola meal-based diet and the impact of different processing methods have not been reported. Thus, this study aimed to evaluate the bioavailability of choline in two types of canola meal: expelled (ECM) and solvent-extracted (SCM), by assessing ileal digestibility. A total of 240-day-old Ross 308 broiler chicks were randomly assigned to six dietary treatments with ECM and SCM at inclusion levels of 5, 10, and 15%, each with eight replications of 5 birds. On day 26, ileal digesta samples were collected and analyzed for choline and sinapine using Liquid Chromatography Mass Spectroscopy (LC-MS). An indigestible marker, celite, was included in the diet to determine digestibility. The sinapine-to-choline ratio was 1.3 and 2.85, and choline levels were 2726 mg/kg and 1238 mg/kg for ECM and SCM, respectively. ECM-based diets had higher choline digestibility compared to SCM, which was 76.15% for ECM versus 44.71% for SCM ($p<0.05$) in diets containing 15% canola meal. There was a negative relationship between sinapine content and its digestibility. The solvent-extracted meal had lower sinapine and choline digestibility, indicating that processing methods affected nutrient digestibility. It was concluded that SCM has lower choline digestibility than

ECM. However, both types of canola meal are a rich source of choline and can be used as a source of choline in chicken diets.

Résumé

Le tourteau de canola, une précieuse source de protéines essentielle dans l'alimentation des volailles produites au Canada, contient une quantité significative de choline, principalement sous forme de sinapine. Cependant, la biodisponibilité de la choline dans un régime à base de tourteau de canola et l'impact des différentes méthodes de transformation n'ont pas été rapportés. Cette étude visait donc à évaluer, par la mesure de la digestibilité iléale, la biodisponibilité de la choline dans deux types de tourteau de canola, soit le tourteau obtenu par expulsion (ECM) et le tourteau résultant de l'extraction par solvant (SCM). Deux cent quarante poussins à griller Ross 308 âgés de un jour ont été aléatoirement assignés à un de six traitements alimentaires contenant 5 %, 10 % ou 15 % d'ECM ou de SCM, chacun prévoyant huit répétions avec cinq oiseaux. Au jour 26, des échantillons de digesta iléal ont été prélevés et analysés pour mesurer leur teneur en choline et sinapine par la méthode du couplage de chromatographie en phase liquide et spectrométrie de masse (LC-MS). Un marqueur non digestible, la célite, a été inclus dans la ration pour déterminer la digestibilité. Le rapport sinapine/choline était de 1,3 pour le tourteau ECM et de 2,85 pour le tourteau SCM, et les concentrations de choline étaient respectivement de 2 726 mg/kg et 1 238 mg/kg. Les régimes à base d'ECM ont présenté une digestibilité de la choline supérieure à celle des régimes à base de SCM, soit 76,15 % pour l'ECM contre 44,71 % pour le SCM ($p<0,05$), dans les rations contenant 15 % de tourteau de canola. Une relation inverse entre la teneur en sinapine et sa digestibilité a été observée. La digestibilité de la sinapine et de la choline a été moindre dans le tourteau traité par extraction par solvant, ce qui indique que les méthodes de traitement influencent la digestibilité des nutriments. Les auteurs ont conclu que la choline était moins digestible dans le cas du SCM que dans celui de l'ECM. Cependant, les deux types de tourteau de canola constituent une riche source de choline et peuvent être utilisés comme source de choline dans l'alimentation des poulets, réduisant ainsi la nécessité d'une supplémentation en choline de synthèse.

Introduction

Poultry diets are supplemented with synthetic choline, as it is required for maintaining cellular integrity, supporting liver lipid metabolism, and functioning as a neurotransmitter and methyl donor (Mohan Lunagariya and Sorathiya 2019). Canola meal, a widely used protein source in Canadian poultry diets, can serve as an important source of choline due to its high choline content. However, its contribution to choline supplementation is often overlooked due to limited information on its bioavailability and the effects of different processing methods. A 1997 study indicated that choline in canola meal had lower bioavailability, however, the specific type of canola meal examined was not mentioned, and the reasons for its reduced bioavailability remained unclear (Emmert and Baker 1997). One possible explanation is that a significant portion of the choline in canola meal is bound as phenolic choline esters, such as sinapine, which constitutes approximately 1% of the canola meal. This compound can be metabolized by microorganisms in the gastrointestinal tract, subsequently releasing choline and sinapic acid. The choline is then converted to trimethylamine before absorption, rendering it unavailable to the chickens (Khajali and Slominski 2012). Thus, this paper aims to evaluate the efficacy of canola meal as a source of choline for broilers, specifically comparing two types of canola meal: expelled canola meal (ECM) and solvent-extracted canola meal (SCM), by assessing ileal digestibility.

Materials and Methods

The protocol for this study was reviewed and approved by the Animal Research Ethics Board at the University of Saskatchewan (approval number 20230088). A 26-day feeding trial was conducted using a 2x3 factorial completely randomized block design (RCBD). The trial involved 240-day-old Ross 308 broiler chicks, which were fed two types of canola meal-based diets (ECM & SCM) with inclusion levels of 5%, 10%, and 15%. There were 8 replicates, with 5 birds in each replicate. The birds were fed a starter diet from days 0 to 14 and a grower diet from days 0 to 26. On day 26, all the birds were euthanized, and pooled ileal samples from each cage were collected. The samples were analyzed for sinapine and choline in both feed and ileal digesta using an Agilent 1100 Liquid Chromatograph paired with a high-resolution mass spectrometer (Q-Star XL, Applied Biosystems, USA), which features an ESI turbo ion source. Sample preparation and analysis were conducted with modifications from (Chen et al. 2019). Digestibility was determined through the indicator method, as celite was incorporated into the grower diet. Statistical analyses of digestibility were performed using analysis of variance (ANOVA) in a randomized complete block design (RCBD) with a factorial arrangement, utilizing the mixed model procedure of SAS. Significant differences between means were determined at $P \leq 0.05$.

Results and Discussions

ECM has higher choline (2726 mg/kg) than SCM (1238 mg/kg), with comparable sinapine content (0.36%). Consequently, ECM has a lower sinapine: choline ratio (1.3) compared to SCM (2.85), as shown in **Table 1**. SCM undergoes 30–40 minutes of desolventization-toasting at 95-120°C, which may lead to choline chloride degradation. This prolonged heating, along with moisture and the canola meal matrix, might have caused more degradation in SCM than in ECM, which uses shorter heating times. Sinapine stability is unaffected by processing.

Table 1. Sinapine and Content in Differently Processed Canola Meal Expelled (ECM) and Solvent-Extracted (SCM)

Meals	Choline (mg/kg)	Sinapine (%)	Sinapine: Choline
ECM	2726	0.37	1.3
SCM	1238	0.36	2.85

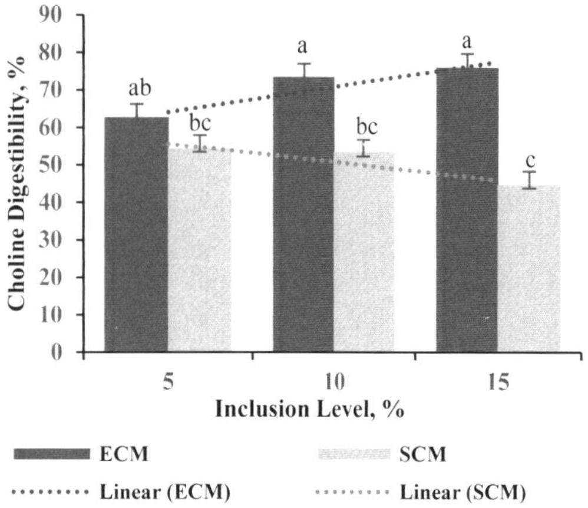

Figure 1: Ileal Choline Digestibility of Two Different Processed Canola Meals: Expelled (ECM) and Solvent-Extracted (ECM) at Inclusion Levels of 5, 10, and 15%. $^{a-c}$ Means Sharing Different Letters Differ Significantly (P<0.05).

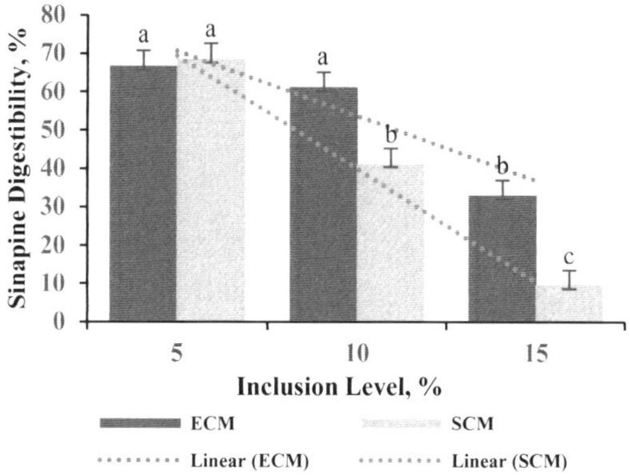

Figure 2: Sinapine Choline Digestibility of Two Different Processed Canola Meals: Expelled (ECM) and Solvent-Extracted (ECM) at Inclusion Levels of 5, 10, and 15%. $^{a-c}$ Means Sharing Different Letters Differ Significantly (P<0.05).

Figures 1 and 2 show contrasting ileal digestibility patterns for choline and sinapine across different canola types (likely ECM and SCM) at different inclusion levels in chickens, which are known to lack or have very low sinapine esterase activity. ECM showed higher digestibility, peaking at 15% inclusion, while SCM digestibility decreased significantly (p<0.05) at the highest level. Conversely, sinapine digestibility was highest at the lowest (5%) inclusion and declined sharply with increasing inclusion for both meal types, reaching the lowest point at 15% for SCM. This decrease in sinapine digestibility with higher inclusion levels in chickens, lacking efficient sinapine esterase, suggests that even limited or non-specific esterase activity might be sufficient to hydrolyze sinapine at lower concentrations, facilitating some choline release and absorption. However, as the sinapine concentration increases with higher inclusion levels, this limited enzymatic capacity becomes saturated, leading to a greater proportion of intact, poorly absorbed sinapine passing through the ileum, thus reducing overall digestibility. The differing trends for free choline digestibility further underscore the impact of the ester linkage in sinapine on its absorption dynamics.

Overall, this study revealed that both ECM and SCM can provide choline in chicken diets, but their effectiveness differs. ECM is likely a better source of digestible choline due to its higher content and more absorbable form. In contrast, choline bound as sinapine in both meals is less bioavailable, as chickens have limited ability to break down this compound, especially at higher inclusion levels; therefore, further studies are required to improve the bioavailability of sinapine-derived choline.

References

Chen, H., Peng, L., Pérez De Nanclares, M., Trudeau, M.P., Yao, D., Cheng, Z., Urriola, P.E., Mydland, L.T., Shurson, G.C., Overland, M., and Chen, C. 2019. Identification of Sinapine-Derived Choline from a Rapeseed Diet as a Source of Serum Trimethylamine N-Oxide in Pigs. J Agric Food Chem 67: 7748–7754. American Chemical Society. doi:10.1021/acs.jafc.9b02950.

Emmert, J.L., and Baker, D.H. 1997. Nutrient Metabolism A Chick Bioassay Approach for Determining the Bioavailable Choline Concentration in Normal and Overheated Soybean Meal, Canola Meal and Peanut Meal 1, 2. J. Nutr. [Online] Available: https://academic.oup.com/jn/article/127/5/745/4724190.

Khajali, F., and Slominski, B.A. 2012. Factors that affect the nutritive value of canola meal for poultry. Oxford University Press. doi:10.3382/ps.2012-02332.

Mohan Lunagariya, P., and Sorathiya, K.K. 2019. An importance of choline chloride for poultry and cattle: an overview. [Online] Available: www.ijset.net.

Blachford
SINCE 1921
Calcium Stearate
For Animal Feed
Anti-Caking • Waterproofing
Improved Flow
Release & Binding Agent
CERTIFIED SAFE FEED SFSF SAFE FOOD FACILITY
ISO 9001:2015 SRI CERTIFIED
SF/SF FSC36 • FSMA Compliant
Organic • Made in Canada
www.blachford.com

SHE'S GIVING YOU HER BEST YEARS.
IT'S UP TO YOU TO MAKE THEM COUNT.
There's nothing special about this cow—except everything. She navigates calving, pregnancy, and dry off with ease, producing high-quality milk along the way. She's a Dia-V NTK cow, leading the pack and built to withstand anything thrown at her—by us or Mother Nature.
Talk to your Diamond V representative to learn more.
Dia-V NTK
Diamond V

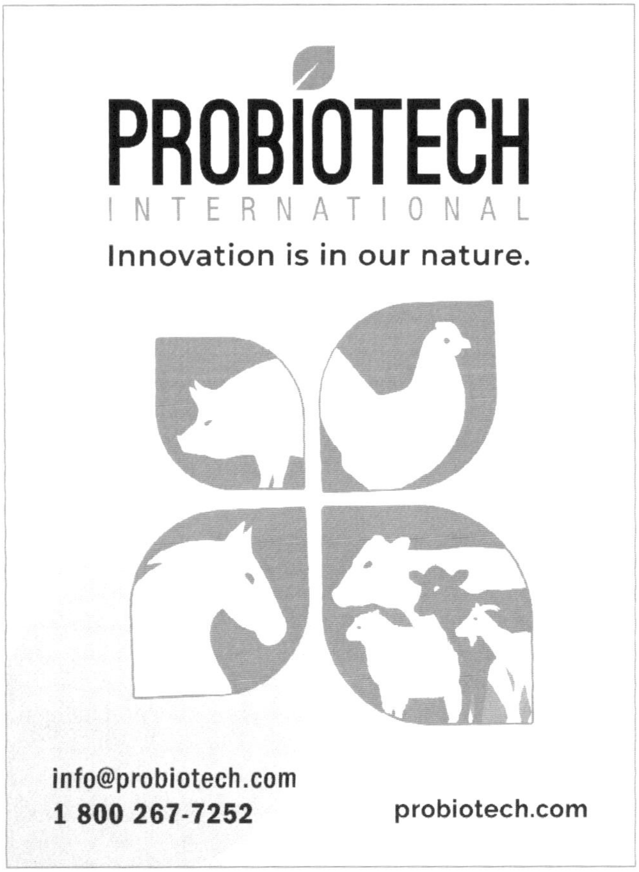
PROBIOTECH
INTERNATIONAL
Innovation is in our nature.
info@probiotech.com
1 800 267-7252
probiotech.com

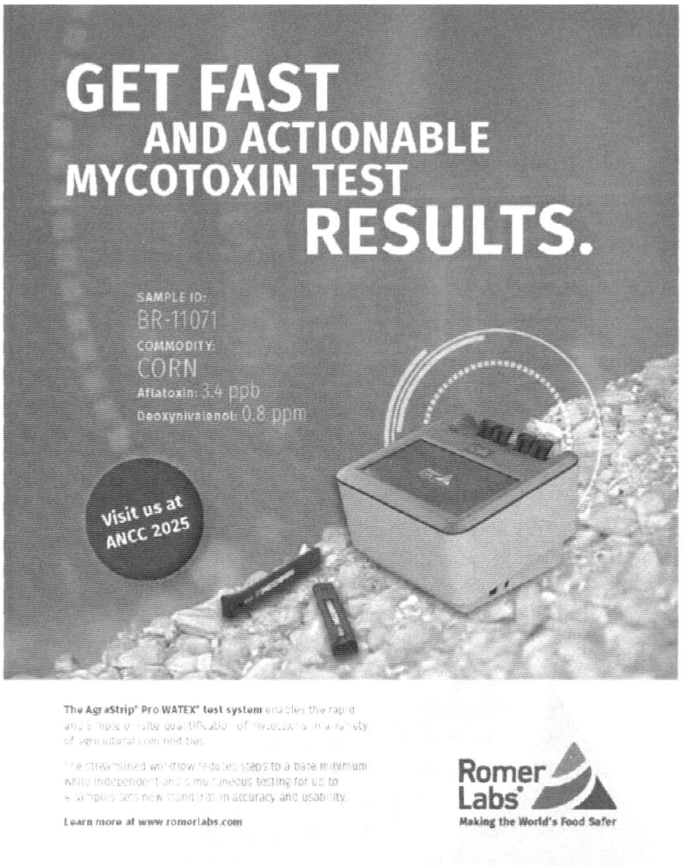
GET FAST AND ACTIONABLE MYCOTOXIN TEST RESULTS.
SAMPLE ID:
BR-11071
COMMODITY:
CORN
Aflatoxin: 3.4 ppb
Deoxynivalenol: 0.8 ppm
Visit us at ANCC 2025
The AgraStrip® Pro WATEX® test system
Learn more at www.romerlabs.com
Romer Labs®
Making the World's Food Safer

Impacts of feeding probiotics and insects on the nutrition and gut microbiome of farmed fish with new insights for other monogastric animals

Impacts de l'alimentation avec des probiotiques et des insectes sur la nutrition et le microbiome intestinal des poissons d'élevage et nouvelles perspectives pour d'autres animaux monogastriques

D. Huyben[1]
[1] *Assistant Professor of Aquaculture,*
Department of Animal Biosciences, University of Guelph, Guelph, ON
huybend@uoguelph.ca

Abstract

Eating fish is a healthy option that provides high levels of protein, omega-3 fatty acids, vitamins and minerals for a growing human population. Majority of the wild fish stocks are overfished, while demand for fish continues to increase. Aquaculture, the farming of finfish, shellfish, crustacean and aquatic plants, provides an alternative method to produce seafood while conserving wild fish. However, taking wild fish to feed fish on a farm in the form of fishmeal is not sustainable, therefore alternative non-fish alternatives need to be researched and developed to support the growth of the aquaculture industry. In addition, these alternatives need to ideally improve the growth, nutrition and gut health of fish to increase the resiliency of fish farm production against common stressors. Recent studies at the University of Guelph aimed to evaluate the effects of feeding insects (black soldier fly), probiotics (live yeast) and prebiotics (autolyzed yeast) on the growth, nutrition and gut microbiome of zebrafish and rainbow trout. 9 and 16 week studies were performed at the Hagen Aqualab and Ontario Aquaculture Research Centre with 0.1g zebrafish at 26°C and 93g rainbow trout at 9°C. Feeding insects improved the growth and FCR of zebrafish and rainbow trout, although not significantly. Nutrient digestibility increased in fish fed the prebiotics, but no effects of the probiotics were found. Insects modulated the gut bacteria the most. These results indicate insects and prebiotic ingredients improve the growth and health of fish and may strengthen the resiliency and growth of the aquaculture industry.

Keywords: aquaculture, prebiotics, rainbow trout, yeast, zebrafish

Résumé

Dans un contexte de croissance démographique, la consommation de poisson est une pratique nutritionnelle bénéfique qui procure des quantités significatives de protéines, d'acides gras oméga-

3, de vitamines et de minéraux essentiels. La forte demande pour le poisson met sous pression la ressource sauvage déjà surexploitée. L'aquaculture, une activité qui consiste à élever des poissons, des mollusques, des crustacés et des plantes aquatiques, constitue une méthode de production de substitution permettant de préserver les stocks de fruits de mer sauvages. Toutefois, transformer des poissons sauvages en farine pour nourrir les poissons d'élevage n'est pas une solution durable. C'est pourquoi il convient de chercher et de créer des aliments issus d'autres sources pour favoriser le développement du secteur de l'aquaculture. En outre, ces substituts doivent idéalement améliorer la croissance, la nutrition et la santé intestinale des poissons afin d'accroître la résilience de la production des piscicultures face aux facteurs de stress habituels. Des études ont récemment été réalisées à l'Université de Guelph pour évaluer les effets de l'utilisation d'insectes (mouche soldat noire), de probiotiques (levure vivante) et de prébiotiques (levure autolysée) sur la croissance, la nutrition et le microbiome intestinal du poisson zèbre et de la truite arc-en-ciel. Des études de 9 et 16 semaines ont été réalisées au Hagen Aqualab et à l'Ontario Aquaculture Research Centre avec 0,1 g de poisson zèbre à 26 ºC et 93 g de truite arc-en-ciel à 9 ºC. L'alimentation avec des insectes a amélioré la croissance et l'ICA du poisson zèbre et de la truite arc-en-ciel, bien que les résultats ne soient pas significatifs. La digestibilité des nutriments a augmenté chez les poissons nourris avec les prébiotiques, mais aucun effet des probiotiques n'a été constaté. Ce sont les insectes qui ont produit le plus d'effet sur les bactéries intestinales. Ces résultats indiquent que les ingrédients à base d'insectes et de prébiotiques améliorent la croissance et la santé des poissons et peuvent favoriser la résilience et le développement de l'industrie de l'aquaculture.

Introduction

Aquaculture is the farming of aquatic plants and animals, which includes algae, molluscs, crustaceans as well as both freshwater and marine fish. Aquaculture is a rapidly expanding agri-food sector since demand for seafood is increasing and production from capture fisheries has stagnated since the 1990's (FAO, 2024). In 2020, per capita consumption of fish and seafood for the world was 20kg out of 63kg (or 32%) compared to poultry, beef and pork, while specific countries varied between 22kg (15%) and 47kg (47%) for USA and Japan, respectively (FAO, 2020). The Canadian aquaculture industry produced 166,000 tonnes in 2022, valued at $1.3 billion CAD (DFO, 2024). This represents a large increase in production overall as in 1991, Canadian aquaculture produced only 50,000 tonnes (valued at $234 million CAD (DFO, 2024). The majority of Canada's aquaculture production is from Atlantic salmon farmed on the coasts, where British Columbia accounted for 58% of total production and the Maritime provinces (NS, NB, PEI and NFLD) accounted for 37% (DFO, 2024). Ontario accounted for 4% or 5,100 tonnes of mainly rainbow trout and most net-pen fish farms are located in Northern Ontario, specifically Georgian Bay, Lake Huron and Manitoulin Island (OARC, 2024). The economic contribution of Ontario's aquaculture industry was estimated to be $98 million CAD annually, and it directly employs 140 people and indirectly employs 150 people full-time (OARC, 2024). The aquaculture industry therefore continues to play an important role in Canada's economy and producing local food for Ontarians.

Farmed rainbow trout that range between 20 to 200 g in size are recommended to be fed a minimum of 40% crude protein on a wet-matter basis (NRC, 2011). Historically, much of the protein requirement of rainbow trout has been provided by feeding fishmeal, although it is not sustainable to feed farmed fish wild fish since this would add pressure onto already overfished wild stocks (FAO, 2024). Following years of research and development the content of marine protein, such as fishmeal, in fish feeds has decreased from 65 to 18% between 1990 and 2013 that has made fish feeds more sustainable (Ytrestøyl et al., 2015). However, the high demand for plant protein, such as soybean meal, from human and other livestock feeds has reduced the affordability and availability of plants as an alternative to fishmeal. In addition, plant proteins commonly possess anti-nutritional factors that lead to gut inflammation and dysbiosis of the gut microbiome, leading to reduced fish growth and health (Francis et al., 2001). Therefore, the purpose of this paper will be to briefly review novel ingredients, such as probiotics and insects, as alternatives to fishmeal as well as functional additives that improve fish growth and health.

Novel ingredients that improve fish nutrition

Fishmeal has been replaced with animal by-products, such as poultry meal, and plant proteins, such as soybean meal, in fish feeds for several decades. However, these protein sources are commonly used to feed humans, livestock and pets as well as use in pharmaceutical industries. Therefore, novel ingredients, such as single cell protein (SCP) and insect meals have a high potential to be included in fish feeds since they are of non-human interest, grown on organic waste stream, culture rapidly and are moderate to high in terms of crude protein (Weththasinghe et al., 2022). Yeast is a SCP that has a crude protein content between 44-55% on a dry matter basis and black soldier fly (BSF) meal is similar, which is much less than fishmeal (69-79%) but equal to soybean meal (46-52%) (Nyman et al., 2017; Huyben et al., 2019). However, other SCP, such as bacterial meals, and insects species, such as cricket meal, have a crude protein level between 66-78%, which is similar to fishmeal and may have higher potential to replace fishmeal in fish feeds (Drosdowech et al., 2024). See Figure 1 for a comparison of crude protein across several feed ingredicnts.

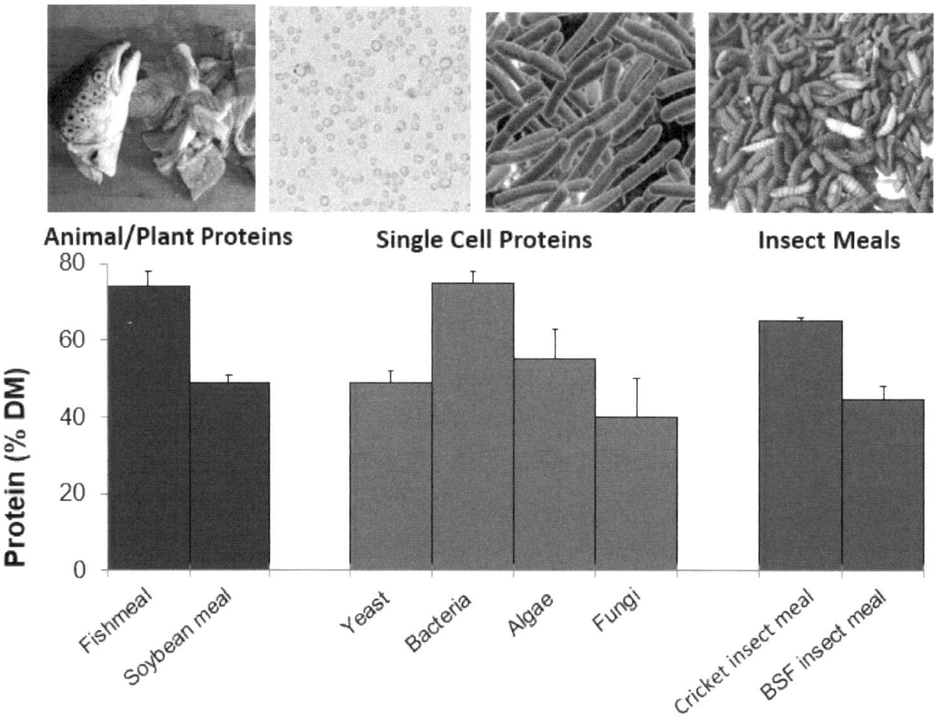

Figure 1. Benefits of the bacteria in the gut of the host. However, most of the research so far has been focused on humans and mice for biomedical research, rather than livestock health and nutrition. Data from Nyman, Huyben et al (2017); Huyben et al (2019); Drosdowech et al (2024).

Novel ingredients that improve the fish gut microbiome

The gut microbiome is the collection of microbes, such as bacteria, fungi and archaea, in the intestinal tract of a host. The gut microbiome is gaining interest for agri-food producers and has been shown to aid in metabolic function, such as fibre digestion, biosynthesis of vitamins and amino acids, production of short chain fatty acids (SCFA) and promotion of fat storage (Laukens et al., 2015). In addition, bacteria and other microbes in the gut aid in protection of epithelial injury, resistance to pathogens, modification of the nervous system, training of the immune system and other roles as mentioned in Figure 2 (Laukens et al., 2015). However, research on the roles of the gut microbiome have mainly focused on humans and mice, and less so on food production animals and even less so for farmed fish. Recently, there has been research on the effects of location (Llewellyn et al., 2014), diet (Huyben et al., 2019; 2020), temperature (Huyben et al., 2018), laboratory methods (Cao et al., 2024) and other factors on the gut microbiome of Atlantic salmon and rainbow trout. Despite these advancements, the effects of feeding probiotics and insects on the gut microbiome of fish and downstream effects on growth performance are still relatively unknown.

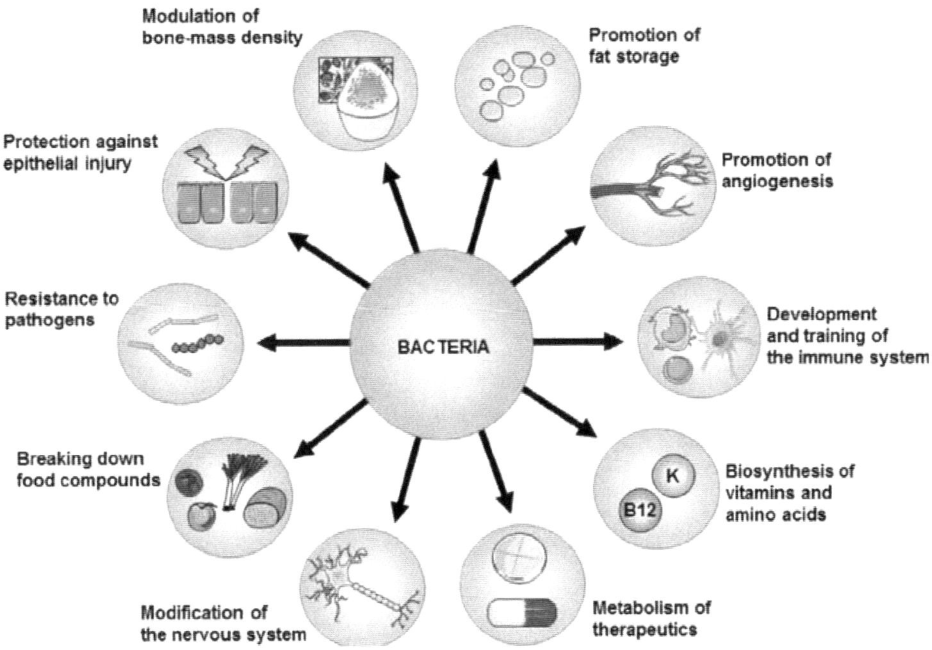

Figure 2. Benefits of the bacteria in the gut of the host. However, most of the research so far has been focused on humans and mice for biomedical research, rather than livestock health and nutrition. Figure from Laukens et al. (2015).

Effects of feeding probiotics and insects on fish growth and gut microbiome

First, a 9-week feeding trial was performed at the Hagen Aqualab (Guelph, ON, Canada) with 0.1g adult zebrafish that were 6-9 months of age. Triplicate tanks of zebrafish at 26°C were hand fed twice per day one of five experimental diets: control (CON), 10% BSF (Enterra), 1% probiotic yeast (Alltech), 0.1% prebiotic yeast (Alltech) and sodium butyrate (SCFA; Sigma). Although no significant effects were found ($p>0.05$), we found that feeding BSF resulted in the highest weight (see Figure 3A), probiotic and BSF resulted in the highest survival rate (89 and 91%) and prebiotic resulted in the highest feed intake (0.4g/fish). BSF, probiotic, prebiotic and SCFA diets maintained diversity of gut microbes and reduced abundance of Vibrio, which is a pathogenic microbe.

Second, a 16-week feeding trial was performed at the Ontario Aquaculture Research Centre (Elora, ON, Canada) with 93g juvenile rainbow trout that were two years of age. Triplicate tanks of trout at 9°C were fed twice per day one of five experimental diets: control (CON), 10% BSF (Enterra), 1% probiotic yeast (Alltech), 0.1% (PR1) and 0.2% (PR2) prebiotic yeast (Alltech). Although no significant effects were found ($p>0.05$), we found that feeding BSF resulted in the highest weight (see Figure 3B), BSF and PR2 resulted in the highest survival rate (100%) and PRO and PR2 resulted in the highest feed intake (355 and 353g/fish). BSF, PRO, PR1 and PR2 diets maintained diversity of gut microbes and reduced abundance of Mycoplasma, which is a pathogenic microbe.

A)

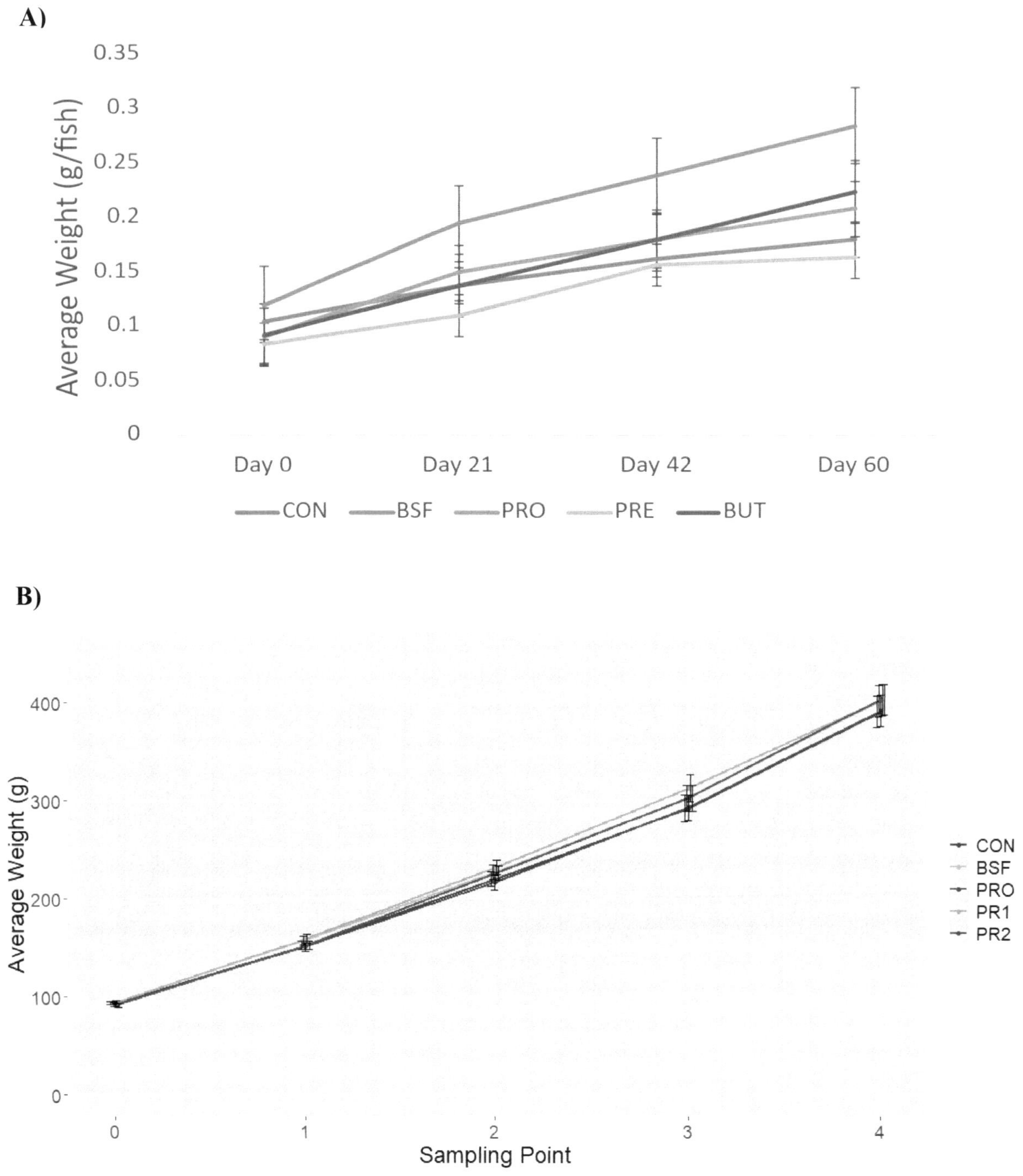

Figure 3. Growth performance (g) of A) zebrafish and B) rainbow trout fed diets of control, black soldier fly (BSF), probiotic yeast and prebiotic yeast.

Conclusions

We used zebrafish as a model for farmed fish and it showed similar results to the later trial on rainbow trout, thus it may be beneficial in the future since trials can be shorter (9 vs 16 weeks) and younger animals can be used (0.5 vs 2-year-olds). Zebrafish may also be applicable to other monogastric animals, such as chickens and pigs, while more research is needed in both fish and other monogastric animals to validate this model. Results from both trials, although not significant, illustrated trends where feeding BSF improved growth and showed that BSF is a viable replacement for fishmeal at 10% inclusion in the diet. Probiotic and prebiotic yeasts were found to increase either feed intake or survival depending on the trial. Lastly, BSF, probiotic and prebiotic diets showed beneficial effects on the gut microbiome and maintained diversity while reducing abundance of pathogenic bacteria. In summary, these results demonstrate probiotics, prebiotics and insects have potential beneficial effects on the growth and gut health of zebrafish and rainbow trout, which indicates potential for other fish species and monogastric animals that may improve production in Canadian and international agri-food sectors.

References

Cao, S., Dicksved, J., Lundh, T., Vidakovic, A., Norouzitallab, P., Huyben, D. 2024. A meta-analysis revealing the technical, environmental, and host-associated factors that shape the gut microbiota of Atlantic salmon and rainbow trout. Rev. Aquac. 16, 1603–1620 https://doi.org/10.1111/raq.12913

Fisheries and Oceans Canada (DFO). 2024. Aquaculture production quantities and value. https://www.dfo-mpo.gc.ca/stats/aqua/aqua22-eng.html (accessed 20 December 2024)

FAO. 2020. The State of World Fisheries and Aquaculture 2020. Sustainability in action. Rome. https://doi.org/10.4060/ca9229en

FAO. 2024. The State of World Fisheries and Aquaculture 2024. Towards Blue Transformation. Rome, FAO. https://doi.org/10.4060/cc0461en

Francis, G., Makkar, P.S., Becker, K. 2001. Antinutritional factors present in plant-derived alternate fish feed ingredients and their effects in fish, Aquacult. 3–4, 197–227. https://doi.org/10.1016/S0044-8486(01)00526-9

Huyben, D., Sun, L., Moccia, R., Kiessling, A., Dicksved, J., Lundh, T. 2018. Dietary live yeast and increased water temperature influence the gut microbiota of rainbow trout. J Appl Microbiol 124, 1377–1392. https://doi.org/10.1111/jam.13738

Huyben, D., Vidaković, A., Hallgren, S. W., Langeland, M. 2019. High-throughput sequencing of gut microbiota in rainbow trout (*Oncorhynchus mykiss*) fed larval and pre-pupae stages of black soldier fly (*Hermetia illucens*). Aquacult. 500, 485-491. https://doi.org/10.1016/j.aquaculture.2018.10.034

Huyben, D., Roehe, B.K., Bekaert, M., Ruyter, B. and Glencross, B. 2020. Dietary lipid: protein ratio and n-3 long-chain polyunsaturated fatty acids alters the gut microbiome of Atlantic salmon under hypoxic and normoxic conditions. Front. Microb. 11, 589898. https://doi.org/10.3389/fmicb.2020.589898

National Research Council (NRC). 2011. Nutrient requirements of fish and shrimp. The National Academies Press. Washington, DC, USA. https://doi.org/10.17226/13039

Ontario Aquaculture Research Centre (OARC). 2024. 'AQUASTATS' Ontario aquacultural production in 2023. University of Guelph. https://www.aquacentre.uoguelph.ca/publications-new/

Ribas, L., & Piferrer, F. 2014. The zebrafish (Danio rerio) as a model organism, with emphasis on applications for finfish aquaculture research. Rev. Aquacult. 6, 4. https://doi.org/10.1111/raq.12041

Sánchez-Muros, M.J., Barroso, F.G., Manzano-Agugliaro, F. 2014. Insect meal as renewable source of food for animal feeding: A review. J Clean Prod. https://doi.org/10.1016/j.jclepro.2013.11.068

Merrifield, D.L., Dimitroglou, A., Foey, A., Davies, S.J., Baker, R.T.M., Bøgwald, J., Castex, M., Ringø, E. 2010. The current status and future focus of probiotic and prebiotic applications for salmonids. Aquacult. https://doi.org/10.1016/j.aquaculture.2010.02.007

Tran, N.T., Li, Z., Wang, S., Zheng, H., Aweya, J.J., Wen, X., Li, S. 2020. Progress and perspectives of short-chain fatty acids in aquaculture. Rev Aquac. https://doi.org/10.1111/raq.12317

Spence, R., Gerlach, G., Lawrence, C., Smith, C. 2008. The behaviour and ecology of the zebrafish, Danio rerio. Biol. Rev. https://doi.org/10.1111/j.1469-185X.2007.00030.x

Zarantoniello, M., Bruni, L., Randazzo, B., Vargas, A., Gioacchini, G., Truzzi, C., Annibaldi, A., Riolo, P., Parisi, G., Cardinaletti, G., Tulli, F., Olivotto, I. 2018. Partial Dietary Inclusion of *Hermetia illucens* (Black Soldier Fly) Full-Fat Prepupae in Zebrafish Feed: Biometric, Histological, Biochemical, and Molecular Implications. Zebrafish 15. https://doi.org/10.1089/zeb.2018.1596

Optimizing gestation feeding strategies to maximize mammary development and milk yield in modern prolific sows

Optimiser les stratégies alimentaires en gestation pour maximiser le développement mammaire et la production laitière des truies modernes hyperprolifiques

C. Farmer[1]

[1] Research scientist in Swine Lactation Biology,
Agriculture and Agri-Food Canada, Sherbrooke Research and Development Centre, Sherbrooke, QC
chantal.farmer@agr.gc.ca

SPONSORED BY / COMMANDITÉ PAR

Sollio Agriculture

Abstract

Sows are expected to produce very large quantities of milk as litter size has kept increasing over the years, although they generally cannot meet the nutritional demands of all their piglets. One avenue to assist sows is to develop feeding strategies maximizing their mammary development in late gestation because the number of milk-synthesizing cells impacts potential milk yield. Overfeeding energy in late gestation can decrease the amount of milk-synthesizing tissue. Crude protein intake (330 versus 216 g/day) in late gestation does not affect mammary development, however, a 40% increase in lysine intake, via greater provision of soybean meal, led to 44% more mammary parenchyma. This effect was present in gilts but not in multiparous sows, suggesting that phase-feeding would be beneficial only in the first gestation. Pregnant gilts are still growing and this may affect the mammary response to supplementary lysine. When providing diets with lysine contents ranging from 70 to 160% of estimated requirements to gilts from days 90 to 110 of gestation, subsequent milk yield was maximized with lysine at 20% above current recommendations. Gilts that are obese or too thin at the end of gestation will have less developed mammary tissue, and consequently lower milk yields. Therefore, the amount of feed provided in gestation should ensure an optimal body condition of gilts, namely a backfat thickness between 16 and 26 mm for Yorkshire Landrace, for maximum milk yields. Current findings demonstrate that feeding of late-pregnant gilts should take into account mammary development in order to maximize future milk yield.

Keywords: feeding, gestation, gilt, lysine, mammary development, milk yield

Résumé

De nos jours, les truies doivent produire de très grandes quantités de lait car la taille des portées n'a cessé d'augmenter au fil des ans. Par conséquent, les truies ne parviennent généralement pas à satisfaire les besoins nutritionnels de tous leurs porcelets. L'une des solutions mises de l'avant pour pallier à cette difficulté est de développer des stratégies d'alimentation visant à stimuler le développement mammaire en fin de gestation, car le nombre de cellules lactifères influence la capacité de production de lait. Un apport énergétique trop important en fin de gestation risque d'entraîner une diminution du tissu mammaire. L'apport en protéines brutes (330 contre 216 g/jour) en fin de gestation n'affecte pas le développement mammaire, mais une augmentation de 40% de la quantité de lysine, en incluant plus de tourteau de soya dans la diète, a augmenté de 44% la masse du parenchyme mammaire. Cet effet a été observé chez les cochettes mais pas chez les truies multipares, ce qui suggère que l'alimentation en phases ne serait bénéfique qu'au cours de la première gestation. Le fait que les cochettes gestantes soient encore en croissance pourrait influencer la réponse de leurs glandes mammaires à un supplément de lysine. Lorsque l'on a servi des régimes à teneur en lysine variant entre 70% et 160% des besoins estimés à des cochettes entre les jours 90 et 110 de gestation, la production laitière subséquente a été maximisée avec une teneur en lysine excédant de 20% les recommandations actuelles. Les cochettes obèses ou trop maigres en fin de gestation auront moins de tissu mammaire et produiront moins de lait. Par conséquent, la quantité d'aliments fournie pendant la gestation doit garantir un état corporel adéquat des cochettes, à savoir une épaisseur de gras dorsal entre 16 et 26 mm pour les Yorkshire Landrace. Les résultats actuels démontrent bien que l'alimentation des cochettes en fin de gestation doit tenir compte du développement mammaire afin de maximiser la future production laitière.

Introduction

At peak lactation, sows produce more milk than cows on a per kilogram body weight basis. Yet, they are still not producing enough to meet the nutritional requirements of all their piglets with the large litters of modern hyperprolific sow genotypes. This is evidenced by the fact that providing supplementary liquid milk replacer to nursing piglets increases their pre-weaning growth rate (Willard et al. 2022). It is therefore imperative to develop novel strategies to assist sows in maximizing their milk output. The number of mammary epithelial cells present at the onset of lactation is a crucial factor impacting potential sow milk yield (Head and Williams, 1991) and it can be affected by nutrition (reviewed by Farmer, 2028). However, the effect of nutrition can only be seen during the periods where there is ongoing mammary development, namely, from 90 days of age to puberty (Sorensen et al. 2002), in the last third of gestation (Sorensen et al. 2002), and during lactation (Kim et al. 1999). Increasing the number of milk-synthesizing cells in the late gestation period holds much promise to improve sow milk yield in the subsequent lactation.

Feeding in late gestation

Energy intake and source: Studies show no beneficial effect of increasing dietary energy in late gestation for mammary development. Increasing dietary energy from approximately 6 to 10.5 Mcal/kg of metabolizable energy/day between days 75 and 105 of gestation either had negative effects on mammary development on day 105 of gestation (Weldon et al. 1991) or had no effect (Howard, 1995). Furthermore, supplementing the diet of gilts with polyunsaturated fatty acids, using 5% soybean oil in isocaloric diets, from days 75 to 105 of gestation did not alter mammary development (Howard, 1995).

Protein and lysine intakes: Increasing protein intake (330 vs. 216 g of crude protein/day) from days 75 to 105 of gestation (Weldon et al. 1991) or providing 4, 8 or 16 g/day of lysine from days 25 to 105 of gestation (Kusina et al. 1999a) had no effect on mammary parenchymal mass or composition at the end of gestation. However, in the latter study, the treatments supplying 4 or 8 g/day of lysine failed to support maximal milk yield in the subsequent lactation (Kusina et al. 1999b), hence suggesting a potential role for additional lysine in late gestation on sow milk production. Recent findings showed that feeding 40% more lysine (26.0 vs 18.6 g/day of standardized ileal digestible (SID) lysine) than estimated requirements (NRC, 2012) by adding soybean meal to the diet of gilts from days 90 to 110 of gestation increased mammary parenchymal mass by 44% (Table 1, Farmer et al. 2022). The same effect was not present in multiparous sows, where feeding a soybean meal-supplemented diet providing 40% more lysine than estimated requirements (20.8 vs 14.8 g/day of SID lysine) did not impact mammary development on day 110 of gestation (Table 1, Farmer et al. 2023). These two studies demonstrate that the nutrient requirements to fully support mammary development differ between parities, being greater in the first gestation, and this could be linked to the fact that pregnant gilts are still growing. The weight of parenchymal mammary tissue in control multiparous sows on day 110 of gestation was 39.5% greater than that of control gilts, in accordance with the greater milk yield of multiparous compared with primiparous sows (review by King, 2000). The greater amount of parenchymal tissue at the end of gestation in multiparous sows likely reflects the presence of more parenchyma at mating post-weaning compared to when mating occurred at puberty. Assuming this is the case, the lysine requirements for mammary growth in late gestation would be reduced in multiparous sows compared with gilts, supporting the absence of any beneficial effect of additional lysine on mammary development in the current study. It is therefore apparent that a phase-feeding regimen to provide more lysine in late gestation via additional soybean meal could be advantageous in the first gestation only.

Table 1. Mammary gland composition on day 110 of gestation for gilts (Farmer et al. 2022) or multiparous sows (parities 2 and 3, Farmer et al. 2023) fed a control diet (CTL) or a diet providing lysine at 40% above estimated requirements (HILYS) from days 90 to 110 of gestation.

	Gilts		Sows	
Mammary variables	CTL	HILYS	CTL	HILYS
Extraparenchymal tissue, g (fat layer around parenchyma)	1691.0	1690.2	2407.0	2348.8
Parenchymal tissue, g (milk-synthesizing tissue)	1437.4[a]	2073.6[b]	2004.7	2099.3
Parenchymal tissue composition				
Dry matter, %	37.8[c]	38.8[d]	29.0	29.2
Fat, %	63.3	60.3	49.4	50.8
Protein, %	32.9[e]	35.9[f]	44.6	44.3
DNA, mg/g (cell number indicator)	6.59	7.06	9.35	9.28
RNA, mg/g (metabolic activity indicator)	7.76	8.38	12.33	12.27

[a,b]Means within a row with different superscripts differ ($P < 0.01$) for gilts.
[c,d]Means within a row with different superscripts differ ($P < 0.05$) for gilts.
[e,f]Means within a row with different superscripts tend to differ ($P < 0.10$) for gilts.

In a more recent study, gilts were fed one of seven diets with lysine contents ranging from 70 to 160% of estimated requirements, using soybean meal to supply the additional lysine, from days 90 to 110 of gestation. Increasing lysine intake 15% above the current recommendations maximized whole body nitrogen retention in late gestation and piglet birth weight, and feeding lysine at 20% above current recommendations maximized milk yield in the subsequent lactation (Kloostra et al. 2025). Providing supplementary lysine in the diet of late-pregnant gilts may therefore improve lactation performance.

Feeding level throughout gestation

The amount of feed offered to gilts in gestation can impact their mammary development via an effect on body condition. An early study showed that when altering body composition of gilts via changes in protein and energy intakes during gestation, obese (36 mm backfat) and leaner gilts (24 mm backfat) had similar weights of mammary tissue at the end of gestation, but the mammary tissue from obese gilts contained three times less DNA, i.e. cell numbers (Head and Williams, 1991). Yet, these body conditions are not representative of what is seen commercially and two projects were carried out to establish if differences in body fat that are commonly seen can lead to changes in mammary development at the end of gestation. When gilts were fed different amounts of the same feed throughout gestation to achieve backfat thicknesses of 12-15 mm (lean), 17-19 mm (medium) or 21-26 mm (fat) on day 109 of gestation, lean gilts had a much reduced parenchymal mass (Farmer et al. 2016a). Being too thin at the end of gestation is therefore detrimental for mammary development. On the other hand, when backfat thicknesses that already differed at mating (lean: 12-15 mm, medium: 17-19 mm, and fat: 21-26 mm) were maintained throughout gestation via varying feeding levels, mammary parenchymal mass was not affected on day 110 of gestation (Farmer et al. 2016b). Together, these findings suggest that feed intake during gestation is more important than body condition at mating to achieve the optimal backfat thickness to support mammary development in late gestation. One should therefore avoid underfeeding to ensure maximal amount of parenchymal tissue mass.

References

Farmer, C. 2018. Nutritional impact on mammary development in pigs: a review. J. Anim. Sci. 96 3748-3756.

Farmer, C., M. Comi, C.R.A. Duarte, M. Vignola, P. Charagu and M.F. Palin. 2016b. Differences in body condition of gilts that are maintained from mating to the end of gestation affect mammary development. J. Anim. Sci. 94 3206-3214.

Farmer, C., C.R.A. Duarte, M. Vignola and M.F. Palin. 2016a. Body condition of gilts at the end of gestation affects their mammary development. J. Anim. Sci. 94 1897-1905.

Farmer, C., C. Gillies, J.C. Johannsen, R.C. Hovey and L.A. Huber. 2023. Dietary supplementation with lysine (protein) in late pregnancy does not enhance mammary development in multiparous sows. J. Anim Sci. 101 1-8.

Farmer, C., M.-F. Palin, R.C. Hovey, T.D. Falt and L.A. Huber. 2022. Dietary supplementation with lysine stimulates mammary development in late-pregnant gilts. J. Anim. Sci. 100 1-11.

Head, R.H. and I.H. Williams. 1991. Mammogenesis is influenced by pregnancy nutrition. In: Batterham, E.S. (ed.) Manipulating Pig Production III. Australasian Pig Science Association, Atwood, Australia, p: 33.

Howard, K.A. 1995. Effect of porcine somatotropin, soybean oil and energy intake on mammary development in gestating gilts. PhD Diss, Dept. Anim. Sci., Michigan State Univ., East Lansing. 72 pp.

Kim, S.W., W.L. Hurley, I.K. Han and R.A. Easter. 1999. Changes in tissue composition associated with mammary gland growth during lactation in sows. J. Anim. Sci. **77** 2510-2516.

King, R. H. 2000. Factors that influence milk production in well-fed sows. J. Anim. Sci. 78 (Suppl. 3) 19-25.

Kloostra, V., C. Farmer and L.A. Huber. 2025. Standardized ileal digestible lysine intake by gilts should be increased in late gestation to maximize whole-body nitrogen retention, piglet birth weight, and subsequent milk yield. Proc. Banff Pork Seminar, Available at: https://www.banffpork.ca/proceedings/search/

Kusina, J., J.E. Pettigrew, A.F. Sower, M.R. Hathaway, M.E. White and B.A. Crooker. 1999a. Effect of protein intake during gestation on mammary development of primiparous sows. J. Anim. Sci. 77 925-930.

Kusina, J., J.E. Pettigrew, A.F. Sower, M.E. White, B.A. Crooker and M.R. Hathaway. 1999b. Effect of protein intake during gestation and lactation on the lactational performance of primiparous sows. J. Anim. Sci. 77 931-941.

NRC. 2012. Nutrient requirements of swine. 11th rev. ed. Natl. Acad. Press, Washington, DC.

Sorensen, M.T., K. Sejrsen and S. Purup. 2002. Mammary gland development in gilts. Livest. Prod. Sci. 75 143-148.

Weldon, W.C., A.J. Thulin, O.A. MacDougald, L.J. Johnston, E.R. Miller and H.A. Tucker. 1991. Effects of increased dietary energy and protein during late gestation on mammary development in gilts. J. Anim. Sci. 69 194-200.

Willard, N.C., N.K. Moest, K.D. Vande Pol, A. Olivio, C.M. Shull, D. McKilligan and M. Ellis. 2022. Effect of litter size and provision of supplementary liquid milk replacer during lactation on piglet pre-weaning performance. J. Anim. Sci. 100 (Suppl. 2) 2.

Up-and-Coming Research Mini Presentations
Courtes présentations sur les recherches en cours et à venir

This presentation category is intended for early-stage graduate students (both master's and PhD level) who do not yet have research results to showcase their project motivation and the benefit their work will have for industry. This is a unique opportunity for graduate students to communicate the value of their work directly to the industry professionals who will apply it.

Cette catégorie de présentation s'adresse aux étudiants en début de cursus (maîtrise et doctorat) qui n'ont pas encore de résultats de recherche pour présenter la motivation de leur projet et les avantages que leur travail apportera à l'industrie. Il s'agit d'une occasion unique pour les étudiants de cycle supérieur de communiquer la valeur de leur travail directement aux professionnels de l'industrie qui l'appliqueront.

Monogastric presentation / *Présentation sur les monogastriques*

Sylvester Mutinda, University of Guelph

Poursuite des recherches sur l'optimisation des concepts de formulation à faible teneur en protéines brutes pour la production de poulets de chair sans antibiotiques

SPONSORED BY / COMMANDITÉ PAR

Alltech®

Further investigations on optimization of low crude protein formulation concepts for antibiotic free broiler production

S. Mutinda[1], N.T. Ferreira[2], S. Torrey[2], M.-P. Létourneau-Montminy[3], E. G. Kiarie[1]
[1] Department of Animal Biosciences, University of Guelph, Guelph, ON
[2] Trouw Nutrition, Puslinch, ON
[3] Département des Sciences Animales, Université Laval, Québec, QC
smutinda@uoguelph.ca

SPONSORED BY / COMMANDITÉ PAR

Alltech®

Brief project summary

Adopting a "less is more" approach in poultry nutrition has gained traction as a sustainable practice in recent years. The low crude protein (LCP) concept involves reducing the inclusion of intact protein feedstuffs and supplementing diets with crystalline amino acids (CAA) to meet the bird's nutritional requirements. This approach is termed "less is more "due to its multiple benefits, including improved nitrogen utilization efficiency, reduced nitrogen excretion and ammonia emissions, enhanced litter quality, and lower feed costs by decreasing reliance on protein-rich ingredients such as soybeans. Despite these advantages, reducing dietary crude protein (CP) by more than 2% in poultry diets under the LCP framework, regardless of CAA supplementation, can lead to unpredictable outcomes in performance, product quality, and yield. These inconsistencies have been linked to factors such as suboptimal essential amino acid (AA) levels and imbalances, disruptions in dietary electrolyte balance, insufficient non-specific nitrogen for the de novo synthesis of non-essential AA, altered essential to non-essential AA rations, and the availability of cost- effective, regulatory-approved CAA. Given increasing consumer demand for high-quality poultry products and regulatory restrictions on antibiotic use in animal agriculture, this study will evaluate LCP diets with CP reductions of -2%, -4%, -6%. A total of 1,200 Ross 708 broilers will be fed these diets with half raised in antibiotic-free system. Optimizing growth performance and nitrogen utilization will provide further insight into the interaction between LCP diet strategies and antibiotic growth promoters (AGP) alternatives in broiler production.

ANCC

2025 ANIMAL NUTRITION
CONFERENCE OF CANADA

CNAC

2025 COLLOQUE DE NUTRITION
ANIMALE DU CANADA

Ruminant Session
Séance sur les ruminants

Effects of low feed intake on gastrointestinal function[1]

Effets d'une prise alimentaire réduite sur la fonction gastro-intestinale

G. Penner, K. Lambert, C. Bertens, and K. Hare
Department of Animal and Poultry Science, University of Saskatchewan
Saskatoon, SK
greg.penner@usask.ca

SPONSORED BY / COMMANDITÉ PAR

O&T Farms
Naturally Better

Abstract

Beef and dairy cattle may experience periods of low feed intake because of weaning, transportation and marketing, parturition, heat stress, or infectious disease. In our research, exposure to low feed intake reduces the concentration of short-chain fatty acids in many regions of the gastrointestinal tract, reduces the weight of the gastrointestinal tract and liver, and decreases absorptive surface area of the gastrointestinal tract. These changes have been reported to reduce short-chain fatty acid absorption across the rumen in some studies, and to compromise permeability, particularly post-ruminal permeability. Increasing the severity and duration of exposure to low feed intake increases the duration of time required to return to voluntary feed intake and consistently increases risks for cattle to have low ruminal pH, highlighting a potential risk factor for ruminal acidosis.

Keywords: absorptive function, barrier function, beef cattle, dairy cattle, permeability

Résumé

Les bovins de boucherie et les bovins laitiers peuvent connaître des périodes de faible prise alimentaire pour diverses raisons: sevrage, transport et commercialisation, vêlage, stress thermique

[1]A version of this paper was previously published in the proceedings for the Southwest Nutrition Conference (2019).

ou maladie infectieuse. Dans nos travaux, l'exposition à une faible prise alimentaire a entraîné une baisse de la concentration d'acides gras à chaîne courte dans de nombreuses régions du tube digestif, une diminution du poids du tube digestif et du foie ainsi qu'une réduction de la surface d'absorption du tube digestif. Selon certaines études, ces changements nuisent à l'absorption des acides gras à chaîne courte dans le rumen et compromettent la perméabilité, en particulier la perméabilité post-ruminale. Une exposition prolongée à une faible prise alimentaire accroît le temps nécessaire pour revenir aux valeurs normales de consommation volontaire et augmente systématiquement les risques de baisse du pH ruminal, un facteur potentiel d'acidose.

Introduction

Ration formulation and evaluation systems have advanced significantly over the past 15 years and now provides a more accurate and detailed prediction for the performance of beef and dairy cattle. Using these models, it is clear that highly fermentable diets are needed to promote milk and milk component yield for dairy cattle and gain:feed (**G:F**) for beef cattle. Feeding highly fermentable diets increases short-chain fatty acid (**SCFA**) production in the rumen and microbial protein growth over diets with a lower fermentability. Thus, this strategy results in a greater supply of metabolizable energy and increases metabolizable protein supply. On the other hand, feeding rapidly fermentable diets can result in excessive fermentation rates that lead to low ruminal pH, high rumen osmolality, and may increase the concentration of antigens within the rumen. Obviously, there is a fine balance between diets that promote rapid fermentation and those that cause deleterious effects.

As feeding strategies for ruminants are implemented to promote SCFA production and metabolizable protein, understanding factors that alter the ability of the gastrointestinal tract (**GIT**) to maintain its critical functions are essential. Historically, the role of the GIT was thought to be limited to facilitating digestion, nutrient absorption, and the transit of feed through the GIT. That said, it has been recognized that the GIT plays a critical role in preventing or limiting the passage of non-desired molecules (e.g. microbial-associated molecular patterns, pathogenic microbes, and other byproducts arising from fermentation [histamine]) from entering portal circulation. This process is referred to as selective permeability and is one critical component of the GIT barrier function. Failure to prevent these molecules from entering portal circulation can result in systemic inflammatory responses. In addition to absorptive and barrier function, the GIT also serves as a communicative organ. As part of this communicative process, the GIT detects nutrients in the lumen and, based on the response, may up-regulate nutrient transport. Moreover, endocrine and paracrine secretions can facilitate communication with other organ systems, including the nervous system. The GIT also communicates with its microbial inhabitants. While much of this research has been conducted in monogastrics, Weimer et al. (2010) have reported that ruminants may also regulate the microbial community structure of the rumen. In that study, they evaluated ruminally cannulated dairy cattle fed the same diet, with similar days in milk, and housed in the same facility. They selected two cattle with the greatest differences in ruminal pH, ruminal SCFA concentrations, and microbial community structure. Subsequently, they completely evacuated the ruminal contents and refilled the rumen with the other cows' contents. Over the next 65 d, they evaluated the microbial community structure and found that the microbial profile returned to a profile that resembled the native community. This study provided new evidence that cattle have the ability to, at least partially, regulate the microbial community structure.

Relevance of Low Feed Intake in Commercial Settings

Beef and dairy cattle are often, albeit inadvertently, exposed to periods of low feed intake (**LFI**). An example of a transient period of LFI is the weaning transition for dairy calves, where the increase in starter intake may not compensate for the reduction of energy intake in milk or milk replacers (Wood et al., 2015). Beef calves also likely reduce DMI when abrupt weaning and separation from the dam are imposed. For beef cattle, it has been reported that walking activity and eating durations are markedly reduced with abrupt weaning (Price et al., 2003; Haley et al., 2005), although weaning methods can have a significant impact on the negative behavioural response (Haley et al., 2005; Wiese et al., 2015). However, no studies have measured calf DMI prior to weaning and post-weaning, and only one study has attempted to measure calf DMI post-weaning and reported no differences among weaning methods (Wiese et al., 2015). Nevertheless, newly received calves certainly experience extended periods of LFI. For example, Hutcheson and Cole (1986) reported that calves in the first, second, and third week after arrival at a feedlot often only consume DM at a rate of 0.5 to 1.5%, 1.5 to 2.5%, and 2.5 to 3.5% of BW, respectively. Thus, newly received cattle experience an extended duration of LFI and variable extents for the magnitude of intake depression (Hutcheson and Cole, 1986; Loerch and Fluharty, 1999). The LFI occurring for newly received cattle is likely related to the complete feed and water deprivation that occurs during transportation (González et al., 2012) in addition to the change in social structure, physical housing methods, and diets that occur upon arrival.

Transition dairy cattle also experience a transient period of LFI around calving. In fact, a literature review reported that a reduction in DMI intake may be as great as 68% on d 1 pre-partum relative to 21 d pre-partum for dairy cattle (Hayirli et al., 2002). Although the range in LFI is large, on average, cattle reduce their feed intake by 33%, with nearly 90% of that reduction occurring in the last week prior to calving. More recent studies have supported the results of Hayirli et al. (2003) as Janovick and Drackley (2010) demonstrated that transition dairy cattle reduced intake by up to 30% as parturition approached, and Penner et al. (2007) reported greater than a 30% reduction in primiparous heifers. The extent of LFI can be exacerbated for transition cows in association with infectious diseases or metabolic and digestive disorders (e.g. displaced abomasum, ketosis; Van Winden et al., 2003; Goldhawk et al., 2009), and numerous studies have been able to identify associations between risk for metabolic disease and LFI pre-partum (Huzzy et al., 2009). Thus, dairy cattle clearly experience a period of LFI prior to calving, and this is coupled with a rapid increase in DMI and diet fermentability following calving.

Environmental conditions can also induce LFI. Heat stress has been reported to lead to marked reductions in DMI (Maust et al., 1972; Knapp and Grummer, 1991; Holter et al., 1996). The magnitude of reduction in DMI differs based on the severity of the heat stress, but under experimental conditions, DMI was reduced by 18 to 30% from exposure to a temperature humidity index of 64 (Wheelock et al., 2010; Baumgard et al., 2011). It appears that the magnitude of DMI depression varies among cattle when exposed to the same severity of heat stress, and it is likely that the depression in intake is exacerbated with greater severity of the thermal heat load.

In addition, infectious disease strongly reduces DMI. Beef cattle with respiratory disease have up to an 88% reduction in DMI for 7 to 10 days (Toaff-Rosenstein et al., 2016). Interestingly, much of the change in feeding behaviour, and presumably intake, occurs prior to the detection of BRD (Wolfger et al., 2015). Dairy cows with mastitis reduce feed intake for up to 5 days prior to diagnosis and treatment and rapidly increase their intake following treatment (Sepulveda-Varas et al., 2016). Moreover, studies using lipopolysaccharide (LPS) as a non-infectious pro-inflammatory challenge have reported marked but short-lived reductions in DMI following an intravenous (Lippolis et al., 2017) or intramammary (Swartz et al., 2023) LPS infusion.

Effects of Low Feed Intake on Function of the Gastrointestinal Tract

Although LFI may be transient, past studies in sheep have demonstrated that periods of complete feed deprivation (FD) can have negative consequences on the absorptive and barrier functions of the rumen epithelium. For example, the transport of Na^+, Cl^-, Mg^{2+}, and SCFA was reduced by approximately 50% for sheep exposed to a 48-h period of FD (Gäbel et al., 1993). With respect to barrier function, Gäbel and Aschenbach (2002) demonstrated that the passive movement of a small hydrophilic molecule (3-O-methyl-α-D-glucose) was increased following FD in sheep. While the previous studies provided evidence to suggest that LFI compromises GIT absorptive and barrier function, the experimental model evaluated 48 h of complete FD; a condition that is not common under industry practices. To address this limitation, we initiated a series of experiments to assess what severities of LFI induce changes in the function of the GIT and to evaluate how the GIT recovers in response to LFI.

In the first study (Zhang et al., 2013a), we assessed the effect of differing severities of LFI by restricting cattle to 75, 50, or 25% of their ad libitum DMI for a 5-d duration. Following exposure to LFI, cattle were provided feed, offered as a total mixed ration, ad libitum and monitored for three consecutive weeks. The diet (30% barley silage, 30% grass hay, 32% rolled barley grain, and 8% of a pelleted barley supplement containing minerals and vitamins) was common among all cattle and experimental periods such that we only evaluated the effect of LFI. Our results showed that in response to LFI, the concentration of SCFA in the rumen decreased in a dose-dependent manner. Heifers that were restricted to 75% of their ad libitum DMI displayed lower SCFA concentrations than those fed ad libitum, while concentrations for heifers restricted to 50% were lower than those at 75%, and heifers restricted to 25% exhibited concentrations lower than those restricted to 50%. The reduction in SCFA concentrations is logical given the decrease in DMI, but it warrants mention as SCFA provide majority of the metabolizable energy supply for ruminants, and the provision of fermentable carbohydrates stimulates metabolizable protein supplies. Corresponding to the reduction in DMI, there was a dose-dependent increase in mean ruminal pH, with greatest increases occurring with the most severe LFI. Increased ruminal pH with LFI is an expected and common outcome (Albornoz et al., 2013a; Lambert et al., unpublished; Bertens et al., 2024).

Low feed intake has been reported to reduce SCFA absorption across the reticulo-rumen and the rate of absorption tended to be reduced as the severity of the LFI increased. However, not all studies report a reduction in SCFA absorption (Lambert et al., unpublished). Moreover, according to Zhang et al. (2013a), permeability of the GIT was increased for heifers exposed to 25% of their ad libitum intake. A second study (Albornoz et al., 2013a) also found that exposure to LFI (25% of ad libitum DMI) reduced SCFA concentration in the rumen, increased ruminal pH, and reduced SCFA absorption across the reticulo-rumen. It should be recognized that not only did exposure to LFI reduce dietary nutrient supply, but the capacity for nutrient absorption was also compromised and permeability was increased. This suggests that exposure to LFI, regardless of the cause (i.e. weaning, heat stress, parturition, transportation, etc.), is likely to reduce metabolizable energy and protein absorption and may predispose cattle to systemic inflammation.

More recently we have evaluated whether the duration of LFI affects GIT responses (Lambert et al., unpublished). In those studies, either sheep or beef cattle were exposed to 30% of their voluntary DMI for 0, 5, or 10 d. Results were consistent with previous studies showing that ruminal pH increased or tended to increase during LFI and the concentration of ruminal SCFA decreased. In a study with lambs, we also observed that digesta pH and SCFA concentrations were linearly affected in other regions of the GIT, highlighting that impacts arising from LFI should not be restricted to the rumen. That said, the absorptive surface area of the ruminal papillae has been reported to be reduced by 59% within 5 d of LFI (Pederzolli et al., 2018) and was linearly reduced by up to 40% as the duration of LFI increased from 0 to 10 d (Lambert et al., unpublished). In addition, weights for other regions of the GIT were also linearly reduced with advancing days of LFI, and a marked reduction in liver weight occurred within 5 d. Reduced liver weight has been confirmed in another study with exposure to 5 d of feed restriction at 30% of their voluntary feed intake (Fehr et al., unpublished).

Corresponding to reduced absorptive function, increased permeability of the GIT to Cr-EDTA was observed for cattle exposed to 25% of their voluntary feed intake for 5 d (Zhang et al., 2013a). Unfortunately, the method to evaluate GIT permeability did not allow for differentiation of the regions affected. We have recently adapted the methodology to characterize post-ruminal and total tract permeability using Cr-EDTA to indicate total tract and Co-EDTA to indicate post-ruminal permeability. Using this model, we have confirmed that LFI increases GIT permeability, with 55 to 82% of the responses occurring due to post-ruminal regions (Bertens et al., 2024a,b, Lambert et al., unpublished).

While our research demonstrated that LFI reduces aspects of GIT function, there was no indication of the recovery time required. To address this limitation, Zhang et al. (2013b) evaluated the recovery response upon return to full feed allocation. A particularly interesting finding was that the severity of LFI alone caused variability in how rapidly heifers returned to ad libitum DMI (Figure 1). For example, heifers exposed to LFI at 25% of their voluntary intake required 3 wk to return to pre-LFI DMI, while those restricted to 75% of their voluntary intake only required 1 wk. Increasing the duration of exposure to LFI from 5 to 10 d also delays recovery of DMI by up to 1 wk (Lambert et al., unpublished). This suggests that in the absence of other challenges (parturition, weaning, heat stress, etc.), the severity and duration of LFI can affect the recovery rate for DMI.

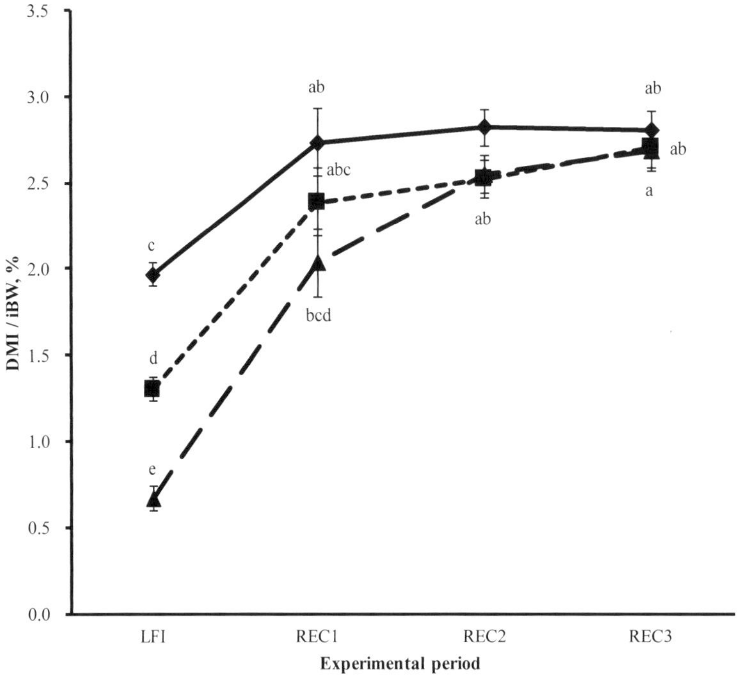

Figure 1. Voluntary recovery of dry matter intake during a 3-wk recovery period (REC1 = 1st week of recovery, REC2 = 2nd week of recovery, and REC3 = 3rd week of recovery) after a 5-d exposure to low feed intake (LFI). For the LFI, heifers were exposed to 75% (solid line with diamonds), 50% (short-dashed line with squares), or 25% (long dashed line with triangles) of their voluntary intake. The DMI was 2.7% of BW prior to LFI. There was a treatment × period interaction ($P < 0.001$) and letters indicate means that differ ($P < 0.05$). Adapted from Zhang et al., (2013b).

Another interesting finding was that despite heifers being fed a diet consisting of 60% forage, return to ad libitum intake resulted in the induction of ruminal acidosis with a dose-dependent response (Zhang et al., 2013b). Thus, heifers exposed to the greatest severity of LFI (25% of ad libitum intake) had the lowest mean ruminal pH (5.88) in the first week of recovery, and those restricted to 75% had the greatest mean ruminal pH, although the mean was still only 6.16. On average, the duration that pH was less than 5.5 was 6 h/d during the first week of recovery, indicating substantial exposure to low pH conditions. Induction of low ruminal pH is a common outcome for cattle as they return to ad libitum feed consumption (Bertens et al., 2024a; Lambert et al., unpublished). Greater risks for low ruminal pH are likely due to the observed reductions in absorptive surface area of the ruminal epithelium (Pederzolli et al., 2018; Lambert et al., unpublished). Furthermore, the severity of LFI also affected SCFA absorption during recovery, with heifers restricted to 50 and 25% requiring more time to recover than those restricted to 75%, and permeability of the GIT not recovering for heifers restricted to 25% of their voluntary intake.

These data suggest that cattle may require up to 3 wk for the GIT function to recover after exposure to LFI, depending on the severity of the event. Given that greater severity of LFI requires more time for recovery, transition dairy cattle that experience metabolic or infectious diseases or feedlot cattle experiencing infectious disease will likely be affected to a greater extent. Moreover, as cattle start to return to their ad libitum intake, they are susceptible to ruminal acidosis, even with low initial feed intake and diets with a substantial forage content.

The data above describe the effects of LFI on GIT function under artificial conditions and industry relevance could be questioned. To address this issue, we evaluated the changes in DMI and total tract permeability of the GIT for dairy calves at weaning (Wood et al., 2015). In this study, 14 Holstein bull calves were used and fed milk replacer at a rate of 15% of their BW (adjusted weekly). The milk replacer was included at 150 g (DM basis)/1 L of water. Calves were exposed to a 42-d milk feeding protocol followed by a 7-d step-down weaning program (WEAN) or were not weaned (CON). The step-down weaning protocol was effective at increasing starter intake as WEAN calves had greater intake than the CON. Calves in the WEAN treatment did not increase BW to the same extent as during the step-down period, further indicating a reduction in nutrient intake associated with weaning. To evaluate GIT permeability, we orally dosed Cr-EDTA and measured its appearance in urine. This approach is the same method used and validated by Zhang et al. (2013). We observed that permeability of the GIT decreased from wk 2 to wk 4 of age, and the CON continued to decrease up to wk 6. However, calves that were weaned followed the same trend until wk 6, where weaning caused a marked increase in permeability of the GIT. This supported the work of Zhang et al. (2013a,b), showing that reductions in DMI and nutrient intake are associated with increased GIT permeability.

Strategies to Minimize the Negative Effect of Low Feed Intake and Accelerate Recovery of the Gastrointestinal Tract

Based on the data above, it appears that LFI consistently decreases aspects of GIT function. However, this data does not indicate what management strategies can be used to mitigate the negative effect caused by LFI or accelerate the recovery response. It should be recognized that there are instances where LFI could be predicted (i.e. weaning, transportation), and hence, dietary scenarios could be put in place to minimize the negative effect of LFI. Alternatively, there are many scenarios where LFI cannot be accurately predicted (metabolic disorders and infectious disease), or the timeline may be difficult to predict accurately (parturition and heat stress). Thus, there is a need to evaluate strategies that can either mitigate the effect of LFI when it is predictable and, more practically, accelerate the recovery following a period of LFI.

For predictable exposure to LFI, we evaluated whether altering the forage:concentrate ratio of the diet could be a mechanism to mitigate the negative effect (Albornoz et al., 2013a,b). In this study heifers were either fed a high forage diet (92% forage with 50% of the forage supplied from barley silage and 50% from grass hay, and 8% of a pelleted barley supplement containing minerals and vitamins; DM basis) or a moderate forage diet (30% barley silage, 30% grass hay, 32% rolled barley grain, and 8% of a pelleted barley supplement containing minerals and vitamins; DM basis).

These treatments were used as feeding a high-forage diet may increase ruminal retention, thus decreasing the severity of the LFI event, while feeding a moderate-forage diet provides a greater nutrient supply per unit of feed intake. Heifers were exposed to a 5-d period of LFI with feed restricted to 25% of voluntary intake. As previously reported, LFI reduced SCFA in the rumen, increased mean pH, and reduced the rate of SCFA absorption. Feeding the high forage diet prior to and during LFI increased serum non-esterified fatty acids (an indicator of adipose tissue mobilization) to a greater extent than the moderate forage diet and reduced the rate of recovery for DMI after returning to voluntary intake. Heifers fed the high forage diet prior to and during the LFI also had a greater risk for ruminal acidosis during the recovery period. Surprisingly, the diet fed prior to and during LFI did not affect the rate of recovery for SCFA absorption. This indicates that diets with a high forage content may have a more deleterious effect than diets with a moderate forage content when fed during a period of LFI, likely because of reduced nutrient density.

On the other hand, most episodes of LFI will not be accurately predicted but can be detected with adequate feed bunk management and behavioural responses. Given that cattle are susceptible to ruminal acidosis during recovery after LFI, we evaluated the effect of feeding a moderate or high forage diet during recovery. Interestingly, heifers fed the high forage diet following LFI resumed voluntary DMI within the first wk of recovery, whereas heifers fed the moderate forage diet required 3 wk. The response of the moderate forage-fed heifers mirrored that reported by Zhang et al. (2013b). Feeding the high-forage diet during recovery also allowed heifers to stabilize their mean ruminal pH, and those fed the high-forage diet did not experience ruminal acidosis. Heifers fed the moderate forage diet had a mean pH below 6 during the first week of recovery, whereas heifers fed the high forage diet had a mean pH above 6.3. The induction of ruminal acidosis during the first week of recovery was also evident as heifers fed the moderate forage diet spent over 4 h below pH 5.5 while those on the high forage diet spent less than 20 min below the same threshold. Although ruminal fermentation parameters were affected, there were no differences in the rate of SCFA absorption between recovery treatment strategies.

References

Albornoz, R.I., J.R. Aschenbach, D.R. Barreda, and G.B. Penner. 2013. Feed restriction reduces short-chain fatty acid absorption across the reticulo-rumen of beef cattle independent of diet. J. Anim. Sci. 91:4730-4738.

Albornoz, R.I., J.R. Aschenbach, D.R. Barreda, and G.B. Penner. 2013. Moderate decreases in the forage-to-concentrate ratio prior to feed restriction and increases thereafter independently improve the recovery from a feed restriction insult in beef cattle. J. Anim. Sci. 91:4739-4749.

Baumgard, L.H., J.B. Wheelock, S.R. Sanders, C.E. Moore, H.B. Green, M.R. Waldron, and R.P. Rhoads. 2011. Postabsorptive carbohydrate adaptations to heat stress and monensin supplementation in lactating dairy cows. J. Dairy Sci. 94:5620-5633.

Bertens, C.A., D.J. Seymour, and G.B. Penner. 2024. Validation of an in vivo dual permeability marker technique to characterize regional gastrointestinal tract permeability in mid lactation Holstein cows during short-term feed restriction. J. Dairy Sci. 107:11708-11727.

Bertens, C.A., C. Stoffel, M. Crombie, P. Vahmani, and G.B. Penner. 2024. The effects of dietary cation-anion difference and dietary buffer for lactating dairy cattle under mild heat stress with night cooling. J. Dairy Sci. 107:10851-10868.

Dengler, F., R. Rackwitz, F. Benesch, H. Pfannkuche, and G. Gabel. 2015. Both butyrate incubation and hypoxia upregulate genes involved in the ruminal transport of SCFA and their metabolites. J. Anim. Physiol. Anim. Nutr. 99:379-390.

Gäbel, G., and J. R. Aschenbach. 2002. Influence of food deprivation on the transport of 3-O-methyl-α-D-glucose across the isolated ruminal epithelium of sheep. J. Anim. Sci. 451 80:2740 2746.

Gäbel, G., M. Marek, and H. Martens. 1993. Influence of food deprivation on SCFA and electrolyte transport across sheep reticulorumen. J. Vet. Med. A. 40:339-344.

Goldhawk, C., N. Chapinal, D. M. Veira, D. M. Weary, and M. A. G. von Keyserlingk. 2009. Prepartum feeding behavior is an early indicator of subclinical ketosis. J. Dairy Sci. 92:4971–4977.

González, L. A., K. S. Schwartzkopf-Genswein, M. Bryan, R. Silasi, and F. Brown. 2012. Benchmarking study of industry practices during commercial long haul transport of cattle in Alberta, Canada. J. Anim Sci. DOI:10.2527/jas.2011-4770.

Gorka, P., Z. M. Kowalski, P. Pietrzak, A. Kotunia, R. Kiljanczyk, J. Flaga, J. J. Holst, P. Guilloteau, and R. Zabielski. 2009. Effect of sodium butyrate supplementation in milk replacer and starter diet on rumen development in calves. J. Physiol. Pharmacol. 60:47–53.

Haley, D. B., D. W. Bailey, and J. M. Stookey. 2005. The effects of weaning beef calves in two stages on their behavior and growth rate. J. Anim. Sci. 83:2205-2214.

Hayirli, A., R. R. Grummer, E. V. Nordheim, and P. M. Crump. 2002. Animal and dietary factors affecting feed intake during the prefresh transition period in Holsteins. J. Dairy Sci. 85:3430-3443.

Holter, J. B., J. W. West, M. L. McGilliard, and A. N. Pell. 1996. Predicting ad libitum dry matter intake and yields of Jersey cows. J. Dairy Sci. 79:912-921.

Hutcheson, D. P. and N. A. Cole. 1986. Management of transit-stress syndrome in cattle: nutritional and environmental effects. J. Anim. Sci. 62:555-560.

Janovick, N.A., and J.K. Drackley. 2010. Prepartum dietary management of energy intake affects postpartum intake and lactation performance by primiparous and multiparous Holstein cows. J. Dairy Sci. 93:3086-3102.

Knapp, D.M., and R.R. Grummer. 1991. Response of lactating dairy cows to fat supplementation during heat stress. J. Dairy Sci. 74:3573-2579.

Kowalski, Z.M., P. Gorka, J. Flaga, A. Barteczko, K. Burakowska, J. Oprzadek, and R. Zabielski. 2015. Effect of microencapsulated sodium butyrate in the close-up diet on performance of dairy cows in the early lactation period. J. Dairy Sci. 98:3284-3291.

Loerch, S. C., and F. L. Fluharty. 1999. Physiological changes and digestive capabilities of newly received feedlot cattle. J. Anim. Sci. 77:1113-1119.

Penner, G. B., K. A. Beauchemin, and T. Mutsvangwa. 2007. Severity of ruminal acidosis in primiparous holstein cows during the periparturient period. J. Dairy Sci. 90:365–375.

Price, E. O., J. E. Harris, R. E. Borgwardt, M. L. Sween, and J. M. Connor. 2003. Fenceline contact of beef calves with their dams at weaning reduces the negative effects of separation on behavior and growth rate. J. Anim. Sci. 81:116–121.

Van Winden, S. C. L., R. Jorritsma, K. E. Muller, and J. P. T. M. Noordhuizen. 2003. Feed intake, milk yield, and metabolic parameters prior to left displaced abomasum in dairy cows. J. Dairy Sci. 86:1465–1471.

Weimer, P.J., D.M. Stevenson, H.C. Mantovani, and S.L. Man. 2010. Host specificity of the ruminal bacterial community in the dairy cow following near-total exchange of ruminal contents. J. Dairy Sci. 93:5902-5912.

Wheelock, J. B., R. P. Rhoads, M. J. VanBaale, S. R. Sanders, and L. H. Baumgard. 2010. Effects of heat stress on energetic metabolism in lactating Holstein cows. J. Dairy Sci. 93:644-655.

Wiese, B.I., S. Hendrick, J.M. Stookey, K.S. Schwartzkopf-Genswein, S. Li, J.C. Plaizier, and G.B. Penner. 2015. Weaning method and timing of weaning on calf behavior, dry matter intake, and reticular pH during a simulated receiving period. Prof. Anim. Sci. 32:229-235.

Wood, K.M., S. Palmer, M.A. Steele, J. Metcalf, and G.B. Penner. 2015. Effect of age and weaning on permeability of the gastrointestinal tract in Holstein bull calves. J. Dairy Sci. 98:7226–7237.

Zhang, S., J.R. Aschenbach, D.R. Barreda, and G.B. Penner. 2013. Recovery of absorptive function of the reticulo-rumen and total tract barrier function in beef cattle after short-term feed restriction. J. Anim. Sci. 91: 1696-1706.

Zhang, S., R.I. Albornoz, J.R. Aschenbach, D.R. Barreda, and G.B. Penner. 2013. Short-term feed restriction impairs the absorptive function of the reticulo-rumen and total tract barrier function in beef cattle. J. Anim. Sci. 91: 1685-1695

Greenhouse gas emissions from beef production systems

Émissions de gaz à effet de serre dans les systèmes de production de viande bovine

K. R. Stackhouse-Lawson[1], P. H.V. Carvalho[1], N. D. Delay[2], J. D. Derner[3], L. A. Kuehn[4], S. E. Place[1], E. J. Raynor[1], J. P. Ritten[2], and J. de J. Vargas[1]
[1] AgNext and Department of Animal Sciences, Colorado State University, Fort Collins, CO, USA
[2] AgNext and Department of Agriculture and Resource Economics, Colorado State University, Fort Collins, CO, USA
[3] Rangeland Resources and Systems Research Unit, US Department of Agriculture−Agricultural Research Service, Cheyenne, WY, USA
[4] Genetics and Animal Breeding Unit-Meat Animal Research Center, US Department of Agriculture−Agricultural Research Service, Clay Center, NE, USA
Kim.stackhouse-lawson@colostate.edu

Abstract

Enteric methane (CH4) emissions from beef cattle contribute to 2.8% of U.S. greenhouse gas emissions. This presentation will briefly highlight the need to better understand enteric CH4 emissions across beef production environments and ecosystems and feature a unique experiment that concluded in January 2025. In this experiment, unique groups of cattle were followed from weaning to harvest and enteric CH4 was measured throughout each life stage. Our goal with this experiment is to provide producers with knowledge about beef cattle emissions on rangeland in comparison to confined settings, introduce an enteric emissions project in the US Central Great Plains co-implemented with producers and researchers, present a current "state of the science" for enteric emissions measurement on working rangelands and confined systems in the western US, and discuss the potential for CH4 reduction inset markets in the beef supply chain. This initiative is a partnership among Crow Valley Livestock Cooperative, Inc., the federal agricultural research service (USDA ARS), and Colorado State University AgNext. Presentation participants will be encouraged to complete a survey that will be integral to the next phase of this work, which is to better understand the potential for an enteric CH_4 reduction inset market in the beef supply chain. Once this experiment is complete, we will extend our findings through multiple communication platforms to learn about fellow agriculturists' interests and needs in enteric CH4 emissions research and help to inform others of our work to expand the adoption of practices that improve the sustainability of beef production systems.

Keywords: beef cattle, enteric methane, sustainability

Résumé

Les émissions de méthane (CH4) entérique des bovins de boucherie comptent pour 2,8 % des gaz à effet de serre produits aux États-Unis. Cette présentation met en lumière la nécessité de mieux comprendre les mécanismes d'émission de CH4 entérique dans les milieux et écosystèmes de production de bovins de boucherie et présente une expérience unique qui s'est conclue en janvier 2025. Au cours de cette étude, des groupes distincts de bovins ont été suivis depuis le sevrage jusqu'à l'abattage, et leurs émissions de CH4 entérique ont été mesurées à chaque étape de leur vie. Cette expérience a pour but d'informer les producteurs sur les émissions des bovins de boucherie gardés au pâturage et en milieu confiné. Elle présente un projet sur les émissions entériques dans les plaines du centre des États-Unis réalisé en collaboration avec les producteurs et les chercheurs. Par ailleurs, l'expérience vise à faire connaître l'état actuel de la science concernant la mesure des émissions entériques au pâturage et dans les systèmes confinés dans l'Ouest des États-Unis et à examiner la possibilité d'intégrer des marchés de réduction du CH4 au sein de la chaîne d'approvisionnement de la viande bovine. Cette initiative est un partenariat entre Crow Valley Livestock Cooperative, Inc, le service de recherche agricole du ministère de l'Agriculture des États-Unis (USDA ARS) et Colorado State University AgNext. Les participants à la présentation seront invités à répondre à un sondage qui fera partie intégrante de la prochaine phase de ce travail, visant à mieux comprendre l'intérêt pour un marché intégré de réduction du CH4 entérique dans la chaîne d'approvisionnement de la viande bovine. À l'issue de cette expérience, nous publierons nos conclusions sur diverses plateformes afin de connaître les intérêts et les besoins de nos collègues spécialistes en agriculture en matière de recherche sur les émissions de CH4 entérique et d'informer un plus large public de notre travail afin de promouvoir l'adoption de pratiques durables au sein des systèmes de production de viande bovine.

Introduction

Developing greenhouse gas (GHG) mitigation strategies in extensive beef cattle production systems is one of the most challenging aspects of creating sustainable production practices across the beef sector (Pelton et al., 2024). The challenging nature of creating such practices is based on the difficulty of measuring emissions of GHG *in vivo* and creating mitigation strategies at-scale that may have widespread adoption in extensive grazing systems. The stocker stage, where growing cattle graze forage before finishing in confinement, has been identified as the portion of an animal's life when most enteric CH_4 emissions are produced (Rotz et al., 2019). Compared to confined settings, this production stage is the most challenging stage for managing enteric CH_4 emissions. However, innovations in livestock production technology incorporated into commercial operations can reduce the beef sector's carbon footprint (Tedeschi et al., 2021).

Currently, management practices that improve the sustainability of beef production systems at scale are mostly oriented toward applications in confined systems (e.g., finishing penned animals on high concentrate diets). An at-scale counterpart for extensive grazing systems is lacking; thus, a need exists to evaluate practices with potential for widespread adoption. Virtual fencing, for example, is an increasingly adopted form of physical management of cattle in grazing systems with the potential for target grazing portions of a pasture that offer resources that may improve production efficiency (Williams et al., 2021). In turn, it is hypothesized that this approach can produce less GHG emissions during an animal's time in extensive production systems (Williams *et al.* 2021).

The US beef production system includes grazing and confined animal systems. Traditionally, beef cattle experience both systems throughout their production cycle. Cow-calf operations are mainly based on grasslands (USDA Economic Research Service, 2023), and weaned calves follow one of two paths: 1) a backgrounding stage, encompassing approximately 60% of the calves, where calves enter a postweaning phase by either being fed a mixed forage-grain ration in a confined feeding setting or grazing on pastures with stockpiled or annual forages (e.g., wheat), with calves then stocked on rangelands/pasturelands during the summer growing season (i.e., the stocker phase) prior to entering feedlots with grain-based diets for the finishing stage, or 2) direct movement of calves to feedlots to be finished with grain-based diet, representative of the approximately 40% of calves nationally (Drouillard, 2018). Thus, it is crucial to understand the effects of each production stage on growth performance and enteric gas flux to reduce uncertainty in enteric emissions prediction models (Hammond et al., 2016; Hristov et al., 2018).

The objectives of this research were to: (1) measure enteric CH_4 emissions and performance of stocker steers managed with and without co-developed practices that may reduce enteric emissions in the stocker and finishing stages of the cattle production systems and (2) determine whether an animal's previous production environment (breeding, genetics, natal management, etc.) before the stocker stage reduces the efficacy of practices in managing enteric CH_4 emissions and animal performance in real-world production systems.

Instrumentation and Strategy Development

Post-weaning stage

Prior to the stocker stage, three groups of 40 yearling steers were evaluated for enteric CH_4 production and animal performance in two dormant-season production settings (Fig. 1). In this initial post-weaning phase, two groups were studied in drylot systems: CSU AgNext Climate-Smart Research Pens and USDA ARS Meat Animal Research Center, while a single group grazed extensive rangeland, at a local cooperating producer operation near Nunn, Colorado. Enteric emissions were measured using an automatic head-chamber system (AHCS; GreenFeed, C-Lock, Inc. Rapid City, SD, USA, Fig. 2).

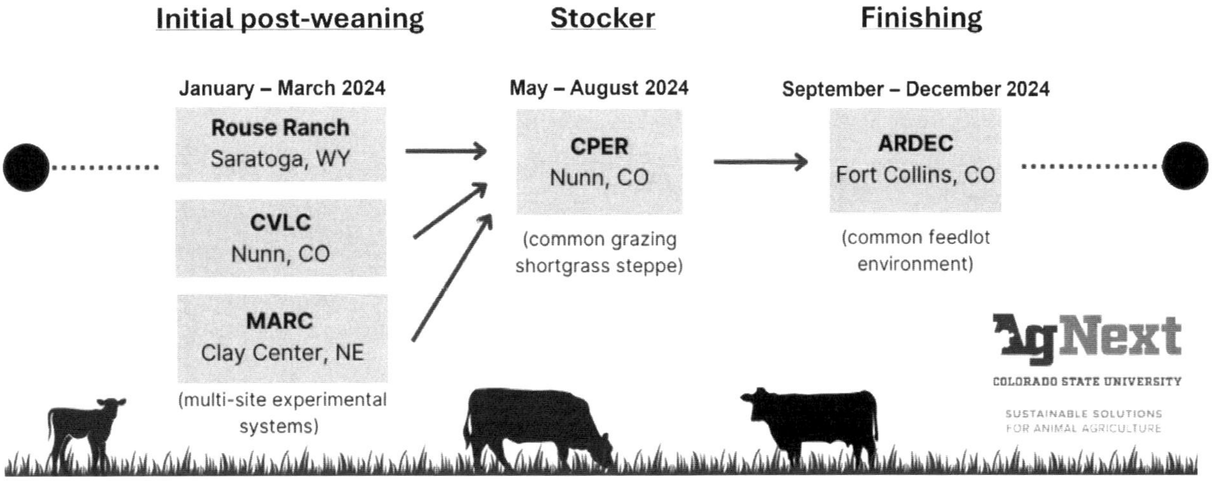

Figure 1. Workflow of cross-production stage experiment performed in collaboration with USDA Agricultural Research Service Central Plains Experimental Range (CPER) and Meat Animal Research Center (MARC), Crow Valley Livestock Cooperative (CVLC), Colorado State University (CSU) Agricultural Research, Development, and Education Center (ARDEC) and CSU AgNext | Department of Animal Sciences.

Stocker stage

Following this post-weaning stage, stocker research was implemented at the USDA ARS Central Plains Experimental Range near Nunn, Colorado where all three groups of steers were managed together in four 130-ha shortgrass steppe pastures from mid-May to early September 2024. Here, two herds of all three groups mixed together were managed with Vence virtual fence technology (Merck Animal Health, Rahway, NJ, USA), and two control herds were managed following traditional rangeland management in the region, continuous season-long stocking. The potential strategy for sustainable grazing management via virtual fencing was co-developed with a stakeholder group making science-informed management decisions within the grazing season. Local knowledge of the spatial availability of forage resources at the study site played a key role in establishing the grazing plan for the stocker stage of the experiment. The plan was set to establish virtual fence boundaries that target the consumption of cool-season grasses at their peak (June), when their digestibility was greatest, i.e., lowering enteric CH_4 emissions. Virtual fence management outcomes for enteric CH_4 emissions and animal performance were then assessed for each group.

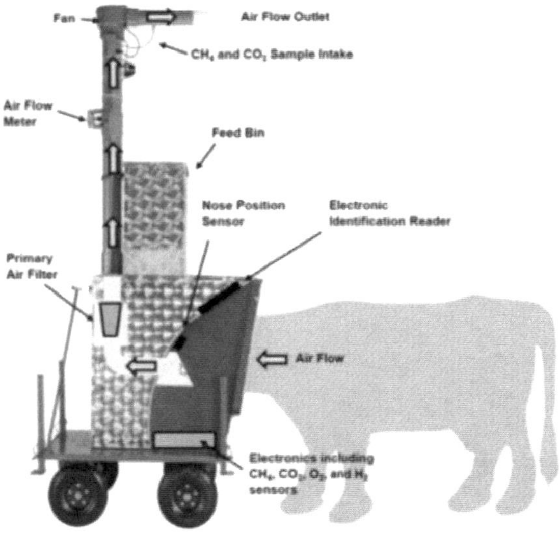

Figure 2. Automated head-chamber system (AHCS; GreenFeed, Rapid City, SD, USA) and components. Modified from Hristov et al. (2015).

Finishing stage

Following the stocker stage, from October to December 2024 the steers were implanted with a growth promoting technology (Synovex Choice®, Zoetis, Parsippany, NJ, USA) and finished in the CSU AgNext Climate-Smart Research Pens (CSRP, 15 m × 43 m), where a protein supplement was on offer daily. Enteric emissions were measured using a drylot pen-based AHCS in each pen. Each CSRP housed 30 animals/pen, equipped with five SmartFeed (SF; C-Lock, Rapid City, SD, USA) bunk systems to measure feed intake, two automatic water tanks, a concrete bunk pad (15 m x 3 m), and a metal roof (15 m × 3 m) covering the SF bunks and approximately 7% of the pen surface. Typical husbandry practices were followed at each facility in this study.

In this preliminary analysis, we calculated the means and standard errors for each group's average daily gain (ADG; kg/d), enteric CH_4 production (g CH_4/d), and CH_4 intensity (g CH_4/ADG) in each production setting. To evaluate whether virtual fence management performed in the stocker stage impacted enteric emissions or performance, we compared emissions and performance outcomes in the stocker stage and any potential legacy effects of VF management carried over to the finishing stage. This comparison utilized a linear mixed effect model with production stage, an animal's previous production environment, and VF treatment as fixed effects and pasture/pen as a random intercept. To achieve adequate estimates of CH_4 emissions for treatment evaluation, emissions data from steers with at least 40 visits at each production stage that met the 2-8 min AHCS visit duration threshold was used for analysis (Beck et al., 2024). Further data-cleaning steps before analysis included removing observations with less than 26 L/s of AHCS airflow and AHCS pipe temperature below -10°C.

Results & Discussion

Preliminary analyses of enteric CH_4 production and performance met our expectations during the initial post-weaning phase: steers managed in confinement gained weight at least twice the rate of steers backgrounded on dormant-season rangeland, confined steers produced 5 to 20 g/d more CH_4 than grazed steers, and confined steers produced substantially less CH_4 per kg gained than grazed steers (Table 1).

Table 1. Means ± SE of enteric emissions measurements and animal performance for each group in each experiment stage. Control = non-virtual fencing (VF) management, VF = virtual fencing management.

Item				Production stage Initial post-weaning	Stocking			Finishing			
	n	Previous production environment		No treatment	Control	VF	P	Control	VF	Trt P	Stage P
Initial body weight, kg	40	Colorado Drylot	-	207 ± 3	390 ± 8	393 ± 8	0.80	551 ± 7	559 ± 8	0.48	0.0001
	40	Nebraska Drylot	-	377 ± 4	445 ± 8	444 ± 8	0.91	585 ± 7	588 ± 7	0.76	0.0001
	40	Colorado Grazing	-	325 ± 4	361 ± 8	358 ± 7	0.85	581 ± 8	554 ± 8	0.01	0.0001
	120	All		303 ± 7	398 ± 6	398 ± 5	0.98	572 ± 6	567 ± 5	0.40	0.0001
Average daily gain, kg/d	40	Colorado Drylot	-	1.23 ± 0.02	0.63 ± 0.10	0.60 ± 0.10	0.85	2.66 ± 0.09	2.68 ± 0.09	0.84	0.0001
	40	Nebraska Drylot	-	1.05 ± 0.03	0.56 ± 0.12	0.47 ± 0.11	0.67	2.71 ± 0.09	2.78 ± 0.09	0.56	0.0001
	40	Colorado Grazing	-	0.23 ± 0.06	1.20 ± 0.10	1.09 ± 0.10	0.41	2.64 ± 0.09	2.46 ± 0.09	0.10	0.0001
	120	All		0.83 ± 0.05	0.82 ± 0.07	0.72 ± 0.08	0.80	2.67 ± 0.07	2.64 ± 0.07	0.98	0.0001
CH_4 production, g CH_4/d	25	Colorado Drylot	-	161 ± 3	219 ± 15	217 ± 15	0.95	200 ± 11	200 ± 11	0.96	0.26
	24	Nebraska Drylot	-	180 ± 3	204 ± 15	200 ± 15	0.86	178 ± 11	189 ± 11	0.21	0.25
	19	Colorado Grazing	-	155 ± 6	247 ± 16	239 ± 16	0.71	219 ± 11	207 ± 11	0.17	0.08
	68	All		167 ± 2	223 ± 14	219 ± 14	0.83	199 ± 10	198 ± 10	0.86	0.16
CH_4 intensity, g CH_4/ADG	25	Colorado Drylot	-	131 ± 3	355 ± 39	384 ± 38	0.61	79 ± 28	77 ± 27	0.88	0.0004
	24	Nebraska Drylot	-	177 ± 5	370 ± 39	407 ± 38	0.53	68 ± 27	70 ± 27	0.89	0.0002
	19	Colorado Grazing	-	631 ± 103	206 ± 39	224 ± 39	0.75	85 ± 27	85 ± 28	0.99	0.02
	68	All		311 ± 31	310 ± 37	339 ± 38	0.61	77 ± 26	77 ± 26	0.98	0.001

In the stocker stage, ADG was greatest for locally grazed Colorado-grazing steers, with local steers managed without VF performing slightly better than those managed with VF, but the difference was not significant. ADG for previously confined steers was about half as much as locally grazed steers, with control steers, again, performing marginally better than VF-managed steers. Likewise, CH_4 production and CH_4 intensity were noticeably lower for previously confined and previously grazed steers that were not managed with VF, although enteric emissions were lower for VF-managed steers when they were bound to a cool-season grass section of the pasture from May to June (*unpublished data*). This outcome suggests that VF management for reducing enteric CH_4 emissions and maintaining animal performance may be possible during peak green-up on semiarid rangeland, yet more work is necessary to determine this technology's potential as an enteric CH_4 reduction tool for extensive grazing systems.

Management and dietary changes in the finishing stage resulted in significantly higher ADG when compared to performance in the previous stages of this experiment. The increase in ADG was not accompanied by a significant reduction in enteric CH_4 production in the finishing stage. However, the amount of CH_4 produced per kg gained (i.e., CH_4 intensity) in this stage was at least 4 times lower than the previous two production stages. Additionally, virtual fencing management in the stocker stage did not impact emissions or performance outcomes in the finishing stage.

Future efforts from this empirical work include providing producers with knowledge about beef cattle emissions on rangeland in comparison to confined settings. Additionally, after more analyses, this data will be used to determine the potential for CH_4 reduction-inset markets in the beef supply chain. In conclusion, our findings contribute to understanding variability in stocker steer performance and GHG emissions at two common stages in the beef production sector. This study adds to the growing body of knowledge on ruminant emissions and suggests areas for further research in sustainable livestock production strategy development.

References

Beck, M. R., L. R. Thompson, Jarret A. Proctor, R. R. Reuter, and S. A. Gunter. 2024. Recommendations on visit duration and sample number requirements for an automated head chamber system. J Anim Sci 102doi: https://doi.org/10.1093/jas/skae158

Drouillard, J. S. 2018. Current situation and future trends for beef production in the United States of America - A review. Asian-Australas J Anim Sci 31(7):1007-1016. doi: https://doi.org/10.5713/ajas.18.0428

Hammond, K. J., L. A. Crompton, A. Bannink, J. Dijkstra, D. R. Yáñez-Ruiz, P. O'Kiely, E. Kebreab, M. A. Eugène, Z. Yu, K. J. Shingfield, A. Schwarm, A. N. Hristov, and C. K. Reynolds. 2016. Review of current in vivo measurement techniques for quantifying enteric methane emission from ruminants. Animal Feed Science and Technology 219:13-30. doi: https://doi.org/10.1016/j.anifeedsci.2016.05.018

Hristov, A. N., E. Kebreab, M. Niu, J. Oh, A. Bannink, A. R. Bayat, T. M. Boland, A. F. Brito, D. P. Casper, and L. A. Crompton. 2018. Symposium review: Uncertainties in enteric methane inventories, measurement techniques, and prediction models. Journal of Dairy Science 101(7):6655-6674. doi: https://doi.org/10.3168/jds.2017-13536

Hristov, A. N., J. Oh, F. Giallongo, T. Frederick, H. Weeks, P. R. Zimmerman, M. T. Harper, R. A. Hristova, R. S. Zimmerman, and A. F. Branco. 2015. The use of an automated system (GreenFeed) to monitor enteric methane and carbon dioxide emissions from ruminant animals. JoVE (Journal of Visualized Experiments) (103):e52904. doi: https://doi.org/10.3791/52904

Pelton, R. E. O., C. E. Kazanski, S. Keerthi, K. A. Racette, S. Gennet, N. Springer, E. Yacobson, M. Wironen, D. Ray, K. Johnson, and J. Schmitt. 2024. Greenhouse gas emissions in US beef production can be reduced by up to 30% with the adoption of selected mitigation measures. Nature Food 5(9):787-797. doi: https://doi.org/10.1038/s43016-024-01031-9

Rotz, C. A., S. Asem-Hiablie, S. Place, and G. Thoma. 2019. Environmental footprints of beef cattle production in the United States. Agricultural Systems 169:1-13. doi: https://doi.org/10.1016/j.agsy.2018.11.005

Tedeschi, L. O., P. L. Greenwood, and I. Halachmi. 2021. Advancements in sensor technology and decision support intelligent tools to assist smart livestock farming. J Anim Sci 99(2):1–11. doi: https://doi.org/10.1093/jas/skab038

USDA Economic Research Service. 2023. Sector at a Glance. https://www.ers.usda.gov/topics/animal-products/cattle-beef/sector-at-a-glance/1/12 (Accessed November 13, 2023 2023).

Williams, T., C. Wilson, P. Wynn, and D. Costa. 2021. Opportunities for precision livestock management in the face of climate change: a focus on extensive systems. Animal Frontiers 11(5):63-68. doi: https://doi.org/10.1093/af/vfab065

Effects of chromium propionate supplementation for beef cows at late gestation on cow performance and skeletal muscle metabolism of the offspring

Effets de l'ajout de propionate de chrome dans l'alimentation des vaches de boucherie en fin de gestation sur les performances des vaches et le métabolisme des muscles squelettiques de la progéniture

E. A. Knapper[1], L. V. Kladt[1,2], I. I. Avalos-Rosario[1], S. Dean[1], W. Silva[1], N.V.L. Serão[3], J. E. Hergenreder[4], E. L. Neto[4], M. Duarte[1]*
[1] Department of Animal Biosciences, University of Guelph, Guelph, ON, Canada
[2] Department of Animal Science, Federal University of Viçosa, Viçosa, MG, Brazil
[3] Department of Animal Science, Federal University of Lavras, Lavras, MG, Brazil
[4] Kemin Industries, Inc., Des Moines, IA, USA
*knappere@uoguelph.ca

SPONSORED BY / COMMANDITÉ PAR

trouw nutrition
a Nutreco company

Abstract

Chromium propionate enhances insulin sensitivity and glucose metabolism, potentially improving maternal energy balance during late gestation. This study evaluated the effects of chromium propionate supplementation on cow-calf performance and offspring muscle growth and metabolism. One hundred late-gestation Angus cows were randomly assigned to two treatments until parturition: control (CON, n = 50), receiving a basal diet, and chromium propionate (CRP, n = 50), receiving the same diet supplemented with 10 g/head/day of 0.04% KemTRACE® chromium. Cow performance was monitored from 56 days prepartum to parturition. From 2-10 days of age, skeletal muscle samples were collected from calves to assess gene expression and protein abundance related to adipogenesis and glucose metabolism. CRP supplementation tended ($P = 0.08$) to interact with calf sex for cow average daily gain (ADG; $P = 0.02$), with CRP cows carrying male calves exhibiting lower ADG compared to CON cows. However, no differences in dry matter intake were observed between treatments when evaluated as kg/d ($P = 0.13$) or as percentage of body weight ($P = 0.29$). Calves from CRP-supplemented dams showed increased skeletal muscle mRNA expression of adipogenic markers *PPARγ* ($P < 0.01$), *ZFP423* ($P < 0.01$), and *DLK1* ($P = 0.02$). Similarly, they exhibited enhanced mRNA expression of *IRS1* ($P = 0.01$) and *GLUT4* ($P = 0.04$) and tended to increase the IRS1 activity ($P = 0.08$). In summary, while CRP had minimal effects on dams, these findings suggest early metabolic programming that may enhance offspring muscle metabolism, increasing their potential for intramuscular fat deposition.

Résumé

Le propionate de chrome accroît la sensibilité à l'insuline et accélère le métabolisme du glucose, ce qui pourrait améliorer l'équilibre énergétique de la mère en fin de gestation. Dans cette étude, nous avons évalué les effets d'une supplémentation en propionate de chrome sur les performances vache-veau et sur la croissance et le métabolisme musculaires de la progéniture. Cent vaches Angus en fin de gestation ont été réparties aléatoirement entre deux groupes jusqu'au moment du vêlage : groupe témoin (CON, n = 50), recevant un régime alimentaire standard, et groupe propionate de chrome (CRP, n = 50), recevant le régime alimentaire standard avec l'ajout de 10 g/jour par tête de KemTRACE^{MD} (0,04 % de chrome). Les performances des vaches ont été contrôlées pendant les 56 derniers jours de la gestation. Des échantillons de muscles squelettiques ont été prélevés sur les veaux des jours 2 à 10 après leur naissance afin d'évaluer l'expression génétique et l'abondance des protéines par rapport à l'adipogenèse et au métabolisme du glucose. Une interaction (*P*=0,08) entre la supplémentation en propionate de chrome et le sexe du veau a été notée pour le gain moyen quotidien (GMQ; *P*=0,02) des vaches, les mère CRP de veaux mâles présentant un GMQ inférieur à celui des mères CON. Cependant, aucune différence de consommation volontaire de matière sèche n'a été observée entre les traitements, telle que mesurée en kg/j (*P*=0,13) ou en pourcentage du poids corporel (*P*=0,29). Les veaux issus de mères ayant reçu le propionate de chrome ont montré une plus forte expression de l'ARNm des marqueurs adipogéniques PPARγ (*P*<0,01), ZFP423 (*P*<0,01) et DLK1 (*P*=0,02) dans les muscles squelettiques. De même, ils ont affiché une meilleure expression de l'ARNm des marqueurs l'IRS1 (*P*=0,01) et GLUT4 (*P*=0,04) et ont eu tendance à montrer une augmentation de l'activité de l'IRS1 (*P*=0,08). En résumé, bien que le propionate de chrome n'ait eu que des effets négligeables sur les mères, ces résultats suggèrent une programmation métabolique précoce susceptible d'améliorer le métabolisme musculaire de la progéniture, augmentant ainsi son aptitude pour le dépôt de graisse intramusculaire.

Introduction

Glucose serves as the primary metabolic fuel for growing fetuses, stimulating insulin release to promote the differentiation of mesenchymal stem cells into preadipocytes involved in skeletal muscle development (Zhu et al., 2006; Lucy et al., 2014). The potential of skeletal muscle capacity and intramuscular adipogenesis in the fetus relies on available energy supplied via the placenta, provided by the maternal diet. Recent studies have indicated that modifying the available energy balance in maternal diets during late gestation can impact intramuscular adipogenesis and development in the offspring (Duarte et al., 2014; Moisá et al., 2015; Costa et al., 2022; Sanglard et al., 2023). Therefore, late gestation nutritional strategies that enhance glucose availability to the fetus may be implemented to improve intramuscular adipogenic differentiation and overall capacity of the skeletal muscle development in the offspring.

With recent biological insights between the connection of maternal nutrition and skeletal muscle metabolism, the organic trace mineral chromium has gained interested as a potential feed additive in late gestation diets. Chromium, particularity in its biologically active form, Glucose Tolerance Factor (GTF), is an essential component to the functional response of the hormone insulin (Mowat, 1996). Chromium is currently understood to bind to the β-subunit of the insulin receptor on the intracellular membrane on insulin-sensitive cells including animal tissues, notably observed in muscle and fat cells, thereby enhancing insulin signaling by activating downstream effectors such as PI3-kinase and AKT (Lewicki et al., 2014; Hua et al., 2013). This activation promotes the translocation of GLUT4, increasing glucose uptake (Hua et al., 2013). As described by Zhu et al. (2006), increased GLUT4 activity enhances glucose uptake by muscle and adipose tissues, supporting protein synthesis and promoting adipogenic differentiation. Therefore, chromium supplementation may represent a strategic approach in cow-calf systems to improve fetal glucose utilization, thereby fostering muscle development, enhancing carcass potential, and increasing overall offspring productivity.

Based on this rationale, we hypothesized that chromium supplementation during late gestation would enhance maternal energy status and performance, while also promoting changes on skeletal muscle growth and energy metabolism in the offspring.

Materials and Methods

A total of 100 late-gestation Angus-cross beef cows, confirmed pregnant via ultrasound with either male (n = 50) or female (n = 50) fetuses, were enrolled in the study. The control group (CON; n = 50) consisted of 25 cows carrying male calves and 25 carrying female calves, all receiving a standard basal diet formulated to meet the National Academies of Sciences, Engineering, and Medicine (NASEM, 2016) requirements for late gestation. The chromium supplementation group (CRP; n = 50) included an identical distribution of fetal sex and was fed the same basal diet, with the addition of 10 g/head/day of a 0.04% chromium propionate premix (KemTRACE® Chromium, Kemin Industries, Inc., Des Moines, IA). Experimental diets were initiated at 56 days prior to expected calving and cows remained on their assigned treatments until parturition. Throughout the supplementation period, individual cow performance was monitored. Daily dry matter intake (DMI) was recorded, and body weight (BW) was measured every 28 days to evaluate performance of the cows. Blood samples were also collected from each cow at the same time of BW recording and analyzed for glucose, insulin, non-esterified fatty acids (NEFA), and beta-hydroxybutyrate (BHB) to assess energy balance. Concurrently, carcass ultrasound images were collected and monitor body condition of the cows throughout the experimental period. At parturition, immediately after calving and prior to nursing, cows were fully milked to measure total colostrum yield, and a colostrum sample was taken to evaluate IgG concentration. Before nursing, calf body weight and blood samples were also collected. After parturition, the cow-calf pair from both treatments were then managed as a single herd and fed the same lactation diet without supplementation of chromium-propionate. Within 10 days of age, skeletal muscle biopsies were obtained from calves to analyze gene expression and protein abundance related to adipogenesis and glucose metabolism.

The analysis of the data included different models depending on the traits and animal type (i.e., cow or calf). A fixed-effect base model, including effects used in all analyses, is described below:

$$Y_{ijkl} = \mu + T_i + PS_j + (T * PS)_{ij} + P_k + iBW_l + iREA_l + iSTF_l + iDPP_l + e_{ijkl}$$

[Eq. 1]

where Y_{ijkl} is the observed value; μ is the intercept; T_i is the effect of maternal dietary treatment; PS_j is the effect of progeny sex; $(T * PS)_{ij}$ is the interaction between maternal dietary treatment and progeny sex; P_k is effect of parity; iBW_l, $iREA_l$, $iSTF_l$, and $iDPP_l$ are the covariates of initial body weight, ribeye area, subcutaneous fat and days pre-partum of the cow; and e_{ijkl} is the residual.

The model in Eq. [1] was used for analysis of the cow data during pregnancy, except for traits measured multiple times within the cow (*i.e.*, blood parameters). For these traits, the model was expanded to include the fixed effect of measurement and its interactions with maternal dietary treatment and progeny sex (one- and two-way interactions), along with the random effect of cow. For post-partum traits, Eq. [1] was expanded to include additional covariates, such as the cow's BW at calving, days post-partum, and cow's ADG post-partum, according to the trait analyzed.

The calf data were analyzed by expanding Eq. [1] to include the random effect of sire of the calf. In addition, according to the trait analyzed, the covariates of colostrum yield of the cow and weaning age were included in the model. Finally, the gene expression and western blot data were analyzed following the procedure described in Steibel et al. (2009), in which data on control and target genes/proteins are analyzed simultaneously in the model. Eq. [1] was expanded to include the fixed effect of control/target gene/protein, along with all one-way interactions with other effects in the model, and the random effect of calf

Prior to final analyses, residuals were evaluated for normality. For each analysis, data points were removed one at a time until absolute Studentized residuals were lower than 3 and had non-significant ($P > 0.01$) Shapiro-Wilk's test for normality. Expected means were generated from the final models when significant ($P < 0.05$). All analyses were performed in SAS Studio 3.81 (Enterprise Edition, SAS Institute Inc., Cary, NC, USA).

Results

Overall, minimal differences in cow performance traits were observed between dietary treatments. There was a trend toward a CRP supplementation and calf sex interaction for cow ADG ($P = 0.08$), with CRP-supplemented cows carrying male calves showing lower ADG than CON cows, while no effects was observed for cows carrying female calves despite treatment. Cow's DMI did not differ between treatments, either in kg/day ($P = 0.13$) or as a percentage of BW ($P = 0.29$). With respect to maternal metabolic markers, concentrations of glucose ($P = 0.96$), and non-esterified fatty acids (NEFA; $P = 0.10$) did not differ between treatment groups. However, there was a trend towards CRP supplementation and days pre-partum (DPP) for insulin ($P = 0.05$), with CRP-supplemented cows having lower blood concentrations approaching parturition then CON cows. There was a treatment × DPP interaction for β-hydroxybutyrate (BHB; $P < 0.01$). At 11 days prepartum (DPP), cows in the CON group had higher BHB concentrations compared to those in the CRP group, with BHB levels increasing from 26 to 11 DPP only in the CON cows. No differences were observed with respect to cow performance after parturition up to the breeding season between treatments ($P > 0.05$).

Differences due to maternal dietary treatments became more evident in the offspring. Although muscle fiber size ($P > 0.53$) and muscle fiber type ($P > 0.05$) did not differ significantly by treatment or calf sex, a significant treatment × calf sex interaction was observed for muscle fiber number ($P = 0.01$). Specifically, female calves born to CON cows had a greater number of muscle fibers compared to those born to CRP-supplemented cows, whereas no differences were observed between male calves from the two dietary groups. No differences were observed between treatment and calf sex for calf serum levels of glucose ($P = 0.10$) and insulin ($P = 0.86$) at birth, neither for calf birth weight ($P = 0.53$), weaning weight ($P = 0.56$), or ADG ($P = 0.41$).

At the molecular level, calves born to CRP-supplemented dams showed increased mRNA expression in skeletal muscle for genes related to mesenchymal progenitor cells - *PDGFRα* ($P < 0.01$), and adipogenesis *DLK1* ($P = 0.02$), *ZFP423* ($P < 0.01$), *PPARγ* ($P < 0.01$), indicating an increase in early intramuscular adipogenesis. We have also observed an increase in mRNA expression glucose transporter 4 (*GLUT4*; $P = 0.04$) and insulin receptor 1 (*IRS1;* P = 0.01), as well as a tendency for an increase of IRS1 activity (pIRS1:IRS1 ratio; $P = 0.08$) in the skeletal muscle of calves born to dams supplemented with CRP compared to those born to dams from the control group. Collectively, these findings suggest that maternal CRP supplementation during late gestation promotes both intramuscular adipogenic capacity and glucose utilization in offspring skeletal muscle.

Conclusions and Further Directions

These results indicate that chromium supplementation during late gestation enhances offspring intramuscular fat deposition potential. Further studies are required to clarify the long-term implications, particularly regarding underlying mechanisms. Additionally, the sex-specific responses observed in some of the output variables assessed highlight the importance of investigating how calf sex influences placental nutrient transport and maternal energy partitioning.

References

Duarte, M. S., M. P. Gionbelli, P. V. R. Paulino, N. V. L. Serão, C. S. Nascimento, M. E. Botelho, T. S. Martins, S. C. V. Filho, M. V. Dodson, S. E. F. Guimarães, and M. Du. 2014. Maternal overnutrition enhances mRNA expression of adipogenic markers and collagen deposition in skeletal muscle of beef cattle fetuses. *J. Anim. Sci.* 92:3846–3854.

Costa, T. C., P. E. C. Lourenço, R. O. Souza, M. M. Lopes, R. D. Araújo, M. M. Santos, L. P. Luciano, J. D. Massensini, L. L. H. Chalfun, L. N. Rennó, C. B. Sampaio, R. Veroneze, P. V. R. Paulino, M. P. Gionbelli, and M. S. Duarte. 2022. Ruminal undegradable protein enriched diet during late gestation of beef cows affects maternal metabolism and offspring's skeletal muscle development. *Anim. Feed. Sci. Tech.* 291:115400.

Hua Y., Clark S., Ren J., and Sreejayan N. 2013. Molecular Mechanisms of Chromium in Alleviating Insulin Resistance. *J Nutr Biochem.* 23(4): 313-319.

Lewicki S., Zdanowski R., Krzyzowska M., Lewicka A., Dębski B., Niemcewicz M., and Goniewicz M. 2014. The role of Chromium III in the organism and its possible use in diabetes and obesity treatment. *An Agr Env Med.* 21(2): 331-335.

Lucy M.C., Butler S.T., and Garverick H.A. 2014. Endocrine and metabolic mechanisms linking postpartum glucose with early embryonic and foetal development in dairy cows. *Anim.* 8(1): 82-90.

Moisá, S. J., D. W. Shike, L. Shoup, S. L. Rodriguez-Zas, and J. J. Loor. 2015. Maternal plane of nutrition during late gestation and weaning age alter Angus × Simmental offspring longissimus muscle transcriptome and intramuscular fat. *Plos One* 10: e0131478.

Mowat, D.N. 1996. Supplemental organic chromium for beef and dairy cattle. Proceedings of Asia-Pacific Lecture Tour. 31.

National Academics of Sciences, Engineering, and Medicine. 2016. *Nutrient Requirements of Beef Cattle: Eighth Revised Edition.* Washington, DC: The National Academies Press.

Steibel J.P., Poletto R., Coussens P.M., and Rosa G.J.M. 2009. A powerful and flexible linear mixed model framework for the analysis of relative quantification RT-PCR data. *Genomics.* 94: 146-152.

Zhu M.J., Ford S.P., Means W.J., Hess B.W., Nathanielsz P.W., and Du M. 2006. Maternal nutrient restriction affects properties of skeletal muscle in offspring. *J Phys.* 575(1): 241-250.

Forage inclusion in finishing feedlot diets: Does source, quantity, and quality matter?

Intégration des fourrages aux rations servies en parc de finition : la source, la quantité et la qualité importent-elles?

G.O. Ribeiro[1], M.M. Dubois[1]
[1]Department of Animal and Poultry Science, University of Saskatchewan, Saskatoon, SK
gabriel.ribeiro@usask.ca

SPONSORED BY / COMMANDITÉ PAR

Abstract

Forage inclusion in finishing feedlot diets is a critical factor influencing cattle growth performance, feed efficiency, health, and the economic and environmental sustainability of beef production. While high-concentrate diets are standard in feedlot operations to maximize energy intake, weight gain, and feed efficiency, forage inclusion is essential for maintaining ruminal function and mitigating the risk of digestive disorders and liver abscess. However, the specific forage or roughage requirements in finishing feedlot diets remain poorly defined. The effects of forage source, quality, and inclusion levels on feed efficiency, growth performance, health, and welfare continue to be subjects of active research and debate. Fiber digestibility is influenced by multiple factors, including plant species, maturity, lignin content, and structural characteristics such as the leaf-to-stem ratio. These factors vary with environmental conditions and management strategies, directly impacting the energy available for animal production. Although the ruminal microbial population is highly effective at breaking down fiber, the process remains incomplete. Strategies to enhance fiber digestion—such as diet adaptation, nutritional supplementation, and physical, chemical, or biological treatments—have shown potential to improve digestibility and boost cattle performance. Additionally, variability in fiber digestion among individual cattle underscores opportunities for selecting animals with greater efficiency. Factors such as feed intake, rumen microbiota composition and activity, digesta passage rate, and chewing behavior contribute to this variability. A deeper understanding of the plant and animal factors affecting forage and fiber digestion can help optimize feedlot diets, enhance feed efficiency, reduce digestive disorders, and minimize the environmental impact of beef production.

Keywords: beef cattle, fibre, forage, digestion, growth performance

Résumé

L'intégration des fourrages dans les rations des bovins en parc de finition est une pratique qui intervient nécessairement sur les performances de croissance des animaux, leur efficacité alimentaire et leur santé ainsi que sur la viabilité économique et environnementale de la production de viande bovine. Bien que les régimes riches en concentré soient couramment employés dans les parcs d'engraissement pour augmenter l'apport énergétique, favoriser la prise de poids et améliorer l'efficacité alimentaire, il demeure essentiel de servir des fourrages pour maintenir la fonction ruminale et atténuer le risque de troubles digestifs et d'abcès hépatique. Cependant, les besoins spécifiques en fourrages dans les rations des bovins en parc de finition sont mal définis. Les effets de la source et de la qualité du fourrage ainsi que du taux d'inclusion sur l'efficacité alimentaire, les performances de croissance, la santé et le bien-être des animaux sont encore largement étudiés et discutés. La digestibilité des fibres est influencée par de multiples facteurs, notamment l'espèce végétale, la maturité, la teneur en lignine et les caractéristiques structurelles, telles que le rapport feuilles-tiges. Ces éléments, qui varient selon les conditions environnementales et les pratiques agronomiques, influencent directement l'énergie disponible pour la production animale. Bien que le microbiote du rumen excelle à décomposer les fibres, le processus demeure incomplet. Certaines stratégies, telles que l'adaptation du régime alimentaire, la supplémentation et les traitements physiques, chimiques ou biologiques, ont montré qu'elles pouvaient améliorer la digestibilité des fibres et stimuler les performances des bovins. En outre, la variabilité au niveau de la digestion des fibres entre les individus ouvre la voie à la sélection d'animaux plus efficaces. Des facteurs tels que la prise alimentaire, la composition et l'activité du microbiote ruminal, la vitesse de transit du digesta et le comportement de mastication contribuent à cette variabilité. Mieux comprendre les phénomènes d'origine végétale et animale qui interviennent dans la digestion des fourrages et des fibres peut aider à optimiser les rations servies en parc d'engraissement, à améliorer l'efficacité alimentaire, à réduire les troubles digestifs et à diminuer l'impact environnemental de la production de viande bovine.

Introduction

Finishing feedlot cattle in North America are typically fed high-concentrate diets with forage or roughage included at levels ranging from 6% to 12% of dry matter (DM; Koenig et al., 2020; Samuelson et al., 2016). While the terms "forage" and "roughage" are often used interchangeably, "roughage" is arguably more appropriate in this context. It is a broader term encompassing both forages (i.e., leaves, stems, and stalks of plants) and fibrous byproducts (e.g., hulls), which are high in crude fibre (>18%), bulky (low weight per unit volume), and lower in energy compared to concentrates.

Roughages are generally more expensive per unit of energy than cereal grains. As a result, they are typically included at minimal levels in finishing diets to reduce the cost of gain, increase average daily gain (ADG), and shorten the time required for cattle to reach slaughter weight (Koenig & Beauchemin, 2011; Koenig et al., 2020; Turgeon et al., 2010). However, a minimum inclusion level is essential to support proper rumen function and health, and to reduce the risk of ruminal and hindgut acidosis, bloat, and liver abscesses (Humer et al., 2018).

Despite extensive research over the past 40 years, the optimal level of roughage inclusion in finishing diets remains poorly defined. This uncertainty likely stems from the wide variability in roughage chemical composition, particle size, and ruminal rate and extent of digestion, as well as interactions with other dietary components such as non-fibrous carbohydrates and fats. This paper reviews recent findings on roughage inclusion in finishing feedlot diets, focusing on effects on growth performance, digestion, and health.

Defining and Assessing Roughage Quality

Before discussing the effects of different roughage inclusion in finishing diets, it is important to understand the factors that may affect its nutritional quality and ability to maintain proper rumen function and health. Factors such as plant species, maturity, structure (leaf-to-stem ratio), and lignin and phenolic acid content are well known to directly affect roughage digestibility (Jung and Allen, 1995; Ribeiro et al., 2016). The digestibility of roughages decreases as the proportion of cell walls or fibre content increases, and their resistance to degradation into small particles during mastication and digestion increases (Wilson & Mertens, 1995).

We routinely assess the fibre content of roughages and their resistance to degradation in the rumen by examining their neutral detergent fibre (NDF), acid detergent fibre (ADF), and lignin content in laboratory feed reports. As NDF, ADF, and lignin increase, the digestibility and the energy content of roughages decrease. In cattle fed high-concentrate diets, the increased ruminal passage rate and the low ruminal pH (<6) further inhibits fibre digestion. Ruminal retention time and access of microbial fibrolytic enzymes to target substrates are considered the main factors affecting fibre degradation in the rumen (Weimer, 1996).

Recent developments in fibre quality metrics include physically effective NDF (peNDF) and undigested NDF (uNDF). The peNDF of a roughage or a diet is a particle-size based measurement calculated by multiplying the proportion of particles that are retained on the 4-, 8- and 19-mm sieves (or 8- and 19-mm sieves only) of the Penn State Particle Separator by the concentration of NDF in that roughage or diet. The peNDF is the portion of the diet that provides the necessary 'scratch factor' that promotes chewing and ruminal contractions. Chewing promotes saliva production and adequate ruminal motility drives nutrient absorption. Both nutrient absorption and saliva are essential to buffer ruminal pH, which will promote ruminal health and effective feed digestion. The chewing activity stimulated by peNDF also promotes a reduction in particle size of the consumed feed, consequently increasing the surface area for ruminal microbial enzymes to act and breakdown the fibre and other feed components.

There is a lack of consensus among scientists and nutritionists regarding which sieves from the Penn State Particle Separator to use in this assessment. Some feeds that may not promote the "scratch factor," such as cereal grains or pellets, may be retained in the 4-mm sieve, while roughages retained in the 4-mm sieve do promote the "scratch factor."

The uNDF is most commonly measured in the lab by incubating samples of the feed in buffered ruminal fluid for 240 hours. The residue remaining is then analyzed for NDF. The uNDF is the portion of the NDF that is unlikely to be digested, regardless of how long it remains in the rumen. The greater the uNDF concentration of a feed or diet, the greater the ruminal filling effect. It has also been shown that the uNDF fraction contributes to the development of a stable rumen mat, stimulating rumen motility, rumination (chewing), and buffering. Due to its indigestible nature, as uNDF increases, diet energy generally decreases.

Current strategies to improve fibre digestibility

While adequate roughage inclusion and particle size are necessary to maintain rumen health, maximizing fibre digestibility is essential to increase growth performance and feed efficiency. This is often accomplished through increasing surface area for microbial attachment and degradation, which often leads to an increase in dry matter intake (DMI) (Soita et al., 2002; Varga & Kolver, 1997). When feeding high grain diets with low roughage inclusion attention should be given to the amount of peNDF in the ration, as it directly affects rumen health through its impact on ruminal buffering (Mertens, 1997).

Several techniques have been implemented by nutritionists and beef producers over the last several years with variable outcomes depending on the procedure utilized and the roughage used in the ration (Soita et al., 2002).

Mechanical processing of roughage is commonly used but impact on intake and digestibility has been inconsistent (Adesogan et al., 2019). Although a reduction in particle size might improve microbial degradation, roughage chopped too finely can pass through the rumen faster and therefore have decreased overall digestion (Soita et al., 2002). Additionally, smaller particle size of roughages (i.e., low peNDF) has been shown to reduce rumination (Beauchemin et al., 1994; Bhandari et al., 2008) leading to an increased risk of acidosis (Mertens, 1997; Zebeli et al., 2012) when used in feedlot finishing diets. Soita et al. (2002) examined the effect of feeding barley silage with short (4.7 mm) compared to long (18.8 mm) chop length to steers fed an all-roughage diet. They found that reducing the chop length increased DMI and nutrient digestibility but reduced ruminal pH (Soita et al., 2002). Conversely, when beef steers were fed a finishing diet containing barley silage (10% of diet DM) with short (10 mm) or long (20 mm) chop length, increasing silage particle size promoted greater DMI and reduced eating rate although no differences in growth performance were detected (Addah et al., 2015). Shredding silage, otherwise known as shredlage, is another form of processing whereby silage is shredded at a longer chop length than average (Ferraretto & Shaver, 2012). Although some studies with dairy cows have found increased starch digestibility with corn shredlage compared to a conventionally processed corn silage, it appears to promote no difference in fibre digestibility (Ferraretto & Shaver, 2012; Vanderwerff et al., 2015). Including corn shredlage in the diet of finishing feedlot steers has been shown to promote greater ADG and lower DMI, resulting in better feed efficiency (4.2% lower) when compared to conventional corn silage (Conroy et al., 2020).

Chemical treatments have also been utilized to improve roughage digestibility, particularly alkali and acid treatments. Alkali treatments can improve fibre digestion by swelling the cellulose molecules and disrupting lignin and hemicellulose cross linkages (Donnelly et al., 2018; Jackson, 1977; Klopfenstein, 1978). Common alkali treatments include calcium oxide (CaO), ammonia (NH_3), sodium hydroxide (NaOH), and calcium hydroxide ($Ca(OH)_2$) (Adesogan et al., 2019; Donnelly et al., 2018; Watson et al., 2015). Improved NDF digestibility has been shown in finishing feedlot cattle fed crop residues treated with CaO (Shreck et al., 2015). Similar results were observed in dairy cows fed corn silage treated with $Ca(OH)_2$ (Cook et al., 2016).

Ammonia has also been used to improve fibre digestibility in low nutritive quality feeds and can be applied through several forms including anhydrous ammonia gas (Thompson et al., 2002), anhydrous ammonia liquid (Horn et al., 1983), urea (Rode et al., 1986), and ammonia fibre expansion (AFEX) (Ribeiro et al., 2020). Each treatment form has been shown to improve NDF digestibility, however, whichever method is chosen by producers depends on equipment availability, the type of feedstuffs, and cost-related factors. Sodium hydroxide has been previously used to treat low quality roughages, however, as noted by Donnelly et al. (2018) and Watson et al. (2015), safety concerns have hindered its utilization. In addition, NaOH may have poor palatability (Horton et al., 1982). Acid treatments, particularly formic acid (CH_2O_2) and sulfuric acid (H_2SO_4), have also been previously utilized to improve fibre digestion (Haigh & Chapple, 1998; Kennedy, 1990), however, health and safety concerns have limited its use (Adesogan et al., 2019).

The addition of exogenous fibrolytic enzymes (EFE) into the diet has been shown to improve NDF digestibility in both beef cattle (Krueger et al., 2008) and dairy cattle (Refat et al., 2018). The alteration of fibre by EFE is achieved via multiple mechanisms of action prior to feeding and in the gastrointestinal system (McAllister et al., 2001). These enzymes in general come from bacterial and/or fungal sources and usually are rich in xylanase and cellulase activity (McAllister et al., 2001). According to Nsereko et al. (2000), EFE improves ruminal fibre degradation through hydrolysis of the NDF and ADF portions of the feed prior to consumption, allowing the rumen microbes to have greater access to nutrients. Additionally, Wang et al. (2001) showed in vitro that EFE promoted the greater microbial attachment to the feed when applied directly to the feed prior to consumption. Successful use of EFE has been reported in feedlot finishing diets, where heifers consuming a total mixed ration (TMR) containing a liquid enzyme experienced greater ADG with no changes in DMI (Beauchemin et al., 1999). Using a different EFE, Eun et al. (2009) reported no differences in performance in feedlot steers fed differing levels of enzymes. Results from research applying EFE to finishing feedlot cattle diets have been quite variable, likely due to the variety of EFE products, as each may vary in their enzyme source, composition, and formulation, inevitability leading to variability in function and performance.

Another alternative feed additive is direct-fed microbials (DFM) which can be used to improve fibre digestibility and enhance performance in cattle (McAllister et al., 2011). Direct-fed microbials contain live microorganisms and are typically composed of bacteria, yeast, and/or fungi, each of which have different effects in ruminants by potentially influencing fermentation patterns, microbial populations, and nutrient absorption (McAllister et al., 2011). Plata et al. (1994) observed increased NDF digestibility in Holstein steers fed *Saccharomyces cerevisiae*, suggesting that the improvement in digestibility may have been due to increased growth of fibrolytic bacterial communities. Other studies have reported similar results (Guedes et al., 2008; Marden et al., 2008), although others found no improvements in NDF digestibility (Arambel & Kent, 1990; Doreau & Jouany, 1998). Variability in efficacy may be attributed to the differences in feed ingredients and proportions of fibre in each of the diets, and different microbial strains in the studies mentioned. There has been great interest on fibrolytic fungi due to their exceptional ability to degrade plant cell wall components (Gruninger et al., 2014). The major genera of anaerobic fungi found in the rumen include *Neocallimastix*, *Piromyces*, and *Orpinomyces*, each of which are able to produce cellulases and hemicellulases needed for the degradation of plant substrates (Gruninger et al., 2014; Nicholson et al., 2005; Wang et al., 2011; Youssef et al., 2013). Improvements in NDF and DM digestibility when anaerobic fibrolytic fungi were fed to buffalo calves has been shown (Kumar et al., 2015; Sehgal et al., 2008). Additional research is needed to examine the effects of rumen fungi administration evaluate the long-term impacts on fibre digestibility.

Impacts of Roughage Inclusion on Feed Intake and Growth Performance

The DMI of finishing feedlot cattle fed high-concentrate/grain diets is not limited by rumen fill and is generally assumed to be controlled by metabolic factors (NASEM, 2016). Consequently, increases in roughage levels (ranging of 0 to 15% of diet DM) usually leads to increased DMI. It is generally accepted that the dilution of diet energy caused by added roughage or increased fibre content is responsible for this increase in DMI, as cattle will try to meet their energy demands. However, other factors such as altered short-chain fatty acids (SCFA) composition, rate of production, and total production, chewing and rumination, ruminal buffering, digesta kinetics, and site and extent of digestion can interfere with DMI making it challenging to predict the outcomes of altering roughage inclusion (NASEM, 2016).

Summarizing data from 11 trials, Galyean and Defoor (2003) showed that the percentage of dietary NDF supplied by roughage explained 92% of the variation in DMI. It is important to note that the roughages (alfalfa hay, corn silage, sorghum silage, cottonseed hulls), byproducts, and cereal grain used in the finishing diets fed to cattle on those trials were quite different from what is typically fed to feedlot cattle in Canada.

Koenig and Beauchemin (2011) fed finishing feedlot steers diets with 3, 6, 9, 12, or 15% barley silage replacing temper-rolled barley grain (DM basis) and observed a linear increase in DMI from 7.19 to 7.75 kg/d as the proportion of silage increased. Steers' ADG was not affected, but gain-to-feed ratio decreased linearly from 0.225 to 0.202 with increasing silage proportion. More recently, Koenig et al. (2020) observed that DMI increased linearly (11.1, 11.3, 11.7, and 11.8 kg/d) as barley silage proportion increased from 0%, 4%, 8%, to 12% replacing dry-rolled barley grain (DM basis) in the diet of finishing feedlot steers. Again, ADG was not affected, but gain-to-feed ration decreased linearly (0.169, 0.164, 0.159, and 0.161). The results from these two studies conducted in Canada suggests that each 1% increase in dietary NDF supplied by roughage (barley silage) promotes a 0.109 kg/d increase in DMI.

Word et al. (2024) observed that increasing ground corn stalks in the diet of feedlot steers from 7, 13 to 19% replacing steam-flaked corn (DM basis) promoted a linear increase in DMI (10.7, 11.1, and 11.3 kg/d), but it also promoted a linear decrease in ADG and gain-to-feed ratio. However, McDaniel et al. (2024) showed that increasing ground alfalfa hay from 5% to 10% replacing steam-flaked corn (DM basis) in the diet of finishing steer promoted an increase in DMI from 9.2 to 9.5 kg/d, but ADG and gain-to-feed ratio were not affected. Differences in growth performance between these two studies with increasing roughage inclusion are likely related to the greater dietary NDF supplied by roughage and its lower ruminal degradability in the study using corn stalks compared to the study using alfalfa hay as roughage.

It has been suggested that dietary NDF supplied by roughage should be around 5 to 7% (DM basis) to optimize DMI, maximize ADG and gain-to-feed ratio, and minimize digestive disorders of finishing feedlot cattle (Benton et al., 2015; Galyean & Defoor, 2003; McDaniel et al., 2024; Swanson et al., 2017). Swanson et al. (2017) showed that finishing steers' growth performance was not influenced by roughage source (alfalfa hay, corn silage, wheat straw, or corn stover) when fed at a similar NDF inclusion level (~6% dietary NDF supplied by roughage) but was negatively impacted by increasing roughage inclusion (5, 10, 15, or 20% DM basis) in high-concentrate finishing diets. However, one should be cautious when substituting roughages in the diet based solely on equivalent dietary NDF supplied by roughage, as chewing and ruminating, ruminal buffering, digesta kinetics, and diet energy may be affected differently. Depending on the grain source, grain processing, inclusion of other feed ingredients such as distillers grains, and the roughage peNDF and uNDF, the optimal inclusion of roughage might change (Gentry et al., 2016; Pereira et al., 2021, 2023a; Quinn et al., 2011; Ran et al., 2021). Pereira et al. (2021) did not observe differences in DMI, ADG, gain-to-feed ratio, and carcass weights when feedlot steers were fed dry rolled barley-based finishing diets with 10% barley silage (4.2% dietary NDF supplied by roughage) or 5% barley silage plus 5% wheat straw (5.8% dietary NDF supplied by roughage), even though dietary NDF supplied by roughage increased 38%.

Roughage Physical Characteristics: peNDF and uNDF

Pereira et al. (2023a) observed that even small increases in peNDF (0% vs. 1.74%) without a change in the roughage type or roughage inclusion rate increased mean ruminal pH, stimulated greater rumen contraction strength, and reduced indicators of systemic inflammation. In a performance study, Pereira et al. (2021) observed that increasing peNDF (peNDF = 3.0 vs. 3.2%) did not affect feed intake, gain, or carcass weight, but increased the dressing percentage, increased the proportion of steers with AAA carcasses, and tended to reduce the proportion of steers with severe liver abscess scores. These results indicate that even small increases in peNDF could help improve rumen health and promote higher quality carcasses. However, it is challenging to increase roughage chop-length much as a means of increasing peNDF. Both Gentry et al. (2016) and Pereira et al. (2021) showed very little change in diet peNDF by increasing chop-length of roughages. Additionally, if the roughage particle size is too long it can alter mixing of the TMR and may make it easier for cattle to sort. This issue is more likely with dry roughages where precise chopping or grinding is more difficult.

Similar to peNDF, uNDF helps to maintain a more stable ruminal environment by stimulating rumen contraction strength, increasing ruminal pH, and reducing indicators of systemic inflammation (Pereira et al., 2023b; Ran et al., 2021). However, increasing uNDF concentration in the diet consistently decreases NDF digestibility with a consequent increase in feed intake and slight reduction in average daily gain (Pereira et al., 2021, 2023b; Ran et al., 2021). Increasing uNDF also promoted a decrease in dressing percentage and may lead to lower carcass weights (Pereira et al., 2021). These results suggest that nutritionists may want to monitor the uNDF concentration ensuring it is below 7% (DM basis) to avoid production losses, which might be particularly important when incorporating fibrous byproducts or novel roughage types.

Roughage Inclusion Strategy

When determining optimal roughage inclusion levels for finishing feedlot cattle, it's essential to consider cattle adaptation—both in terms of intake and ruminal SCFA absorption—as well as the ruminal microbiome's response to high-concentrate diets. By gradually acclimating cattle to these highly fermentable diets and implementing a feed and bunk management strategy that minimizes competition for feed and prevents instances of feed restriction, it is possible to maintain lower roughage levels successfully and cost-effectively in finishing diets.

In recent research aimed at reducing ruminal acidosis and liver abscesses—and minimizing the need for in-feed antimicrobials like tylosin—while maintaining or enhancing growth performance, we explored different roughage inclusion strategies for finishing steers (Paterson et al., 2024). While it has long been established that higher levels of roughage (over 15% of diet DM) can decrease liver abscess occurrences (Foster & Woods, 1970), typical feedlot finishing diets in North America often contain lower roughage levels (8%–10% of diet DM; Samuelson et al., 2016) due to improved feed efficiency (Koenig & Beauchemin, 2011; Koenig et al., 2020). This study investigated diet variations with both decreasing and increasing roughage levels—without tylosin—compared to a conventional diet with constant roughage levels throughout the finishing period, evaluating impacts on growth performance, carcass characteristics, and liver abscess rates (Paterson et al., 2024).

The study's treatments included: (1) a positive control (+CTRL) fed a diet with 7.5% roughage and tylosin; (2) a negative control (−CTRL) with no tylosin; (3) a decreasing roughage diet (DECR), where roughage levels were reduced every 42 days, remaining static for the last 84 days, comprising 15%, 9%, 3%, and 3% of DM, respectively; and (4) an increasing roughage diet (INCR), where roughage levels began at 3% and increased to 15% of DM over the same period. The aim was to mirror total roughage amounts typically found in finishing diets, thereby preventing losses in feed efficiency while supplying more roughage during critical periods when cattle are more susceptible to developing liver abscesses. Our findings indicated that incorporating greater dietary roughage concentrations earlier in the finishing phase—with a subsequent decline—could effectively reduce the incidence of minor liver abscesses, comparable to typical finishing diets that include tylosin, without adversely impacting growth performance or carcass quality.

Animal Variation in Fibre Utilization and Methane Mitigation

The variability in individual cattle's fibre digestion reveals opportunities for selecting animals with superior capacities for eating and digesting roughages. The rumen microbiome plays a crucial role in breaking down complex fibres found in roughages. However, enhanced fibre digestion often promotes fermentation pathways that increase acetate production, leading to greater methane emissions per unit of organic matter digested. Nevertheless, recent studies have shown that efficient fibre-digesting cattle may produce lower methane emissions per kilogram of fibre digested (Payne et al., 2023; Terranova et al., 2024).

We have recently studied characteristics linked to improved intake of digestible fibre—specifically, digestible neutral detergent fibre (NDF) intake (DFI, g/kg $BW^{0.75}$)—which involved factors such as DMI, rumen microbial population and diversity, SCFA and ammonia-nitrogen (NH_3-N) production, rumen passage rates, and chewing behaviours, examining 16 beef heifers selected from a larger group of 64. Our initial findings confirmed that beef heifers selected for their enhanced ability to consume and digest fibre showed increased ADG while producing similar levels of methane per day, but 16% less methane per kilogram of DMI (Payne et al., 2023). These results corroborate a recent study indicating efficient fibre-digesting cattle generate lower methane levels per kilogram of fibre digested (Terranova et al., 2024).

Rumen passage rates are particularly significant in this context; faster digesta passage has been associated with reduced methanogenesis and a shift towards propionate production, which decreases hydrogen availability that could otherwise contribute to methane formation (Hammond et al., 2014; Terranova et al., 2024). Our recent research demonstrated that cattle could achieve higher passage rates without compromising NDF digestibility (Delver et al., 2023). Additionally, the feeding behaviours of cattle, often overlooked, significantly influence fibre digestion. Cattle exhibit distinct chewing mechanisms that facilitate roughage digestion during eating and rumination, stimulating saliva production and reducing particle size for effective microbial digestion of fibre (NASEM, 2016). Increased rumination has been shown to affect rumen passage rates, microbial metabolism, and microbial protein production, all of which can inhibit methanogenesis (Pfau et al., 2021). Data from our prior research indicates that high DFI heifers consume feed faster and spend more time ruminating than those with low DFI (Dubois et al., 2024).

Final remarks

Roughage inclusion in finishing feedlot diets remains a complex but essential component of modern beef production. While high-concentrate diets offer improved feed efficiency, strategic roughage inclusion is critical for maintaining rumen function, animal health, and reducing reliance on antimicrobial interventions.

- Optimal roughage inclusion ranges between 5–7% dietary NDF from roughage for maximizing performance and minimizing digestive disorders.
- Roughage source matters: Highly digestible roughages (e.g., alfalfa, corn silage) may have different effects than low-quality roughages (e.g., wheat straw), even when NDF levels are similar.
- Physically effective fibre (peNDF) enhances chewing, rumen pH, and carcass quality, while excessive uNDF can impair performance. However, increasing diet peNDF by increasing roughage particle size can be challenging.
- Dynamic roughage inclusion strategy, such as increasing roughage early in the finishing period but reducing it later on, may reduce liver abscess incidence without sacrificing performance.
- Individual animal variation in fibre digestion and behaviour represents a promising avenue for selection, potentially improving growth performance and lowering environmental impact through reduced methane emissions per unit of fibre digested.

References

Addah, W., Baah, J., Okine, E. K., & McAllister, T. A. (2015). Effect of barley silage chop length and inoculation on growth performance, feeding behavior, and ruminal acidosis in finishing feedlot steers. *J. Anim. Sci. 93*(5), 2309-2321. doi: 10.2527/jas.2014-8247

Adesogan, A. T., Arriola, K. G., Jiang, Y., Oyebade, A., Paula, E. M., Pech-Cervantes, A. A., Romero, J. J., Ferraretto, L. F., & Vyas, D. (2019). Symposium review: Technologies for improving fiber utilization. *J. Dairy Sci. 102*(6), 5726-5755. doi: 10.3168/jds.2018-15334

Arambel, M. J., & Kent, B. A. (1990). Effect of Yeast Culture on Nutrient Digestibility and Milk Yield Response in Early- to Midlactation Dairy Cows. *J. Dairy Sci. 73*(6), 1560-1563. doi: 10.3168/jds.S0022-0302(90)78825-X

Beauchemin, K. A., Farr, B. I., Rode, L. M., & Schaalje, G. B. (1994). Effects of Alfalfa Silage Chop Length and Supplementary Long Hay on Chewing and Milk Production of Dairy Cows. *J. Dairy Sci. 77*(5), 1326-1339. doi: 10.3168/jds.S0022-0302(94)77072-7

Beauchemin, K. A., Rode, L. M., & Karren, D. (1999). Use of feed enzymes in feedlot finishing diets. *Can. J. Anim. Sci. 72*(2), 243-246. doi: 10.4141/A98-124

Benton, J. R., Watson, A. K., Erickson, G. E., Klopfenstein, T. J., Vander Pol, K. J., Meyer, N. F., & Greenquist, M. A. (2015). Effects of roughage source and inclusion in beef finishing diets containing corn wet distillers' grains plus solubles. *J. Anim. Sci. 93*(9), 4358-4367. doi: 10.2527/jas.2015-9211

Bhandari, S. K., Li, S., Ominski, K. H., Wittenberg, K. M., & Plaizier, J. C. (2008). Effects of the Chop Lengths of Alfalfa Silage and Oat Silage on Feed Intake, Milk Production, Feeding Behavior, and Rumen Fermentation of Dairy Cows. *J. Dairy Sci. 91*(5), 1942-1958. doi: 10.3168/jds.2007-0358

Conroy, B., Jaynes, M., Pritchard, R., & Jenkins, K. H. (2020). Comparing SHREDLAGE® and conventional silage as a roughage component in steamflaked corn diets for finishing cattle. *Nebraska Beef Cattle Rep.* (1089), 89-90.

Cook, D. E., Bender, R. W., Shinners, K. J., & Combs, D. K. (2016). The effects of calcium hydroxide-treated whole-plant and fractionated corn silage on intake, digestion, and lactation performance in dairy cows. *J. Dairy Sci. 99*(7), 5385-5393. doi: 10.3168/jds.2015-10402

Delver, J. J., Lardner, H. A., McKinnon, J. J., Ribeiro, G. O., Asai-Coakwell, M., & Penner, G. B. (2023). Development of an efficiency ranking system for beef cows and effects on feed intake, ruminal fermentation, NDF turnover, and apparent total tract digestibility. *Can. J. Anim. Sci. 103*(4), 338-354. doi: 10.1139/cjas-2023-0004

Donnelly, D. M., de Resende, L. C., Cook, D. E., Atalla, R. H., & Combs, D. K. (2018). Technical note: A comparison of alkali treatment methods to improve neutral detergent fiber digestibility of corn stover. *J. Dairy Sci. 101*(10), 9058-9064. doi: 10.3168/jds.2017-14317

Doreau, M., & Jouany, J. P. (1998). Effect of a Saccharomyces cerevisiae culture on nutrient digestion in lactating dairy cows. *J. Dairy Sci. 81*(12), 3214-3221. doi: 10.3168/jds.S0022-0302(98)75885-0

Dubois, M. M., Penner, G. B., Lardner, H. A., Abbott, D. W., Gruninger, R. J., & Ribeiro, G. O. (2024). *Impact of Selecting Beef Heifers with Divergent Digestible Fibre Intake when Fed Four Different Diets.* Animal Nutrition Conference of Canada (ANCC), May 14-16, 2024, Winnipeg, MB, Canada

Eun, J. -S., ZoBell, D. R., Dschaak, C. M., Diaz, D. E., & Tricarico, J. M. (2009). Effects of Supplementing a Fibrolytic Feed Enzyme on the Growth Performance and Carcass Characteristics of Beef Steers. *Prof. Anim. Sci. 25*(3), 382-387. doi: 10.15232/S1080-7446(15)30729-4

Ferraretto, L. F., & Shaver, R. D. (2012). Effects of Corn Shredlage on lactation performance and total tract starch digestibility by dairy cows. *Prof. Anim. Sci. 28*(6), 639-647. doi: 10.15232/S1080-7446(15)30423-X

Foster, L., & Woods, W. (1970). Influence of liver abscesses on animal performance. *J. Anim. Sci. 31*, 241.

Galyean, M. L., & Defoor, P. J. (2003). Effects of roughage source and level on intake by feedlot cattle. *J. Anim. Sci. 81*(Suppl. 2), E8–E16. doi: 10.2527/2003.8114_suppl_2E8x

Gentry, W. W., Weiss, C. P., Meredith, C. M., McCollum, F. T., Cole, N. A., & Jennings, J. S. (2016). Effects of roughage inclusion and particle size on performance and rumination behavior of finishing beef steers. *J. Anim. Sci. 94*(11), 4759–4770. doi: 10.2527/jas.2016-0734

Gruninger, R. J., Puniya, A. K., Callaghan, T. M., Edwards, J. E., Youssef, N., Dagar, S. S., Fliegerova, K., Griffith, G. W., Forster, R., Tsang, A., McAllister, T., & Elshahed, M. S. (2014). Anaerobic fungi (phylum Neocallimastigomycota): advances in understanding their taxonomy, life cycle, ecology, role and biotechnological potential. *FEMS Microbiol. Ecol. 90*(1), 1-17. doi: 10.1111/1574-6941.12383

Guedes, C. M., Gonçalves, D., Rodrigues, M. A. M., & Dias-da-Silva, A. (2008). Effects of Saccharomyces cerevisiae yeast on ruminal fermentation and fibre degradation of maize silages in cows. *Anim. Feed. Sci. Technol. 145*(1-4), 27-40. doi: 10.1016/j.anifeedsci.2007.06.037

Haigh, P. M., & Chapple, D. G. (1998). The Effect of Formic Acid, Formic Acid Salt and Formic Acid with Formalin on Silage Fermentation, Digestibility and Intake, and on Liveweight Change of Young Cattle. *J. Agric. Eng. Res. 69*(3), 267-271. doi: 10.1006/jaer.1997.0231

Hammond, K. J., Humphries, D. J., Westbury, D. B., Thompson, A., Crompton, L. A., Kirton, P., Green, C., & Reynolds, C. K. (2014). The inclusion of forage mixtures in the diet of growing dairy heifers: Impacts on digestion, energy utilisation, and methane emissions. *Agric. Ecosyst. Environ. 197*, 88-95. doi: 10.1016/j.agee.2014.07.016

Horn, G. W., Batchelder, D. G., Manor, G., Streeter, C. L., & McLaughlin, G. L. (1983). Ammoniation of wheat straw and native grass hay during baling of large round bales. *Anim. Feed. Sci. Technol. 8*(1), 35-46. doi: 10.1016/0377-8401(83)90041-X

Horton, G. M. J., Nicholson, H. H., & Christensen, D. A. (1982). Ammonia and sodium hydroxide treatment of wheat straw in diets for fattening steers. *Anim. Feed. Sci. Technol. 7*(1), 1-10. doi: 10.1016/0377-8401(82)90031-1

Humer, E., Petri, R. M., Aschenbach, J. R., Bradford, B. J., Penner, G. B., Tafaj, M., Südekum, K. H., & Zebeli, Q. (2018). Invited Review: Practical feeding management recommendations to mitigate the risk of subacute ruminal acidosis in dairy cattle. *J. Dairy Sci. 101(*2), 872–888. doi:10.3168/jds.2017-13191

Jackson, M. G. (1977). Review article: The alkali treatment of straws. *Anim. Feed. Sci. Technol. 2*(2), 105-130. doi: 10.1016/0377-8401(77)90013-X

Jung, H. G., & Allen, M. S. (1995). Characteristics of Plant Cell Walls Affecting Intake and Digestibility of Forages by Ruminants. *J. Anim. Sci. 73*(9), 2774-2790. doi: 10.2527/1995.7392774x

Kennedy, S. J. (1990). Comparison of the fermentation quality and nutritive value of sulphuric and formic acid-treated silages fed to beef cattle. *Grass Forage Sci. 45*(1), 17-28. doi: 10.1111/j.1365-2494.1990.tb02178.x

Klopfenstein, T. (1978). Chemical Treatment of Crop Residues. *J. Anim. Sci. 46*(3), 841-848. doi: 10.2527/jas1978.463841x

Koenig, K. M., & Beauchemin, K. A. (2011). Optimum extent of barley grain processing and barley silage proportion in feedlot cattle diets: growth, feed efficiency, and fecal characteristics. *Can. J. Anim. Sci. 91*, 411–422. doi: 10.1139/CJAS2010-03

Koenig, K. M., Chibisa, G. E., Penner, G. B., & Beauchemin, K. A. (2020). Optimum roughage proportion in barley-based feedlot cattle diets: growth performance, feeding behavior, and carcass traits. *J. Anim. Sci. 98*(10), skaa299. doi: 10.1093/jas/skaa299

Koenig, K. M., McGinn, S. M., & Beauchemin, K. A. (2013). Ammonia emissions and performance of backgrounding and finishing beef feedlot cattle fed barley-based diets varying in dietary crude protein concentration and rumen degradability. *J. Anim. Sci. 91*(5), 2278-2294. doi: 10.2527/jas.2012-5651

Krueger, N. A., Adesogan, A. T., Staples, C. R., Krueger, W. K., Kim, S. C., Littell, R. C., & Sollenberger, L. E. (2008). Effect of method of applying fibrolytic enzymes or ammonia to Bermudagrass hay on feed intake, digestion, and growth of beef steers. *J. Anim. Sci. 86*(4), 882-889. doi: 10.2527/jas.2006-717

Kumar, S., Sehgal, J. P., Puniya, A. K., & Kumari, R. (2015). Growth performance and fibre utilization of Murrah male buffalo calves fed wheat straw based complete feed blocks incorporated with superior anaerobic fungal zoospores (*Neocallimastix* sp. GR-1). *Indian. J. Anim. Sci. 85*(3), 275-281. doi: 10.56093/ijans.v85i3.47323

Marden, J. P., Julien, C., Monteils, V., Auclair, E., Moncoulon, R., & Bayourthe, C. (2008). How does live yeast differ from sodium bicarbonate to stabilize ruminal pH in high-yielding dairy cows? *J. Dairy Sci. 91*(9), 3528-3535. doi: 10.3168/jds.2007-0889

McAllister, T. A., Beauchemin, K. A., Alazzeh, A. Y., Baah, J., Teather, R. M., & Stanford, K. (2011). Review: The use of direct fed microbials to mitigate pathogens and enhance production in cattle. *Can. J. Anim. Sci. 91*(2), 193-211. doi: 10.4141/cjas10047

McAllister, T. A., Hristov, A. N., Beauchemin, K. A., Rode, L. M., & Cheng, K. J. (2001). Enzymes in ruminant diets. In M. R. Bedford, & G. G. Partridge (Eds.), *Enzymes in farm animal nutrition* (273-298). CABI Publishing, Wallingford, Oxon, UK.

McDaniel, Z. S., Galyean, M. L., Broadway, P. R., Carroll, J. A., Burdick Sanchez, N. C., Hanratty, A. N., Dornbach, C. W., Line, D. J., Smock, T. M., Manahan, J. L., & Hales, K. E. (2024). Effects of increasing the concentration of neutral detergent fiber in roughage and bulk density of steam-flaked corn on growth performance, carcass characteristics, and liver abscesses of finishing beef steers fed diets without tylosin phosphate. *Appl. Anim. Sci. 40*(3), 269-278. doi: 10.15232/aas.2023-02484

Mertens, D. R. (1997). Creating a System for Meeting the Fiber Requirements of Dairy Cows. *J. Dairy Sci. 80*(7), 1463-1481. doi: 10.3168/jds.S0022-0302(97)76075-2

Nicholson, M. J., Theodorou, M. K., & Brookman, J. L. (2005). Molecular analysis of the anaerobic rumen fungus *Orpinomyces* – insights into an AT-rich genome. *Microbiology. 151*(1), 121-133. doi: 10.1099/mic.0.27353-0

Nsereko, V. L., Morgavi, D. P., Rode, L. M., Beauchemin, K. A., & McAllister, T. A. (2000). Effects of fungal enzyme preparations on hydrolysis and subsequent degradation of alfalfa hay fiber by mixed rumen microorganisms in vitro. *Anim. Feed. Sci. Technol. 88*(3), 153-170. doi: 10.1016/S0377-8401(00)00225-X

Paterson, T. J., O'Hara, E., Gruninger, R. J., Penner, G. B., Lardner, H. A., Stephens, E., Yang, W., Beauchemin, K. A., McAllister, T. A., & Ribeiro, G. O. (2024). Assessment of forage inclusion strategies as a means of reducing liver abscesses in finishing feedlot cattle. *Appl. Anim. Sci. 40*(3), 279-296. doi: 10.15232/aas.2023-02481

Payne, N. A., Penner, G. B., Lardner, H. A., Abbott, W. Gruninger, R. J., & Ribeiro, G. O. (2023). 211 Evaluation of Beef Heifer Variability in the Ability to Eat and Digest a High Forage Diet. *J. Anim. Sci. 101*(Suppl. 3), 128-129. doi: 10.1093/jas/skad281.157

Pereira, M. C. S., Beauchemin, K. A., McAllister, T. A., Yang, W. Z., Wood, K. M., & Penner, G. B. (2023a). Effect of physically effective neutral detergent fiber and undigested neutral detergent fiber on eating behavior, ruminal fermentation and motility, barrier function, blood metabolites, and total tract digestibility in finishing cattle. *J. Anim. Sci. 101*, skad023. doi: 10.1093/jas/skad023

Pereira, M. C. S., Yang, W. Z., Beauchemin, K. A., McAllister, T. A., Wood, K. M., & Penner, G. B. (2021). Effect of silage source, physically effective neutral detergent fiber, and undigested neutral detergent fiber concentrations on performance and carcass characteristics of finishing steers. *Trans. Anim. Sci. 5*(1), txaa236. doi: 10.1093/tas/txaa236

Pereira, M. C. S., Yang, W. Z., Beauchemin, K. A., McAllister, T. A., Wood, K. M., & Penner, G. B. (2023b). Effect of forage types differing in undigested neutral detergent fiber concentration and forage inclusion rate on reticulo-ruminal motility and fermentation, total tract barrier function, and blood metabolites of finishing beef heifers. *J. Anim. Sci. 101*, skad043. doi: 10.1093/jas/skad043

Pfau, F., Hünerberg, M., Südekum, K., Breves, G., Clauss, M., & Hummel, J. (2021). Effects of Dilution Rate on Fermentation Characteristics of Feeds with Different Carbohydrate Composition Incubated in the Rumen Simulation Technique (RUSITEC). *Front. Anim. Sci. 2*, 715142. doi: 10.3389/fanim.2021.715142

Plata, F. P., Mendoza, G. D., Bárcena-Gama, J. R., & González, S. (1994). Effect of a yeast culture (Saccharomyces cerevisiae) on neutral detergent fiber digestion in steers fed oat straw based diets. *Anim. Feed. Sci. Technol. 49*(3-4), 203-210. doi: 10.1016/0377-8401(94)90046-9

Quinn, M. J., May, M. L., DiLorenzo, N., Ponce, C. H., Smith, D. R., Parr, S. L., & Galyean, M. L. (2011). Effects of roughage source and distillers grain concentration on beef cattle finishing performance, carcass characteristics, and in vitro fermentation. *J. Anim. Sci. 89*(8), 2631–2642. doi: 10.2527/jas.2010-3563

Ran, T., Saleem, A. M., Beauchemin, K. A., Penner, G. B., & Yang, W. (2021). Processing index of barley grain and dietary undigested neutral detergent fiber concentration affected chewing behavior, ruminal pH, and total tract nutrient digestibility of heifers fed a high-grain diet. *J. Anim. Sci. 99*(1), skab011. doi: 10.1093/jas/skab011

Refat, B., Christensen, D. A., McKinnon, J. J., Yang, W., Beattie, A. D., McAllister, T. A., Eun, J. S., Abdel-Rahman, G. A., & Yu, P. (2018). Effect of fibrolytic enzymes on lactational performance, feeding behavior, and digestibility in high-producing dairy cows fed a barley silage-based diet. *J. Dairy Sci. 101*(9), 7971-7979. doi: 10.3168/jds.2017-14203

Ribeiro, G. O., Gruninger, R. J., Badhan, A., & McAllister, T. A. (2016). Mining the rumen for fibrolytic feed enzymes. *Anim. Front. 6*(2), 20-26. doi: 10.2527/af.2016-0019

Ribeiro, G. O., Gruninger, R. J., Jones, D. R., Beauchemin, K. A., Yang, W. Z., Wang, Y., Abbott, D. W., Tsang, A., & McAllister, T. A. (2020). Effect of ammonia fiber expansion-treated wheat straw and a recombinant fibrolytic enzyme on rumen microbiota and fermentation parameters, total tract digestibility, and performance of lambs. *J. Anim. Sci. 98*(5), skaa116. doi: 10.1093/jas/skaa116

Rode, L. M., Cheng, K. J., & Costerton, J. W. (1986). Digestion by cattle of urea-treated, ammonia-treated, or rolled high-moisture barley. *Can. J. Anim. Sci. 66*(3), 711-721. doi: 10.4141/cjas86-078

Samuelson, K. L., Hubbert, M. E., Galyean, M. L., & Löest, C. A. (2016). Nutritional recommendations of feedlot consulting nutritionists: The 2015 New Mexico State and Texas Tech University survey. *J. Anim. Sci. 94*(6), 2648-2663. doi: 10.2527/jas.2016-0282

Sehgal, J. P., Jit, D., Puniya, A. K., & Singh, K. (2008). Influence of anaerobic fungal administration on growth, rumen fermentation and nutrient digestion in female buffalo calves. *J. Anim. Feed. Sci. 17*(4), 510-518. doi: 10.22358/jafs/66678/2008

Shreck, A. L., Nuttelman, B. L., Harding, J. L., Griffin, W. A., Erickson, G. E., Klopfenstein, T. J., & Cecava, M. J. (2015). Digestibility and performance of steers fed low-quality crop residues treated with calcium oxide to partially replace corn in distillers grains finishing diets. *J. Anim. Sci. 93*(2), 661-671. doi: 10.2527/jas.2013-7194

Soita, H. W., Christensen, D. A., McKinnon, J. J., & Mustafa, A. F. (2002). Effects of barley silage of different theoretical cut length on digestion kinetics in ruminants. *Can. J. Anim. Sci. 82*(2), 207-213. doi: 10.4141/A01-064

Swanson, K. C., Carlson, Z. E., Ruch, M. C., Gilbery, T. C., Underdahl, S. R., Keomanivong, F. E., Bauer, M. L., & Islas, A. (2017). Influence of forage source and forage inclusion level on growth performance, feeding behavior, and carcass characteristics in finishing steers. *J. Anim. Sci. 95*(3), 1325-1334. doi: 10.2527/jas.2016.1157

Terranova, M., Kreuzer, M., Li, Y., Schwarm, A., & Clauss, M. (2024). Indications for a lower methane yield from digested fibre in ruminants digesting fibre more efficiently. *Anim. Feed Sci. Technol. 307*, 115834. doi: 10.1016/j.anifeedsci.2023.115834

The National Academies of Sciences, Engineering, and Medicine (NASEM). (2016). *Nutrient Requirements of Beef Cattle*. 8th ed. National Academies Press; Washington, DC, USA

Thompson, R. K., McKinnon, J. J., Mustafa, A. F., Maenz, D. D., Racz, V. J., & Christensen, D. A. (2002). Chemical composition, ruminal kinetic parameters, and nutrient digestibility of ammonia treated oat hulls. *Can. J. Anim. Sci. 82*(1), 103-109. doi: 10.4141/A01-036

Turgeon, O. A., Szasz, J. I., Koers, W. C., Davis, M. S., & Vander Pol, K. J. (2010). Manipulating grain processing method and roughage level to improve feed efficiency in feedlot cattle. *J. Anim. Sci. 88*(1), 284-295. doi: 10.2527/jas.2009-1859

Vanderwerff, L. M., Ferraretto, L. F., & Shaver, R. D. (2015). Brown midrib corn shredlage in diets for high-producing dairy cows. *J. Dairy Sci. 98*(8), 5642-5652. doi: 10.3168/jds.2015-9543

Varga, G. A., & Kolver, E. S. (1997). Microbial and Animal Limitations to Fiber Digestion and Utilization. *J. Nutr. 127*(5), 819S-823S. doi: 10.1093/jn/127.5.819S

Wang, T. Y., Chen, H. L., Lu, M. Y. J., Chen, Y. C., Sung, H. M., Mao, C. T., Cho, H. Y., Ke, H. M., Hwa, T. Y., Ruan, S. K., Hung, K. Y., Chen, C. K., Li, J. Y., Wu, Y. C., Chen, Y. H., Chou, S. P., Tsai, Y. W., Chu, T. C., Shih, C. C. A., Li, W. H., Shih, M. C. (2011). Functional characterization of cellulases identified from the cow rumen fungus *Neocallimastix patriciarum* W5 by transcriptomic and secretomic analyses. *Biotechnol. Biofuels. 4*(1), 24. doi: 10.1186/1754-6834-4-24

Wang, Y., McAllister, T. A., Rode, L. M., Beauchemin, K. A., Morgavi, D. P., Nsereko, V. L., Iwaasa, A. D., & Yang, W. (2001). Effects of an exogenous enzyme preparation on microbial protein synthesis, enzyme activity and attachment to feed in the Rumen Simulation Technique (Rusitec). *Br. J. Nutr. 85*(3), 325-332. doi: 10.1079/BJN2000277

Watson, A. K., MacDonald, J. C., Erickson, G. E., Kononoff, P. J., & Klopfenstein, T. J. (2015). FORAGES AND PASTURES SYMPOSIUM: Optimizing the use of fibrous residues in beef and dairy diets. *J. Anim. Sci. 93*(6), 2616-2625. doi: 10.2527/jas.2014-8780

Weimer, P. J. (1996). Why Don't Ruminal Bacteria Digest Cellulose Faster? *J. Dairy Sci. 79*(8), 1496-1502. doi: 10.3168/jds.S0022-0302(96)76509-8

Wilson, J. R., & Mertens, D. R. (1995). Cell Wall Accessibility and Cell Structure Limitations to Microbial Digestion of Forage. *Crop Sci. 35*(1), 251-259. doi: 10.2135/cropsci1995.0011183X003500010046x

Word, A. B., Karr, K. J., Holland, B. P., Maxwell, C. L., Linneen, S. K., & Defoor, P. J. (2024). Removing tylosin phosphate from finishing diets with increasing roughage concentrations affects growth performance, carcass characteristics, and prevalence of liver abscesses of finishing steers. *Appl. Anim. Sci. 40*(3), 260-268. doi: 10.15232/aas.2023-02489

Youssef, N. H., Couger, M. B., Struchtemeyer, C. G., Liggenstoffer, A. S., Prade, R. A., Najar, F. Z., Atiyeh, H. K., Wilkins, M. R., & Elshahed, M. S. (2013). The Genome of the Anaerobic Fungus Orpinomyces sp. Strain C1A Reveals the Unique Evolutionary History of a

Remarkable Plant Biomass Degrader. *Appl. Environ. Microbiol. 79*(15), 4620-4634. doi: 10.1128/AEM.00821-13

Zebeli, Q., Aschenbach, J. R., Tafaj, M., Boguhn, J., Ametaj, B. N., & Drochner, W. (2012). Invited review: Role of physically effective fiber and estimation of dietary fiber adequacy in high-producing dairy cattle. *J. Dairy Sci. 95*(3), 1041-1056. doi: 10.3168/jds.2011-4421

Strategies related to calcium homeostasis across the transition period

Stratégies liées à l'homéostasie du calcium pendant la période de transition

L. L. Hernandez[1]
[1] *University of Wisconsin-Madison, Madison, WI, USA*
llhernandez@wisc.edu

Abstract

The periparturient period is characterized by increased demand for calcium. For this reason, different nutritional strategies have been adopted for use to mitigate hypocalcemia. It is well established that the mammary gland contributes to the homeostatic and homeorhetic regulation of maternal and lactational physiology during the periparturient period when nutrients, such as calcium, must be partitioned to the conceptus and the mammary gland. As a result, mammals have adapted multiple mechanisms to ensure support of both maternal physiology and milk production when normal mechanisms of support may fail. Dietary calcium is insufficient during this period to replenish maternal calcium lost to colostrum and milk and does not achieve neutrality until post-peak lactation, resulting in activation of additional adaptive mechanisms beyond those utilized for regulation of calcium metabolism during non-lactating and late-lactation states. Cows and other mammals use the rich calcium stores in maternal bone to augment maternal calcium concentrations. A subsequent decline in blood calcium concentrations often occurs at the onset of lactation and is believed to be a key trigger in activating supporting negative feedback mechanisms. Without adequate negative feedback, ensuing endocrine mechanisms cannot be sufficiently stimulated, thus resulting in a failure in calcium mobilization from bone results. This failure in bone resorption may lead to severe decreases in blood calcium concentrations at the onset of lactation, further prompting a loss in rumen contractility, suppressing ruminal calcium absorption, and further straining transition cow calcium metabolism. Therefore, making informed decisions on calcium management during the transition period is critical.

Résumé

La période péripartum se caractérise par une demande accrue en calcium. C'est pourquoi différentes stratégies nutritionnelles ont été adoptées pour combattre l'hypocalcémie. On sait que la glande mammaire joue un rôle dans la régulation homéostatique et homéorétique de la physiologie maternelle et de la lactation pendant la période péripartum, au cours de laquelle les nutriments, tels que le calcium, doivent être répartis entre le conceptus et la glande mammaire. En conséquence, de multiples mécanismes ont été élaborés par les mammifères pour que la physiologie de la mère et la production de lait soient prises en charge lorsque les mécanismes de soutien normaux font défaut. Le calcium d'origine alimentaire ne suffit pas pendant cette période pour remplacer celui utilisé par la mère pour produire le colostrum et le lait – l'équilibre n'est atteint qu'après le pic de lactation – ce qui entraîne l'activation de mécanismes adaptatifs autres que ceux utilisés normalement pour la régulation du métabolisme calcique pendant la période non productive et en fin de lactation. Les bovins et d'autres mammifères utilisent les riches réserves de calcium de leurs os pour augmenter les quantités de calcium disponible chez les mères. Une baisse subséquente de la concentration de calcium dans le sang se produit souvent au début de la lactation, et on pense qu'il s'agit d'un important élément déclencheur des mécanismes de contre réaction. En l'absence de contre-réaction adéquate, les mécanismes endocriniens consécutifs ne peuvent pas être suffisamment stimulés, ce qui entraîne un arrêt de la mobilisation du calcium à partir des os. Cet obstacle à la résorption osseuse peut provoquer une sérieuse diminution de la calcémie en début de lactation, ce qui entraîne une perte de contractilité ruminale, qui gêne l'absorption du calcium dans le rumen, et perturbe encore davantage le métabolisme calcique des vaches en période de transition. Il est donc essentiel de prendre des décisions éclairées par rapport à la gestion du calcium pendant la période de transition.

Introduction

Due to the rapid and robust demand for Ca at the onset of lactation by the mammary gland, a dynamic change in Ca metabolism ensues as the mammary gland's demand for Ca from the blood and extracellular pools outpaces the rate at which pools can be maintained. This can result in Ca dysregulation due to the sudden shift from calcemic nadir in late gestation to extreme Ca deficit in early lactation (Horst et al., 2005). Reinhardt and colleagues demonstrated that a large proportion of cows may be experiencing more severe Ca perturbations during this timeframe than anticipated (Reinhardt et al., 2011). Data indicated ~50% of multiparous dairy cows were experiencing subclinical hypocalcemia (SCH; Figure 1) (Reinhardt et al., 2011).

Figure 1. Incidence of hypocalcemia in US Dairy Herds(Reinhardt et al., 2011).

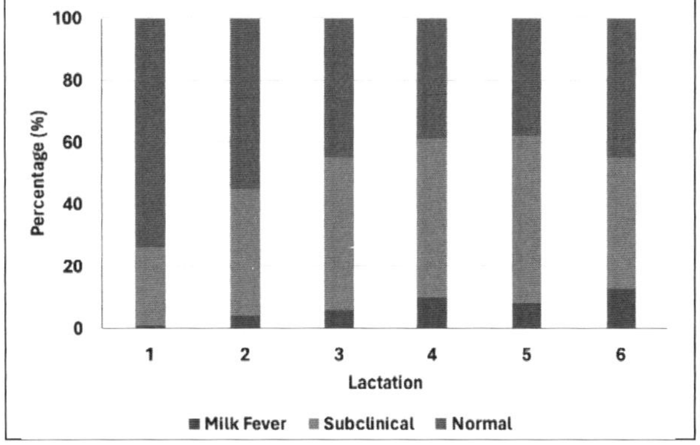

Interestingly, the prevalence of hypocalcemia was observed despite an increase in the active form of Vitamin D, 1α,25-(OH)2D3. This finding suggests that homeostatic responses to maintain Ca concentrations may be limited as animals age and are a factor in increased incidences of hypocalcemia in older cows (Reinhardt et al., 2011). Congruently, older cows excrete more Ca/d into milk and experience greater Ca balance disruptions during the peripartal period when compared to nulliparous animals (Rodney et al., 2018), supporting the increased risk of poor Ca status in older peripartal cows. The rapid change from calcemic nadir to secreting ~30-50 grams of Ca per day into milk during lactation results in a challenge not easily overcome(Horst et al., 2005).

A subsequent decline in blood Ca concentrations often occurs at the onset of lactation and is believed to be a key trigger in activating supporting negative feedback mechanisms. Without adequate negative feedback, ensuing endocrine mechanisms cannot be sufficiently stimulated, thus a failure in Ca mobilization from bone results. This failure in bone resorption may lead to severe decreases in blood Ca concentrations at the onset of lactation, further prompting a loss in rumen contractility, suppressing ruminal Ca absorption, and further straining transition cow Ca metabolism (Hyde and Fraser, 2014). In the initial 6-8 weeks after parturition cows experience a negative Ca balance, with some cows losing up to 13% of their skeletal Ca reservoir to meet the massive peripheral and mammary Ca requirements(Horst et al., 2005) and may need additional factors beyond classical Ca homeostatic mechanisms to maintain Ca pools. Hence, stimulation of additional supporting physiological mechanisms to prompt calciotropic feedback loops for the maintenance of blood Ca concentrations is critical for an integrated maternal metabolism.

Management strategies for optimization of calcium homeostasis during the peripartal period

Negative DCAD: A common strategy to prevent hypocalcemia involves feeding an acidogenic diet during the prepartum period. This dietary approach, known as negative dietary cation-anion difference (DCAD), induces metabolic acidosis, which triggers a response to increased blood Ca concentrations postpartum (Block, 1984). Metabolic acidosis prompts a process where Ca is released from bones to elevate the cation levels in the blood, counteracting the increased acidity (Horst et al., 2005). Evidence suggests that metabolic acidosis can disrupt Ca balance by enhancing the excretion of Ca through the urine, and the Ca lost in the urine can be recovered under the influence of parathyroid hormone (PTH) (Lemann J, 1967, Bushinsky et al., 1993, Schonewille JT, 1999). Metabolic acidosis inhibits collagen synthesis by osteoblasts and stimulates osteoclastic activity, leading to increased bone resorption (Bushinsky et al., 1993). Acute metabolic acidosis stimulates PTH secretion. Interestingly, cows on a negative DCAD diet exhibit higher $1,25(OH)_2D$

and Ca concentrations in response to PTH treatment than those on an alkaline diet ((Goff, 2014). Further, metabolic acidosis may increase ionized Ca concentration due to a decrease in the binding of calcium to albumin (Lopez et al., 2002).

In a recent study, researchers investigated how preventing transient hypocalcemia at parturition, combined with prepartum dietary DCAD, would impact blood Ca dynamics ((Connelly et al., 2024). Cows at 251 days of gestation were allocated to receive positive DCAD (+120 mEq/kg) or negative DCAD (-120 mEq/kg) (Connelly et al., 2024). Immediately upon calving, cows received a continuous 24-hour infusion of either Ca gluconate or dextrose, resulting in four groups: positive DCAD and Ca gluconate (PCa), positive DCAD and dextrose (PDex), negative DCAD and Ca gluconate (NCa), and negative DCAD and dextrose (NDex) (Connelly et al., 2024). The study found that cows in the NCa group required lower infusion rates to maintain normocalcemia compared to the PCa group, indicating that cows receiving a negative DCAD diet prepartum required less exogenous Ca to maintain normal Ca concentration, due to the acidogenic state that increases Ca flux (Connelly et al., 2024).

Studies have shown that feeding multiparous cows for less than 14 days tended to increase culling risk, but extending this management to 28 days, regardless of parity, led to increased milk production and decreased culling risk (Venjakob et al., 2021, Vieira-Neto et al., 2021). (Vieira-Neto et al., 2020) detected an increased culling risk for cows receiving negative DCAD for more than 30 days, while (Venjakob et al., 2021) did not observe an association between an extended length on the close-up diet and increased culling risk, stillbirth risk, or reproductive performance. Due to variations in gestation length, the authors suggested moving cows from the far-off to the close-up group at 254 days of gestation, but when cows are expecting twins, they should be moved at 250 days of gestation due to their shorter gestation length (Venjakob et al., 2021).

In summary, the negative DCAD diet-induced metabolic acidosis has complex effects on Ca regulation, bone health, hormonal responses, and ionized Ca concentrations in cows, all contributing to preventing hypocalcemia during early lactation.

<u>Zeolite clays</u>: An additional prepartum dietary strategy to prevent hypocalcemia is the addition of a synthetic zeolite clay, which was thought to act primarily as a Ca binder. Zeolite clays are able to bind 2+ cations (Thilsing et al., 2007; Kerwin et al., 2019). Recent work has demonstrated that zeolite largely binds P, leading to mild hypophosphatemia, which improves Ca homeostasis (Kerwin et al., 2019)). Hypophosphatemia potentially increases Ca concentrations by elevating vitamin D synthesis. Higher blood P concentrations inhibited the conversion of 25-hydroxyvitamin D to 1,25$(OH)_2$D, decreasing Ca absorption in the gut (Horst et al., 2005). However, P-feed restriction during the final 4 weeks of gestation in cows led to increased Ca concentration without a concurrent increase in PTH and 1,25$(OH)_2$D concentrations (Wachter et al., 2022b). Nevertheless, cows undergoing P-restriction exhibit increased concentrations of carboxy-terminal collagen crosslinks (CTX), a marker of increased bone resorption (Wachter et al., 2022b).

It has been suggested that the relationship between hypophosphatemia and bone resorption occurs through P regulating the hormone fibroblast growth factor (FGF23). During hyperphosphatemia, osteocytes synthesize FGF23 to regulate P concentrations by increasing renal P excretion and decreasing 1-α-hydroxylase synthesis, thereby decreasing $1,25(OH)_2D$ concentrations (Kuro-o, 2009). It is hypothesized that the decrease in blood P induced by zeolite A supplementation might decrease FGF23 expression, potentially increasing bone resorption and subsequently increasing Ca concentration. However, there is no data at this time to support this finding (Frizzarini et al., 2024).

In summary, using Ca binders like synthetic zeolite A in the prepartum diet involves intricate connections between P regulation, vitamin D synthesis, FGF23-mediated processes, and their combined effects on Ca homeostasis and bone health in cows. It is imperative to exercise caution regarding the duration of zeolite supplementation in dairy cows. Industry recommendations suggest feeding zeolites for a span of 14 to 21 days. However, this regimen presents challenges due to the inherent difficulty in accurately predicting the timing of calving. Further work should also be done exploring the amount of zeolite to be fed.

<u>Low P diets</u>: Research findings indicate that rats maintained on a low P diet and supplemented with $25(OH)D_3$ exhibit higher intestinal Ca transport compared to those on an adequate P diet and supplemented with $25(OH)D_3$, due to an increase in $1,25(OH)_2D_3$ biosynthesis (Tanaka Y., 1973). Additionally, studies in rats have demonstrated that P deficiency enhances both the number and resorptive activity of osteoclasts (Thompson et al., 1975). In dairy cows, (Cohrs et al., 2018) investigated the effects of feeding a low-P or adequate-P diet during the last 4 weeks of gestation until 10 days postpartum, observing elevated plasma Ca and CrossLaps (CTX, a bone resorption marker) concentrations alongside reduced plasma PTH concentrations in the low-P group. This suggests that early lactation bone resorption and increased Ca concentration may occur independently of the classical PTH pathway (Cohrs et al., 2018). However, a more recent study reported only a statistical increase in CTX concentration on day 1 post-calving, possibly due to less severe hypophosphatemia observed, indicating a milder activation of osteoclast activity (Wachter et al., 2022b).

Despite evidence of increased bone mobilization during P deficiency, PTH secretion is typically decreased or unaffected in these animals (Kichura et al., 1982, Cohrs et al., 2018, Wachter et al., 2022a), suggesting an incomplete understanding of the exact mechanisms underlying increased bone resorption due to consumption of a low P diet. A recent study hypothesized that reduced dietary P availability leading to decreased blood P concentrations might downregulate FGF23 expression in the bone, potentially resulting in increased bone resorption and consequent Ca concentration (Frizzarini et al., 2024).

Despite improvements in Ca homeostasis observed when reducing dietary P, it is extremely difficult to formulate a dry cow diet with low P due to the common feed ingredients used in Europe and the Americas, making it hard to implement this management in dairy farms (Wachter et al., 2022b).

Postpartum calcium boluses: Oral Ca bolus therapy is a widely employed method to counteract the common Ca nadir observed in early lactation, and it is typically administered within hours postpartum. The administration of multiple Ca boluses effectively elevates blood Ca concentration by enhancing the passive diffusion in the rumen and passive or active absorption in the small intestines (Goff, 1999, Horst et al., 2005, Wachter et al., 2022a). There are different types of Ca salts used in these boluses, and each one has different solubility and absorption properties, and this information should be considered before applying this management (McArt and Oetzel, 2023). The efficacy of oral Ca boluses administration in elevating blood Ca concentrations relies on dosage and frequency of administration (McArt and Oetzel, 2023).

Recent prospective research has revealed intriguing findings regarding the adaptation of cows experiencing transient hypocalcemia (tSCH; [Ca] ≤ 1.77 at 1 day in milk (DIM) and > 2.20 mmol/L at 4 DIM, for multiparous cows) compared to those with persistent (pSCH; [Ca] ≤ 1.77 at 1 DIM and ≤ 2.20 mmol/L at 4 DIM, for multiparous cows) or delayed (dSCH; [Ca] > 1.77 at 1 DIM and ≤ 2.20 mmol/L at 4 DIM, for multiparous cows) hypocalcemia (McArt and Neves, 2020). Cows in the tSCH group exhibited increased milk production compared to those in the pSCH, dSCH, and normocalcemic (NC; [Ca] > 1.77 at 1 DIM and 2.20 mmol/L at 4 DIM, for multiparous cows) groups, indicating their efficient adaptation to the heightened Ca demand during early lactation (McArt and Neves, 2020). This finding suggests that cows experiencing tSCH are high milk producers and effectively adapt to the heightened Ca demand during early lactation ((McArt and Neves, 2020). This adaptation is evident in their substantial decrease in plasma Ca concentration by 1 DIM followed by a rapid rebound by 4 DIM (McArt and Neves, 2020)). In a different prospective study, cows classified into the NC and tSCH cohorts exhibited numerically lower PTH concentrations and elevated blood Ca concentrations relative to those in the dSCH and pSCH categories ((Seely et al., 2023). Additionally, cows in dSCH cohort, had higher PTH concentration one-week prepartum (Seely et al., 2023). These findings lend credence to the notion that cows affected by dSCH and pSCH faced an additional metabolic perturbation, culminating in disruption of Ca homeostasis (Seely et al., 2023).

The expanding body of literature indicates that transient SCH occurring at 1 DIM correlates with increased milk production, reduced incidence of health disorders, and improved reproductive outcomes, while SCH at 4 DIM is associated with diminished milk yield, feed intake, and reproductive performance ((Neves et al., 2018, McArt and Neves, 2020, Seely et al., 2021, Venjakob et al., 2021, Seely et al., 2023). Therefore, a recent investigation explored the impact of delayed oral Ca supplementation to 2 and 3 DIM on milk production, creating 3 distinct treatment groups: (1) control (CON; no Ca supplementation); (2) conventional bolus (BOL-C; one Ca bolus at parturition and another 24 h later); and (3) delayed bolus (BOL-D; one Ca bolus at 48 and another at 72 h postpartum) (Seely et al., 2023). Notably, the timing of Ca bolus administration exerts disparate effects across parity groups. While no discernible disparities in milk production were observed in parity 2 and 4 cows, parity 3 cows subjected to BOL-D treatment exhibited augmented milk production compared to CON and BOL-C counterparts (Seely et al., 2023). Intriguingly, no discrepancies in blood Ca concentrations were detected among the treatments in parity 3 cows, indicating that this group utilized Ca supplementation to support mammary gland function for milk synthesis rather than elevating blood Ca concentrations (Seely et al., 2023).

Hypocalcemia, serving as a trigger for the secretion of PTH and parathyroid related protein (PTHLH), initiates Ca homeostasis ((Horst, 1986, Horst RL, 1994, Wysolmerski, 2012, Goff, 2014). The intense milk production leads to a rapid decline in blood Ca concentration as Ca is secreted into colostrum and milk (Ramberg CF, 1967). fact that cows naturally restore normal Ca concentrations within a few days without treatment indicates an appropriate homeostatic response. Consequently, caution should be exercised in the use of oral Ca boluses to avoid inhibiting the Ca negative feedback loop responsible for maintaining Ca homeostasis.

Conclusion

A transient decline in blood Ca at the onset of lactation is critical for stimulating vital adaptive responses that allow for a successful adaptation to lactation. (McArt and Neves, 2020), found that cows experiencing transient hypocalcemia at the onset of lactation produce more milk than prolonged-, delayed-, and normo-calcemic counterparts. These cows also experience fewer incidences of early lactation diseases. (Seely et al., 2021) further supported these findings as cows experiencing transient hypocalcemia produced more milk while also consuming similar amounts of feed postpartum. This research has prompted considerations when discussing the classification and timing of hypocalcemia thresholds, as well as the regulation of calcemic status. Further, the homeorhetic coordination may differ between cows dependent on their calcemic status and the timing of blood sampling. In tandem, commonly used oral and intravenous Ca intervention strategies to improve blood Ca postpartum have been re-evaluated as calcemic interventions to introduce the potential to perturb normal homeostasis and homeorhetic adaptations that are stimulated by transient hypocalcemia or other pre-calving nutritional and management strategies ((Seely et al., 2021, Connelly et al., 2024). Research has demonstrated that bolus and intravenous Ca treatment shortly after parturition successfully increases blood Ca acutely and provides benefits for certain types of cows but can disrupt blood Ca dynamics and impact the production and health of cows ((Blanc et al., 2014, Martinez et al., 2016). The benefits of transient prepartum hypocalcemia must be considered when developing treatments, deciding types of treatments (i.e., bolus, gel), and the timing of administration of treatments when formulating diets and intervention plans.

References

Blanc, C. D., M. Van der List, S. S. Aly, H. A. Rossow, and N. Silva-del-Rio. 2014. Blood calcium dynamics after prophylactic treatment of subclinical hypocalcemia with oral or intravenous calcium. J Dairy Sci 97(11):6901-6906.

Block, E. 1984. Manipulation of dietary cation anion differene on nutritionally related production diseases, productivity, and metabolic responses of dairy cows. J Dairy Sci 77:1437-1450.

Bushinsky, D. A., W. Wolbach, N. E. Sessler, R. Mogilevsky, and R. Levi-Setti. 1993. Physicochemical effects of acidosis on bone calcium flux and surface ion composition. J Bone Miner Res 8(1):93-102.

Cohrs, I., M. R. Wilkens, and W. Grunberg. 2018. Short communication: Effect of dietary phosphorus deprivation in late gestation and early lactation on the calcium homeostasis of periparturient dairy cows. J Dairy Sci 101(10):9591-9598.

Connelly, M. K., R. M. R. Harris, J. Kuehnl, J. P. N. Andrade, F. S. Andrade, S. Henschel, E. Block, I. J. Lean, and L. L. Hernandez. 2024. Influence of prepartum dietary cation-anion difference and the magnitude of calcium decline at the onset of lactation on mineral metabolism and physiological responses. Journal of Dairy Science 107(2):1228-1243.

Frizzarini, W. S., P. L. J. Monteiro, J. P. Campolina, A. L. Vang, O. Soudah, L. R. Lewandowski, M. K. Connelly, S. I. Arriola Apelo, and L. L. Hernandez. 2024. Mechanisms by which feeding synthetic zeolite A and dietary cation-anion difference diets affect mineral metabolism in multiparous Holstein cows: Part I. Journal of Dairy Science 107(7):5204-5221.

Goff, J. P. 1999. Treatment of Calcium, Phosphorus, and Magnesium Balance Disorders. Veterinary Clinics of North America: Food Animal Practice 15(3):619-639.

Goff, J. P. 2014. Calcium and magnesium disorders. Vet Clin North Am Food Anim Pract 30(2):359-381, vi.

Horst, R. L. 1986. Regulation of calcium and phosphorous homeostasis in the dairy cow. J. Dairy Sci. (69):604-616.

Horst, R. L., G. J., Reinhardt, T.A. 1994. Calcium and vitamin D metabolism in the dairy cow. J. Dairy Sci. 77:1936-1951.

Horst, R. L., J. P. Goff, and T. A. Reinhardt. 2005. Adapting to the transition between gestation and lactation: differences between rat, human and dairy cow. J Mammary Gland Biol Neoplasia 10(2):141-156.

Hyde, M. L. and D. R. Fraser. 2014. In vivomeasurement of the absorption of strontium in the rumen and small intestine of sheep as an index of calcium absorption capacity. British Journal of Nutrition 112(5):718-724.

Kerwin, A. L., C. M. Ryan, B. M. Leno, M. Jakobsen, P. Theilgaard, D. M. Barbano, and T. R. Overton. 2019. Effects of feeding synthetic zeolite A during the prepartum period on serum mineral concentration, oxidant status, and performance of multiparous Holstein cows. J Dairy Sci 102(6):5191-5207.

Kichura, T. S., R. L. Horst, D. C. Beitz, and E. T. Littledike. 1982. Relationships between prepartal dietary calcium and phosphorus, vitamin D metabolism, and parturient paresis in dairy cows. J Nutr 112(3):480-487.

Kuro-o, M. 2009. Overview of the FGF23-Klotho axis. Pediatric Nephrology 25(4):583-590. Lemann J, L. J., Lennon EJ. 1967. Studies of the mechanism by which chronic metabolic acidosis augments urinary calcium excretion in man. J. Clin. Invest. 46(8):1318-1328.

Lopez, I., E. Aguilera-Tejero, A. J. Felsenfeld, J. C. Estepa, and M. Rodriguez. 2002. Direct effect of acute metabolic and respiratory acidosis on parathyroid hormone secretion in the dog. J Bone Miner Res 17(9):1691-1700.

Martinez, N., L. D. P. Sinedino, R. S. Bisinotto, R. Daetz, C. Lopera, C. A. Risco, K. N. Galvao, W. W. Thatcher, and J. E. P. Santos. 2016. Effects of oral calcium supplementation on mineral and acid-base status, energy metabolites, and health of postpartum dairy cows. J Dairy Sci 99(10):8397-8416.

McArt, J. A. A. and R. C. Neves. 2020. Association of transient, persistent, or delayed subclinical hypocalcemia with early lactation disease, removal, and milk yield in Holstein cows. J Dairy Sci 103(1):690-701.

McArt, J. A. A. and G. R. Oetzel. 2023. Considerations in the Diagnosis and Treatment of Early Lactation Calcium Disturbances. Veterinary Clinics of North America: Food Animal Practice 39(2):241-259.

Neves, R. C., B. M. Leno, M. D. Curler, M. J. Thomas, T. R. Overton, and J. A. A. McArt. 2018. Association of immediate postpartum plasma calcium concentration with early-lactation clinical diseases, culling, reproduction, and milk production in Holstein cows. J Dairy Sci 101(1):547-555.

Ramberg CF, M. G., Kronfeld DS, Aurbach GD, Sherwood LM, Potts JT. 1967. Plasma calcium and parathyroid hormone responses to EDTA infusion in the cow. Amer. J. Physiol. 213(4):878-882.

Reinhardt, T. A., J. D. Lippolis, B. J. McCluskey, J. P. Goff, and R. L. Horst. 2011. Prevalence of subclinical hypocalcemia in dairy herds. The Veterinary Journal 188(1):122-124.

Rodney, R. M., N. Martinez, E. Block, L. L. Hernandez, P. Celi, C. D. Nelson, J. E. P. Santos, and I. J. Lean. 2018. Effects of prepartum dietary cation-anion difference and source of vitamin D in dairy cows: Vitamin D, mineral, and bone metabolism. J Dairy Sci 101(3):2519-2543.

Schonewille JT, V. t. K. A., Wouterse H, Beynen AC. 1999. Hypocalcemia induced by intravenous administration of disodium ethylenediaminotetraacetate and its effects on excretion of calcium in urine of cows fed a high chloride diet. J. Dairy Sci. 82:1317-1324.

Seely, C. R., B. M. Leno, A. L. Kerwin, T. R. Overton, and J. A. A. McArt. 2021. Association of subclinical hypocalcemia dynamics with dry matter intake, milk yield, and blood minerals during the periparturient period. J Dairy Sci 104(4):4692-4702.

Seely, C. R., C. N. Wilbur, K. M. Fang, and J. A. A. McArt. 2023. Impact of timing of oral calcium administration on milk production in high producing early lactation Holstein cows. J Dairy Sci.

Tanaka Y., F. H., DeLuca H.F. 1973. Intestinal calcium transport: stimulation by low phosphorous diets. Science 181(4099):564-566.

Venjakob, P. L., R. Staufenbiel, W. Heuwieser, and S. Borchardt. 2021. Association between serum calcium dynamics around parturition and common postpartum diseases in dairy cows. J Dairy Sci 104(2):2243-2253.

Vieira-Neto, A., I. M. R. Leao, J. G. Prim, A. C. M. Silva, M. Nehme Marinho, R. Zimpel, S. Etheve, C. D. Nelson, and J. E. P. Santos. 2020. Effect of duration of exposure to diets differing in dietary cation-anion difference on Ca metabolism after a parathyroid hormone challenge in dairy cows. J Dairy Sci.

Vieira-Neto, A., G. Negro, R. Zimpel, M. Poindexter, F. Lopes, Jr., W. W. Thatcher, C. D. Nelson, and J. E. P. Santos. 2021. Effects of injectable calcitriol on mineral metabolism and postpartum health and performance in dairy cows. J Dairy Sci 104(1):683-701.

Wachter, S., I. Cohrs, L. Golbeck, T. Scheu, K. Eder, and W. Grunberg. 2022a. Effects of restricted dietary phosphorus supply during the dry period on productivity and metabolism in dairy cows. J Dairy Sci 105(5):4370-4392.

Wachter, S., I. Cohrs, L. Golbeck, M. R. Wilkens, and W. Grunberg. 2022b. Effects of restricted dietary phosphorus supply to dry cows on periparturient calcium status. J Dairy Sci 105(1):748-760.

Wysolmerski, J. J. 2012. Osteocytic osteolysis: time for a second look? Bonekey Rep 1:229.

Up-and-Coming Research Mini Presentations
Courtes présentations sur les recherches en cours et à venir

This presentation category is intended for early-stage graduate students (both master's and PhD level) who do not yet have research results to showcase their project motivation and the benefit their work will have for industry. This is a unique opportunity for graduate students to communicate the value of their work directly to the industry professionals who will apply it.

Cette catégorie de présentation s'adresse aux étudiants en début de cursus (maîtrise et doctorat) qui n'ont pas encore de résultats de recherche pour présenter la motivation de leur projet et les avantages que leur travail apportera à l'industrie. Il s'agit d'une occasion unique pour les étudiants de cycle supérieur de communiquer la valeur de leur travail directement aux professionnels de l'industrie qui l'appliqueront.

Ruminant presentations / *Les présentations sur les ruminants*

Katie Kroeze, University of Guelph
Methane Reduction Strategies for Grazing Beef Cows
Stratégies de réduction des émissions de méthane pour les vaches de boucherie en pâturage

Tatum Schooley, University of Guelph
Reduce, reuse, recycle: Using milk processing by-products as an alternative for corn in dairy cow diets
Réduire, réutiliser, recycler : Utilisation des sous-produits de la transformation du lait comme alternative au maïs dans l'alimentation des vaches laitières

Kylie Grimes, University of Saskatchewan
Maximizing wheat straw use in gestating beef cow diets with oilseed fines supplementation
Optimisation de l'utilisation de la paille de blé dans l'alimentation des vaches de boucherie en gestation avec un apport complémentaire de fines de graines oléagineuses

Gus Halibert, University of Saskatchewan
A Benchmark Study of the Canadian Feedlot Industry and an Evaluation of Best Management Practices to Improve the Sustainability of Feedlots
Une étude comparative de l'industrie canadienne des parcs d'engraissement et une évaluation des meilleures pratiques de gestion pour en améliorer la durabilité

SPONSORED BY / COMMANDITÉ PAR

Alltech®

Methane reduction strategies for grazing beef cows

K. Kroeze[1], G. Rebeiro[2], K. Wood[3]
[1] Ph.D. Student, Department of Animal Biosciences, University of Guelph, Guelph, ON
[2] Assistant Professor, Department of Animal and Poultry Science, University of Saskatchewan, Saskatoon, SK
[3] Associate Professor, Department of Animal Biosciences, University of Guelph, Guelph, ON
kkroeze@uoguelph.ca

SPONSORED BY / COMMANDITÉ PAR

Alltech®

Brief project summary

Reducing greenhouse gas emissions is a goal for Canada's beef industry. However, there is a paucity of research investigating how grazing strategies can help to reduce methane emissions in Canadian beef production. Experiment 1 investigates two summer pasture management strategies: 3-day rotational grazing or continuous grazing. The hypothesis for experiment 1 is that the increase in more consistent grazing of the field in smaller portions will allow for consistent regrowth and therefore lower fibre content, thus be more easily digestible, requiring less fermentation, therefore the cows on the 3-day rotation treatment will produce less enteric methane. In experiment 2, cows will late season graze either high or low birdsfoot trefoil content over the Fall. The hypothesis for experiment 2 is that the increased birdsfoot trefoil content will have a higher amount of tannins and therefore reduce the enteric methane emissions from the cows. These two experiments will measure animal production metrics as well as enteric emission data collected using GreenFeed trailers (C-Lock Inc.), and pasture composition and biomass using quadrat methods. To improve methane prediction from *in vitro* methane measurements, experiment 3 will evaluate *in vivo* data from experiments 1 and 2 and compare to *in vitro* methane production using an artificial rumen, Rusitec. These data will create a mathematical model to better predict *in vivo* methane emissions from the *in vitro* experiments. These experiments could provide beef producers with pasture management strategies to reduce enteric methane emissions and provide a better understanding of emission predictions from *in vitro* experiments.

Reduce, reuse, recycle: Using milk processing by-products as an alternative for corn in dairy cow diets

T.E. Schooley[1], A. Heebner[2], B.E. Abbassi[2], P.F. Wu[2], M.A. Steele[1], J.P. Cant[1], T.C. Wright[3], K.M. Wood[1]
[1] Department of Animal Biosciences, University of Guelph, Guelph, ON
[2] School of Engineering, University of Guelph, Guelph, ON
[3] Ontario Ministry of Agriculture, Food and Agribusiness, Guelph, ON

SPONSORED BY / COMMANDITÉ PAR

Alltech®

Brief project summary

Milk processing results in a large volume of byproducts, including solids non-fat (SNF) and cheese whey. Half of these are processed into food or feed, but disposal of the surplus presents environmental concerns. Separating and dewatering whey byproducts produces separated lactose (SEP-LAC) and whey protein plus bentonite (WPB) to replace HMC in a dairy cow TMR. Replacing up to 12% of HMC with lactose and WPB could improve rumen health by increasing ruminal pH and butyrate concentrations. Objectives include estimating the nutritive value of lactose and WPB relative to HMC and observing the effects on rumen health and milk production. Cheese whey was mixed with 30g/L of sodium bentonite clay and separated into SEP-LAC and WPB. Synthetic lactose (SYN-LAC) was mixed at 150g/L to simulate SEP-LAC DM% after nanofiltration. Treatment diets were formulated from a lactating cow TMR as a control with SYN-LAC added at 4, 8, 12, and 8% + 2% WPB. Diets will be fed to ten multiparous mid-lactation Holstein cows in a replicated 5x5 Latin square, with five 21d periods consisting of 2.5w adaptation and 3d sampling, and two cows on each diet. The removal of proteins with bentonite before nanofiltration replaces pre-filtration steps and extends membrane lifespan, reducing costs of de-watering. Revalorizing milk byproducts could reduce food waste and provide a sustainable method for the disposal of whey and SNF surplus. Additionally, replacing HMC with lactose sugar could prevent sub-acute ruminal acidosis caused by high starch diets in early lactation cows and feedlot steers.

Maximizing wheat straw use in gestating beef cow diets with oilseed fines supplementation

K. Grimes[1], K. Larson[2], G. B. Penner[1], H. A. Lardner[1], G. O. Ribeiro[1]
[1] Department of Animal and Poultry Science, University of Saskatchewan, Saskatchewan, SK
[2] Department of Agricultural and Resource Economics, University of Saskatchewan, Saskatchewan, SK

SPONSORED BY / COMMANDITÉ PAR

Alltech®

Brief project summary

With a growing global population and increasing demand for food, sustainable agricultural strategies are essential. Incorporating byproducts like oilseed fines and wheat straw into beef rations could improve sustainability, mitigate feed shortages during droughts, and reduce costs. Additionally, the higher fat content in oilseed fines may lower methane emissions from ruminal fermentation of low-quality forages. The objective is to evaluate the effects of a wheat straw-based diet supplemented with high- or low-fat oilseed fines (6% fat on a diet dry matter basis) on feed intake, weight change, body condition score, ruminal fermentation, enteric methane emissions, reproductive efficiency, and production costs. We hypothesize that cows fed these alternative diets will show similar outcomes to those on a standard western Canadian barley silage/grass hay diet. To test this, 72 pregnant cows will be fed one of three diets during their second and third trimesters of gestation: (1) a control diet of barley silage and hay, (2) wheat straw with low-fat oilseed fines, or (3) wheat straw with high-fat oilseed fines. Methane emissions will be measured in vivo using the GreenFeed system, while in vitro ruminal fermentation parameters, methane production, and microbial protein synthesis will be analyzed using the rumen simulation technique (RUSITEC). Findings from this research will help optimize the use of low-quality crop residues and byproducts in western Canada, enhancing the sustainability of cow-calf operations.

A benchmark study of the Canadian feedlot industry and an evaluation of best management practices to improve the sustainability of feedlots

K. A. Halibert[1], M. St. Martin[2], G. B. Penner[1], K. H. Ominski[3], S. A. Terry[4], M. Van Schaik[5], B. Grant[6], M. L. Swift[1], J. F. Peters[1], T. A. McAllister[4], K. M. Wood[2], and G. O. Ribeiro[1]

[1] Department of Animal and Poultry Science, University of Saskatchewan, Saskatoon, SK

[2] Department of Animal Biosciences, University of Guelph, Guelph, ON

[3] Department of Animal Science, University of Manitoba, Winnipeg, MB

[4] Lethbridge Research and Development Center, Agriculture and Agri-Food Canada, Lethbridge, AB

[5] Ontario Ministry of Agriculture, Food, and Rural Affairs, Guelph, ON

[6] Canfax Research Services, Calgary, AB

gus.halibert@usask.ca

SPONSORED BY / COMMANDITÉ PAR

Alltech®

Brief project summary

Canadian beef production sustainability assessments have been a topic of great interest from the government and consumers, with many estimates of greenhouse gases (GHG) conducted in the last decade. These estimates are based on common practices used by livestock operations across the country; however, operations are very diverse. Feedlot practices in Canada have been underrepresented in previous beef cattle surveys and key components are missing. To increase accuracy of sustainability assessments and to identify best management practices (BMPs) for beef production, accurate data must be collected. Further evaluation of industry practices is also necessary to identify BMPs. Fecal near infrared spectroscopy (NIRS) has been identified as a rapid analysis tool with promising results. Fecal NIRS assessments may allow beef producers to identify issues in a timely manner and make changes to their practices, consequently improving production and reducing environmental impact. The present research will first develop and conduct a national survey for backgrounding and finishing feedlot producers to collect accurate data on current production, feeding, and management practices. The second objective of this research is to develop accurate fecal NIRS calibrations applicable to commercial feedlot production by sampling twenty feedlots across Canada over 2 years. This work will provide accurate information on backgrounding and finishing feedlot practices not collected in previous surveys, for more accurate GHG emission and other sustainability assessments, and will support the development of effective fecal NIRS calibrations enabling the fine tuning of current practices increasing the efficiency and sustainability of beef production in Canada.

ANCC

2025 ANIMAL NUTRITION CONFERENCE OF CANADA

CNAC

2025 COLLOQUE DE NUTRITION ANIMALE DU CANADA

Closing Plenary
Plénière de clôture

Can AI truly transform livestock and feed industries, or are we overpromising?

L'IA est-elle en mesure de révolutionner les secteurs de l'élevage et de l'alimentation animale, ou nos attentes sont-elles excessives?

J. R. R. Dorea[1,2], G. L. Menezes[1], R. E. P. Ferreira[1], P. S. Lamas[1,3]
[1] Department of Animal and Dairy Sciences, University of Wisconsin-Madison, Madison, WI, USA
[2]Department of Biological Systems Engineering, University of Wisconsin-Madison, Madison, WI, USA
[3] Escola Superior de Agricultura Luiz de Queiroz - University of Sao Paulo
joao.dorea@wisc.edu

Abstract

The advance of AI systems in different fields of science has created incredible opportunities for the new generation of students and scientists to answer research questions that would not have been possible before the recent progress toward more intelligent systems. AI technologies such as computer vision, natural language processing, and robotics have become integral to our daily lives through applications like facial recognition, speech-to-text, drones, and virtual reality. The agricultural sector has leveraged AI advancements from other scientific domains, and livestock systems are increasingly integrating modern solutions to address critical challenges related to animal health and welfare monitoring, greenhouse gas emissions, animal traceability, and labor shortages. Beyond these applications, the future of the livestock industry is poised to be transformed by digital technologies and predictive analytics, enabling proactive decision-making and precision management at large scale. AI has the potential to reshape the feed industry by optimizing formulations, improving efficiency, and enhancing overall sustainability. Machine learning models can analyze vast datasets, including ingredient costs, nutritional profiles, and animal performance metrics, to create precise, cost-effective feed formulations that maximize productivity while minimizing waste. In this talk, we will explore key AI technologies, such as computer vision and natural language processing, with the potential to positively impact livestock systems and the feeding industry in the coming decades. We will discuss how these innovations align with real-world challenges faced by farmers, industry stakeholders, and the scientific community.

Résumé

Les progrès réalisés par les systèmes d'IA dans différents domaines scientifiques ont créé des opportunités incroyables pour la nouvelle génération d'étudiants et de chercheurs qui peuvent désormais résoudre des problèmes auparavant impossibles à surmonter. Les technologies d'IA comme la vision par ordinateur, le traitement automatique du langage naturel et la robotique sont désormais omniprésentes dans notre quotidien, notamment avec les applications de reconnaissance faciale, de reconnaissance de la parole, de drones et de réalité virtuelle. Le secteur agricole a tiré parti des avancées de l'IA dans d'autres domaines scientifiques, et les systèmes d'élevage intègrent de plus en plus de solutions modernes pour relever les défis importants liés à la surveillance de la santé et du bien-être des animaux, aux émissions de gaz à effet de serre, à la traçabilité des animaux et à la pénurie de main-d'œuvre. En outre, l'avenir de l'industrie de l'élevage est en passe d'être révolutionné par les technologies numériques et l'analyse prédictive, ces outils permettant une prise de décision proactive ainsi qu'une gestion de précision à grande échelle. L'IA a le potentiel de remodeler l'industrie de l'alimentation animale par l'optimisation du travail de formulation, l'amélioration de l'efficacité et l'accroissement de la pérennité en général. Les modèles d'apprentissage automatique peuvent analyser de vastes ensembles de données, notamment les coûts des ingrédients, les profils nutritionnels et les mesures de performance des animaux, ce qui permet d'améliorer la précision et la rentabilité de la formulation des aliments et, ainsi, de maximiser la productivité tout en réduisant les déchets. Dans cet exposé, nous explorerons les principales technologies d'IA, telles que la vision par ordinateur et le traitement du langage naturel, susceptibles d'avoir un impact positif sur les systèmes d'élevage et l'industrie de l'alimentation animale au cours des prochaines décennies. Nous aborderons la question de la pertinence de ces innovations face aux défis concrets auxquels sont confrontés les producteurs agricoles, les acteurs de l'industrie et la communauté scientifique.

1. Introduction

Artificial intelligence (AI) is increasingly being promoted as a transformative force in the livestock and feed industries, with promises of enhanced efficiency, sustainability, and precision. Technologies such as computer vision systems (CVS), large language models (LLMs), and precision livestock farming tools have already made significant progress in certain areas, such as automated animal identification, health monitoring, and feed optimization. However, while AI-driven innovations continue to be developed and refined, widespread deployment remains debatable due to concerns over scalability, cost, and practical implementation within the agricultural sector.

Some AI applications, such as CVS for automated body condition scoring and open-set cattle identification, have demonstrated success in controlled settings. Similarly, machine learning models have been effectively used to optimize feed formulations by analyzing nutritional profiles, ingredient costs, and animal performance metrics. Large language models offer promise in integrating multimodal farm data, supporting predictive analytics, and serving as virtual assistants for farm management. Retrieval-Augmented Generation (RAG) is a technique that integrates a retrieval mechanism with a generative language model to enhance response accuracy. In this approach, relevant documents or text chunks are first retrieved from an external knowledge base using embedding-based similarity or traditional search methods. These retrieved text chunks are then provided as context to the language model, which conditions its output on both the user question and the retrieved information. RAG improves performance on tasks requiring up-to-date or domain-specific knowledge by reducing hallucinations and enabling grounded generation.

Despite these advancements, many challenges hinder immediate large-scale implementation in livestock and feed industries. The agricultural sector has historically faced slower adoption of digital technologies due to infrastructure limitations, connectivity constraints, and financial barriers. Unlike crop monitoring, which benefits from well-established satellite imagery and remote sensing, livestock monitoring is inherently more complex due to the dynamic nature of animal behavior and environmental variability. Furthermore, issues related to data quality, integration across diverse farm management systems, and the need for multidisciplinary expertise raise concerns about scalability and economic feasibility.

While AI has the potential to reshape livestock and feed industries, there remains a significant gap between technological capabilities and real-world deployment. The hype surrounding AI's role in agriculture often overlooks the practical hurdles that must be addressed before these technologies become universally accessible and commercially viable. This paper critically evaluates the progress AI has made in specific areas, the ongoing developments in others, and the key barriers, including deployment challenges, scalability limitations, and cost considerations, that may preclude immediate adoption in modern livestock production.

2. Computer Vision Systems in Livestock

Computer vision systems for animal identification provide an automated approach to individual monitoring without requiring external accessories such as RFID tags, reducing labor and improving animal welfare (Chapa et al., 2020; Johnston & Edwards, 1996). Identification methods within CVS can be categorized by the anatomical region used, image modality, and whether the system is designed for closed-set or open-set identification (Hossain et al., 2022; Qiao et al., 2021). While closed-set models function within a predefined set of animals, open-set models can integrate new individuals dynamically, improving scalability (Andrew et al., 2021; Wang et al., 2024; Sharma et al., 2025). Various methods leverage unique features such as the muzzle, retina, iris, and coat pattern, with RGB, RGB-D, and depth images supporting identification (Hossain et al., 2022; Qiao et al., 2021)

Achieving high-quality facial images is difficult, necessitating alternative imaging strategies. Side and multiple-angle imaging have been explored (Bhole et al., 2019; Shen et al., 2020; Hu et al., 2020), but occlusion remains a challenge. Body coat pattern-based identification is advantageous due to its practicality in large-scale operations (Phyo et al., 2018; Zin et al., 2018). More specifically, top-down imaging, leveraging convolutional neural networks (CNNs), mitigates occlusion issues and allows concurrent phenotype assessments such as body weight (Cominotte et al., 2020; Manzanilla-Pech et al., 2023), body condition score (Qiao et al., 2021), ribeye area and circularity (Caffarini et al., 2022), behavior (Tsai & Huang, 2014), and individual feed intake (Manzanilla-Pech et al., 2023). Additionally, combining CNNs with recurrent neural networks has improved identification accuracy by incorporating temporal information (Andrew et al., 2017; Qiao et al., 2019, 2020). However, video-based methods may be impractical due to farm hardware and connectivity limitations (Rumelhart et al., 1986).

For cattle with similar coat colors, alternative approaches include depth imaging and point cloud analysis, allowing three-dimensional (3D) surface-based identification (Ferreira et al., 2022; Sharma et al., 2025). Optical character recognition techniques for reading ear tags and collars offer breed-independent solutions but suffer from occlusion, wear, and loss (Bezen et al., 2020; Ilestrand, 2017; Smink et al., 2024; Velez et al., 2013; Zin, Misawa, et al., 2020). Despite their applicability to large-scale systems, these methods raise concerns regarding management reliability and animal welfare (Johnston & Edwards, 1996).

Most existing methods require retraining to accommodate new animals, limiting their scalability and practicality in commercial livestock settings. To address this, open-set identification should be adopted to enable reliable animal recognition under dynamic conditions, such as frequent animal movement, a common feature in livestock production systems. Open-set identification leverages embeddings extracted from images (Figure 1) or point clouds to classify animals dynamically without retraining (Bergamini et al., 2018; Andrew et al., 2021; Wang et al., 2024; Sharma et al., 2025). By mapping similar data points within an embedding space, new animals can be grouped for identification without retraining. Studies have demonstrated the viability of this approach for commercial herds, ensuring scalability. However, long-term tracking remains a challenge as animal morphology changes over time. While biometric features such as muzzle prints and retina scans provide consistent identification (Allen et al., 2008), methodologies relying on coat patterns require periodic updates. In livestock systems, where long-term animal identification is essential for effective traceability, the need to continuously update models with recent images poses a significant challenge, particularly when relying on manual image labeling. To reduce the intensive annotation burden associated with CVS, semi-supervised learning techniques such as pseudo-labeling have been proposed as a practical alternative as proposed by Ferreira et al. (2023). By leveraging previously trained models to annotate new images, dataset diversity and model robustness improve significantly, enhancing performance (Qiao et al., 2021).

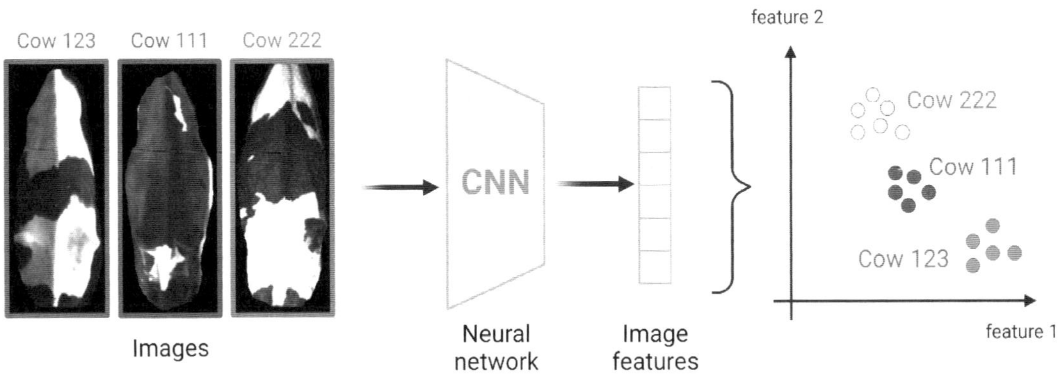

Figure 1. Feature extraction procedure for images of cows with different coat color patterns. In the feature space, each data point is an image, and images of the same cow are grouped together and positioned close to each other. (source: Ferreira and Dorea, 2025)

For body condition score (BCS) evaluation, CVS automates the traditionally subjective and labor-intensive process (Evans, 1978). Techniques employing two-dimensional (2D) and 3D imaging, including RGB and depth images, enable precise assessments (Qiao et al., 2021). CNNs and vision transformers have demonstrated high accuracy (Dosovitskiy et al., 2021; Wu et al., 2021) but concerns regarding dataset redundancy and generalizability persist (Bresolin & Dórea, 2020). Advanced methodologies incorporate automated keypoint detection and shape descriptors, enhancing predictive reliability (Higaki, Menezes, et al., 2024; Sun et al., 2013; Ferreira et al., 2024). Despite these advancements, BCS remains a subjective measurement. Future research should explore quantitative metrics derived from CVS for direct predictions of cow health and performance, reducing reliance on traditional scoring systems (Barletta et al., 2017).

3. Large Language Models

LLMs offer significant potential for dairy farming by leveraging NLP techniques to extract information from text, support dimensionality reduction, integrate diverse datasets, and enhance knowledge retrieval. These models are based on advanced language modeling techniques, capturing the structure, grammar, and context of language for various applications, including text generation, translation, and understanding (Brown et al., 2020; Devlin, 2018). Transformers, the foundational deep learning architecture of LLMs, enable long-range contextual understanding, making them suitable for multimodal tasks spanning text and vision (Vaswani et al., 2017; Radford et al., 2021).

Prompt engineering is a critical technique for optimizing LLM performance by structuring queries to improve accuracy and contextual relevance (Brown et al., 2020). Effective prompt engineering enables zero-shot and few-shot learning, allowing models to generalize across tasks with minimal training data. In dairy applications, structured prompts incorporating herd data, such as body weight, days in milk, and diet composition, can enhance predictive capabilities for nutrition and management decision-making.

Text embeddings play a pivotal role in data integration by transforming unstructured text into high-dimensional vector representations, facilitating semantic similarity analysis across multiple data sources (Figure 2). These embeddings enable efficient multimodal data fusion, integrating information from text, images, and sensor outputs to support decision-making in dairy systems (Radford et al., 2021). For example, embeddings allow heterogeneous datasets, such as wearable sensor outputs and farm management records, to be standardized for predictive modeling (Cabrera et al., 2020; van der Voort et al., 2021). LLMs also improve knowledge accessibility in dairy science by summarizing and retrieving information from extensive scientific literature. In transition cow management, deep learning models have demonstrated the ability to extract insights from domain-specific research and answer applied questions with expert-level accuracy (Zhu et al., 2023). This capability enhances decision support for veterinarians, animal scientists, and farmers.

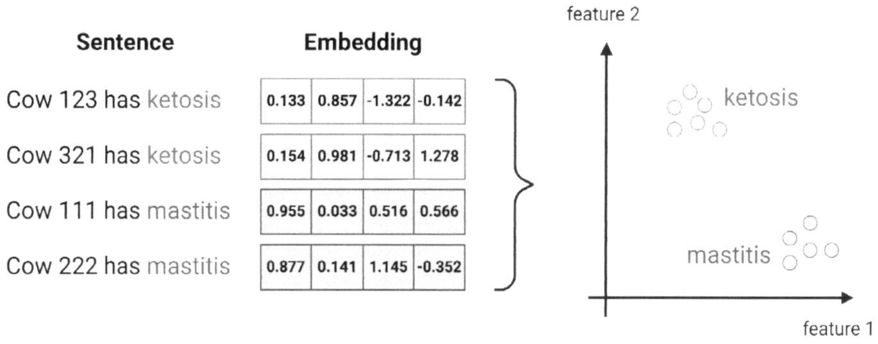

Figure 2. Embeddings are extracted from sentences, representing semantics in the form of a vector of numbers. These vectors belong to an embedding space, which groups together similar sentences and concepts. (source: Ferreira and Dorea, 2025)

Ensuring the factual reliability of LLM outputs remains a challenge, as these models can generate responses based on incomplete or outdated training data. Retrieval-Augmented Generation mitigates this issue by combining generative models with real-time information retrieval, improving accuracy and trustworthiness in domains where precision is critical, such as dairy health and nutrition (Lewis et al., 2020). RAG enhances trust in AI-generated content by integrating external information, ensuring responses are factually grounded and transparent. This capability is critical for applications requiring both fluency and reliability, such as virtual dairy farm assistants. By allowing models to cite sources, RAG increases user confidence in AI-driven insights. Future research should focus on optimizing LLM applications for dairy systems, emphasizing robust integration of multimodal datasets and real-time decision support tools.

3.1. Can Large Language Models Replicate Meta-Analysis Findings?

In the following sections (3.1.1 and 3.1.2), we present the application of RAG for data retrieval, aiming to assess whether applying RAG to a text corpus composed of studies used in a meta-analytical approach would yield conclusions consistent with those of the original meta-analyses. Although RAG typically relies on pre-trained models to extract embeddings for text retrieval, we fine-tuned the embedding model using question–answer pairs generated by the GPT-3.5 model. The objective was to enhance the retrieval of relevant information from the knowledge base to more accurately address specific questions. The GPT-3.5 model was integrated into the framework for two key tasks: (1) generating questions from text chunks within the corpus (question-answer pairs), and (2) answering those questions in natural language format.

3.1.1. Case-Study 1: Dorea and Armentano (2017)

In this case study, we utilized the complete set of articles reviewed in the meta-analysis by Dorea and Armentano (2017) to fine-tune a large language model (LLM) specifically for embedding extraction. The primary goal was to fine-tune the model using the identical corpus of literature employed in the original meta-analysis, subsequently querying the model with the same research questions addressed by Dorea and Armentano (2017) and evaluating whether the generated responses were consistent with their meta-analytic findings. The referenced meta-analysis articulated its objective as follows: "*The objective of the present article was to summarize the effects of five common dietary fatty acids (C16:0, C18:0, C18:1, C18:2, and C18:3) on the major milk fat groups (<C16, C16, and C18).*" To accomplish this objective, two distinct analytical approaches were undertaken in the original meta-analysis. The first analysis examined how different dietary fat supplements, categorized by their predominant fatty acid composition, influenced the milk yield of fatty acid groups <C16, C16, and C18. The second analysis employed regression techniques to determine how total dietary concentration or intake of specific fatty acid groups (<C16, C16, and C18) impacted the secretion of these same fatty acids in milk.

To assess the efficacy of RAG, an experimental evaluation was conducted employing a RAG-based methodology aimed at enhancing response accuracy concerning the meta-analytical outcomes reported by Dorea and Armentano (2017). The developed framework integrated a retrieval component alongside a generative language model, applied to a knowledge base derived directly from the papers used in Dorea and Armentano (2017) dataset. The constructed corpus consisted of 40 parsed PDF documents used by Dorea and Armentano (2017), which were further segmented into 1,147 text snippets, referred to as "nodes". Additionally, a validation corpus comprising 11 additional articles (PDF documents) was parsed, resulting in 346 additional nodes. The validation dataset was used in the RAG framework to evaluate the quality and relevance of the retrieved text in response to a given question. Specifically, it enabled comparison between the model's retrieved answer and a ground-truth reference answer, providing a means to assess whether the retrieval component is successfully identifying contextually appropriate and semantically accurate information. A large language model (BAAI/bge-large-en) was fine-tuned utilizing the LlamaIndex framework, leveraging GPT-3.5 Turbo for generating a total of 2,293 question-answer pairs. Model training consisted of two epochs, each containing 519 iterations, optimizing for accurate embedding extraction. Query questions were transformed into embeddings and compared to the corpus (original text) using cosine similarity metrics to identify and retrieve the most relevant context.

For the generative component, the Azure OpenAI GPT-3.5 Turbo model was prompted with both the retrieved contextual data and the query itself, with key parameters adjusted to optimize response precision and relevance (temperature set at 0.7 and maximum token length capped at 400). In large language models (LLMs), tokens are the basic units of text that the model processes. A token can be a word, part of a word, or even just a character like a punctuation mark. For example, the sentence "Cows eat silage." might be split into the tokens "Cows", "eat", and "silage", ".". Models like GPT read and generate text one token at a time. Comparative evaluations were performed against the responses found the meta-analysis. Below are the results of the case study, where questions were constructed to align with the objectives of the meta-analysis:

Question 1: Which unsaturated fatty acid has the greatest effect on milk fat depression?

RAG Response 1: Based on the information provided, diets containing higher concentrations of polyunsaturated fatty acids (PUFA) are associated with milk fat depression. Among unsaturated fatty acids, it is generally observed that increasing concentrations of PUFA, such as C18:2 and C18:3, tend to depress milk fat synthesis the most. This is due to the inhibition of milk fat synthesis by fatty acid intermediates formed in the rumen during the biohydrogenation of unsaturated fatty acids.

Dorea and Armentano (2017) - Response 1: *"When total milk fatty acid yield is modelled, the slope for added linoleic acid is numerically more negative than the slopes for oleic acid and linolenic acid."*

Question 2: Rank oleic, linoleic, and linolenic acid based on their impact on milk fat depression

RAG - Response 2: Based on the information provided, we can rank linoleic, oleic, and linolenic acid based on their power on milk fat depression as follows:

1. Linoleic acid - This oil had a significant effect on milk fat depression compared to other oils (3 = OSAF vs. LNSD), indicating a strong impact on reducing milk fat content.

2. Oleic acid - Though oleic acid was not directly compared in the given context, it can be inferred that oleic acid may have a moderate effect on milk fat depression based on the comparison of different types of safflower oils.
3. Linolenic acid - Linolenic acid had the least significant effect on milk fat depression compared to other oils, as it did not show a significant difference when compared to high-linoleic safflower oil or high-oleic safflower oil.

Therefore, linoleic acid would be ranked highest in terms of power on milk fat depression, followed by oleic acid, and then linolenic acid.

Dorea and Armentano (2017) - Response 2: *"Once again, linoleic acid had the most negative coefficient, and considering the three unsaturated fatty acids individually..."*

Question 3: Based on all the literature you have reviewed, how do dietary C16:0, C18:1, C18:2, and C18:3 fat supplements affect milk yield and the composition of <C16, C16, and C18 fatty acids?

Response 3: Based on the literature provided, the effects of dietary C16:0, C18:1, C18:2, and C18:3 fat supplements on milk yield of <C16, C16, and C18 fatty acids can be summarized as follows:

1. C16:0: Diets supplemented with primarily C16:0 resulted in a negligible change in the milk yield of both <C16 and C18 fatty acids.
2. C18:1: Diets supplemented with C18:1 led to a decrease in the milk yield of <C16 fatty acids and an increase in the yield of C18 fatty acids. The magnitude of the decrease in <C16 fatty acids was relatively moderate.
3. C18:2: Diets supplemented with C18:2 resulted in a significant decrease in the milk yield of <C16 fatty acids, with the most severe decreases observed compared to other supplements. These diets also increased the milk yield of C18 fatty acids.
4. C18:3: The effects of C18:3 fatty acid supplements were not specifically mentioned in the provided literature excerpt.

In general, the type of dietary fat supplement can impact the milk yield of different fatty acids in varying ways. The specific fatty acid composition of the supplement can influence the synthesis and incorporation of fatty acids into milk fat, leading to changes in the yield of specific fatty acid components in the milk.

Dorea and Armentano (2017) - Response 3: *Item 1: "The C16:0 and C16:0 + C18:0 fat supplements are the only supplements that did not decrease* *secretion of de novo fatty acids or increase milk C18 yield." Items 2, 3, and 4: "It is clear from these data that the three unsaturated C18 dietary fatty acids decrease the yield of milk <C16 and C16 and increase the yield of milk C18."*

Question 4: On average, if I increase C18:2 by one percent, how much of a reduction in fat yield (in grams per day) can be expected?

Response 4: To determine the expected reduction in fat yield in grams per day when increasing one percent of C18:2, we need to analyze the relationship between C18:2 and fat yield based on the information provided in the text. From the text, we can see that the C18:2 data is not directly provided in the tables, but we can infer some general trends related to the changes in fatty acid composition and fat yield. Increasing C18:2 (linoleic acid) content in the diet may lead to changes in milk fat composition and yield, although the exact quantitative relationship is not explicitly given. In general, an increase in polyunsaturated fatty acids (PUFAs) like C18:2 could potentially lead to a decrease in milk fat yield, as observed by changes in other fatty acid concentrations in the study. However, the specific magnitude of the reduction in fat yield per one percent increase in C18:2 is not provided in the text. To accurately determine the expected reduction in fat yield in grams per day when increasing one percent of C18:2, a detailed quantitative analysis or specific data related to the direct impact of C18:2 on fat yield would be required.

Dorea and Armentano (2017) - Response 4: *"For each 1% diet DM inclusion of C18:2, a reduction in milk fatty acids of 95.7 g/d is expected" (extracted from Table 7 of the paper).*

These findings demonstrate the capability of the RAG-enhanced model to effectively retrieve and synthesize domain-specific knowledge. The alignment of the model's responses with the findings of the original meta-analysis by Dorea and Armentano (2017) illustrates the potential of AI-driven approaches for enhancing scientific data retrieval and interpretation in livestock and nutritional sciences. Questions 1 and 2, which were qualitative in nature, yielded responses consistent with the meta-analysis findings. Question 3 also produced results closely aligned with those of the meta-analysis, although the model encountered difficulty specifically addressing item 4 of this question. The model incorrectly indicated that the effects of C18:3 fatty acid supplements were not covered by the provided literature excerpt, despite this information being clearly present within the corpus. This discrepancy highlights the critical influence of prompt design on response quality, emphasizing that the model's performance depends significantly on the prompts used for information retrieval. In this case study, the prompt consisted of the question and a basic context directive, stating {"role": "system", "content": "You are an AI assistant providing responses for scientific questions based on all the papers used in the knowledge base."}. To enhance response accuracy and prevent incorrect assertions such as those experienced in this scenario, the prompt could be refined to provide more detailed instructions. Additionally, adjustments to the model's generation parameters, specifically tuning the "temperature," could further optimize the precision and reliability of generated responses by controlling their randomness and creativity.

In Question 4, the model did not adequately address the query. This question was deliberately quantitative to evaluate the model's capability in summarizing numerical data extracted from the literature. The current model and employed approach were limited in their ability to effectively extract and analyze quantitative information from the source documents utilized in the meta-analysis. While this limitation was anticipated due to the simplicity of the approach and prompts used, it shows an inherent constraint associated with certain models and their implementation. Nevertheless, recent advancements in multimodal RAG techniques represent an area of significant research interest, offering potential for more sophisticated interpretation of tables, figures, and complex quantitative data embedded within textual documents. These advanced methodologies in combination with prompt engineering could significantly enhance data extraction accuracy and efficiency. A notable challenge associated with many advanced models, like multimodal LLMs, is their substantial computational demands, including large model sizes and the corresponding GPU memory requirements for both fine-tuning and inference. Recent efforts have explored smaller-scale models and techniques such as quantization, aimed at reducing computational requirements and facilitating broader deployment.

3.1.2. Case-Study 2: Pupo et al. (2025)

A subsequent case study was conducted utilizing the same open-source model (BAAI/bge-large-en) as described earlier. In this second investigation, we constructed a more detailed and comprehensive prompt specifically designed to improve the extraction and interpretation of quantitative data. This second study utilized a total of 15 articles reviewed by Pupo et al. (2025). The documents were processed into a corpus comprising 497 individual nodes for training. The model was fine-tuned utilizing the LlamaIndex framework, leveraging GPT-3.5 Turbo for generating a total of 993 question-answer pairs. Model training consisted of 10 epochs, each containing 5,000 iterations. Query questions were transformed into embeddings and compared to the corpus using cosine similarity metrics to identify and retrieve the most relevant context. Additionally, a validation corpus of 5 research articles (PDF documents) was processed, resulting in 70 parsed nodes. The study published by Pupo et al. (2025) specifically evaluated the impact of 3-NOP supplementation on methane emissions, dry matter intake, dairy cow performance, and associated income over feed costs.

A more advanced and integrated data extraction and synthesis pipeline was developed for this second case study to enable a more quantitative and contextual interpretation of findings from peer-reviewed articles. Similar to the first case study, all articles were stored in PDF format within a designated directory and dynamically loaded through a recursive search script that identified ".pdf" files across all subdirectories. The textual content of each file was extracted using the PyMuPDF library (version 1.25.5). Files with no readable content were excluded from the pipeline.

Full text from each document was split into individual sentences, and only those containing numerical expressions (e.g., digits, units, decimal values) were retained using regular expression-based filtering. The filtered content was segmented into batches of 5,000 characters to ensure compatibility with the token limit of GPT models. Token counting and truncation were handled using the "tiktoken" library with the "cl100k_base` encoding, optimized for GPT-3.5 Turbo. Each batch was submitted to an Azure OpenAI GPT-3.5 Turbo model with a structured system prompt that instructed the model to extract only numerical results and categorize them under four core topics: (1) Dry Matter Intake, (2) Milk Yield and Composition, (3) Rumen Metabolism, and (4) Methane Production and Emissions. The model returned tabulated outputs with six standardized columns: Topic, Variable, Value, Units, Treatment Name, and Study Reference. Outputs were aggregated and parsed into a unified structured dataset, where duplicate and inconsistent entries were removed, and numeric coercion was applied to ensure data integrity.

To support downstream synthesis and interpretation, each study's structured numerical findings were embedded using the fine-tuned model previously described. A parallel corpus of raw full-text paragraphs was also embedded. For each user query, semantic similarity was computed between the query and all embedded rows (structured numerical and corpus). The top 50 most relevant structured entries were selected based on cosine similarity between their embeddings and the question embedding. Additionally, the most semantically similar unstructured text snippets were retrieved and integrated with the structured context to support contextual analysis.

The final step involved constructing a hybrid prompt consisting of both structured tabular evidence and raw textual excerpts. This composite context was passed to GPT-3.5 Turbo to generate a concise, domain-aware summary. The prompt instructed the model to synthesize the overall direction of the effect across studies, emphasize consistency or variability, and mention outliers only when relevant, while omitting unnecessary statistical, dietary, or methodological detail. The temperature was set to 0.2. The result was a compact expert-level interpretation that directly answered scientific questions, as detailed below:

Question 1: What is the effect of 3-NOP on dry matter intake?

Response 1: Overall, the studies indicate a consistent reduction in dry matter intake (DMI) when cows were fed 3-NOP. This decrease in DMI was observed across multiple studies, with no significant interactions with fat supplementation noted. However, a notable exception was seen in one study where DMI was not affected by 3-NOP. For example, Haisan et al. (2015) reported an increase in DMI, but this was an outlier compared to the general trend observed in other studies.

Pupo et al. (2025) - Response 1: *"Dairy cows fed 3-NOP had 0.5 kg/d lower (P = 0.004; Figure 2) DMI compared with control cows. Dose of 3-NOP was negatively associated with the 3-NOP effect on DMI response."*

Question 2: What is the effect of 3-NOP on milk yield?

Response 2: Overall, the effect of 3-NOP on milk yield across multiple studies shows a consistent decrease in milk yield. Most studies reported a reduction in milk yield when cows were fed 3-NOP, with one notable exception showing a decrease in milk yield when fat supplementation was combined with 3-NOP. For example, Van Gastelen et al. (2020) reported a decrease in milk yield with 3-NOP supplementation, while Vyas et al. (2018) observed a 7.0% decrease in milk yield. Despite some variations in the magnitude of the effect, the general trend of reduced milk yield with 3-NOP supplementation is consistent across the studies.

Pupo et al. (2025) - Response 2: *"Similarly, milk yield was 0.7 kg/d lower (P = 0.01; Figure 3) when cows were supplemented with 3-NOP."*

Question 3: What is the effect of 3-NOP on methane emission?

Response 3: Overall, the studies consistently show that 3-NOP supplementation leads to a reduction in methane emissions in dairy cows. For example, various studies reported reductions from 7% to 60% in methane emissions when cows were fed diets supplemented with 3-NOP. Additionally, the efficacy of 3-NOP in reducing methane production was maintained over different lactation stages, with some variations observed in specific diets. Despite some fluctuations in mitigation potential over time within specific diets, the overall trend indicates a reduction in methane emissions with 3-NOP supplementation across the studies.

Pupo et al. (2025) - Response 3: *"Typically, studies evaluating the effects of 3-NOP supplementation mixed the feed additive into the TMR, and reductions* *in methane emissions were on average 30% with a similar target concentration of 3-NOP in dairy diets as reported in this study."*

Question 4: What is the average reduction on methane emission when 3-NOP is supplemented?

Response 4: The supplementation of 3-nitrooxypropanol (3-NOP) shows a consistent trend towards reducing methane emissions across multiple studies. For instance, studies reported methane reduction percentages ranging from 11% to 42%, with some studies showing specific reductions like 15%, 18%, or 33%. Although there are variations in the exact reduction values reported, the overall direction of the effect is a decrease in methane emissions with 3-NOP supplementation. This trend is further supported by findings showing reductions in methane yield, methane production, and methane emission across different experimental conditions and diets.

Pupo et al. (2025) - Response 4: *"The mixed effects model detected a reduction of 27.9% (P = 0.0001; Figure 8) for CH4 production (g/d) at an average 3-NOP dose of 80.3 mg/kg DM."*

The model accurately responded to the questions posed, using the findings from Pupo et al. (2025) as a reference point. The answers to questions 1, 2, and 3 were consistent with the conclusions presented in that meta-analysis. These initial questions were framed to elicit qualitative interpretations of the effect of 3-NOP on dry matter intake, milk yield, and methane emissions, without requiring the model to compute averages or summarize descriptive statistics. In contrast, question 4 explicitly asked for a numerical estimate of the average reduction in methane emissions. As observed in Case Study 1, the model did not fully resolve the quantitative component of the question. However, following improvements in prompt engineering added to this case study, it was able to provide a range of numerical values. These values aligned well with the results reported by Pupo et al. (2025), who found a 27.9% reduction in methane production at an average 3-NOP dose of 80.3 mg/kg DM. The range provided by the model (11 - 42%) encompassed this value, reflecting a reasonable approximation. While these findings demonstrate the potential of large language models (LLMs) to approximate expert-derived conclusions, it is important to emphasize that the outputs are not derived from rigorous statistical analysis. Unlike formal meta-analytical approaches, which incorporate factors such as study design, standard error, sample size, covariance structures, random effects, among others, the LLM operates without structured weighting or model-based inference. Thus, the examples presented here are not intended to suggest that LLMs replace meta-analytical techniques. Rather, they illustrate how LLMs, when paired with a curated knowledge base can approximate conclusions typically reached through statistically rigorous synthesis. As LLM technology evolves, it is anticipated that these models will increasingly mirror the outcomes of analytical frameworks like meta-analysis, especially when provided with well-structured, domain-specific data.

The objective of these two case studies is to illustrate the potential impact of large language models (LLMs) on the field of livestock management, encompassing domains such as animal nutrition, reproduction, physiology, and beyond. We foresee a fundamental shift in the roles of technical consulting and on-farm decision-making as end-users, including farmers, industry professionals, and students, become increasingly familiar with and gain easier access to these AI-driven tools. Given the widespread availability of open-source LLMs, the abundance of domain-relevant text datasets, and the growing accessibility of computational resources, the development and deployment of these technologies is less a question of feasibility and more one of strategic vision and community initiative. Within this evolving landscape, artificial intelligence has the potential to reshape the livestock sector profoundly over the coming decade.

4. Future Directions

One of the most promising areas of development is open-set cattle identification, which ensures accurate and scalable animal recognition as herds grow and change. While depth imaging has demonstrated potential for identifying individual animals within uniform-colored breeds, further research is needed to develop long-term identification strategies, particularly for growing animals. Additionally, transitioning from subjective health assessments to AI-based objective monitoring will be transformative. Computer vision techniques can analyze quantitative traits such as body shape, gait patterns, and movement metrics to enhance animal health monitoring, reducing reliance on subjective evaluations such as body condition and locomotion scoring.

The use of LLMs in dairy farming holds vast potential by enabling seamless integration of structured and unstructured data sources. These models can facilitate real-time decision-making, improve operational efficiency, and serve as virtual farm assistants, translating complex scientific insights into actionable recommendations for farmers and industry stakeholders. Multimodal AI models, which process images, text, and tabular data simultaneously, could transform data-driven decision-making, particularly in areas such as feed optimization, disease prevention, and precision nutrition. Education and multidisciplinary collaboration are key to ensuring that AI innovations are successfully implemented in commercial dairy operations. Academic and industry partnerships can drive the development of user-friendly, cost-effective AI solutions while ensuring that farmers and industry professionals are equipped with the necessary knowledge to adopt these technologies effectively. Furthermore, increased funding in digital agriculture is essential for accelerating innovation and fostering entrepreneurship in livestock technology.

References

Allen, A., Golden, B., Taylor, M., Patterson, D., Henriksen, D., & Skuce, R. (2008). Evaluation of retinal imaging technology for the biometric identification of bovine animals in Northern Ireland. Livestock Science, 116(1), 42–52. https://doi.org/10.1016/j.livsci.2007.08.018

Andrew, W., Greatwood, C., & Burghardt, T. (2017). Visual localisation and individual identification of Holstein Friesian cattle via deep learning. 2017 IEEE International Conference on Computer Vision Workshops (ICCVW), 2850–2859. https://doi.org/10.1109/ICCVW.2017.336

Andrew, W., Gao, J., Mullan, S., Campbell, N., Dowsey, A. W., & Burghardt, T. (2021). Visual identification of individual Holstein-Friesian cattle via deep metric learning. Computers and Electronics in Agriculture, 185, 106133. https://doi.org/10.1016/j.compag.2021.106133

Barletta, R. V., Maturana Filho, M., Carvalho, P. D., Del Valle, T. A., Netto, A. S., Rennó, F. P., & Wiltbank, M. C. (2017). Association of changes among body condition score during the transition period with NEFA and BHBA concentrations, milk production, fertility, and health of Holstein cows. Theriogenology, 104, 30–36. https://doi.org/10.1016/j.theriogenology.2017.07.030

Bergamini, L., Porrello, A., Dondona, A. C., Del Negro, E., Mattioli, M., D'Alterio, N., & Calderara, S. (2018). Multi-views embedding for cattle re-identification. 2018 14th International Conference on Signal-Image Technology & Internet-Based Systems (SITIS), 184–191. https://doi.org/10.1109/SITIS.2018.00036

Bezen, R., Edan, Y., & Halachmi, I. (2020). Computer vision system for measuring individual cow feed intake using RGB-D camera and deep learning algorithms. Computers and Electronics in Agriculture, 172, 105345. https://doi.org/10.1016/j.compag.2020.105345

Bhole, A., Falzon, O., Biehl, M., & Azzopardi, G. (2019). A computer vision pipeline that uses thermal and RGB images for the recognition of Holstein cattle. In M. Vento & G. Percannella (Eds.), Computer Analysis of Images and Patterns (pp. 108–119). Springer International Publishing.

Bresolin, T., & Dórea, J. R. R. (2020). Infrared spectrometry as a high-throughput phenotyping technology to predict complex traits in livestock systems. Frontiers in Genetics, 11. https://doi.org/10.3389/fgene.2020.00923

Brown, T. B., Mann, B., Ryder, N., Subbiah, M., Kaplan, J., Dhariwal, P., Neelakantan, A., Shyam, P., Sastry, G., Askell, A., Agarwal, S., Herbert-Voss, A., Krueger, G., Henighan, T., Child, R., Ramesh, A., Ziegler, D. M., Wu, J., Winter, C., Hesse, C., Chen, M., Sigler, E., Litwin, M., Gray, S., Chess, B., Clark, J.,

Berner, C., McCandlish, S., Radford, A., Sutskever, I., & Amodei, D. (2020). Language models are few-shot learners. In Advances in Neural Information Processing Systems (Vol. 33, pp. 1877–1901). Curran Associates, Inc. https://proceedings.neurips.cc/paper_files/paper/2020/file/1457c0d6bfcb4967418bfb8ac142f64a-Paper.pdf

Cabrera, V. E., Barrientos-Blanco, J. A., Delgado, H., & Fadul-Pacheco, L. (2020). Symposium review: Real-time continuous decision making using big data on dairy farms. Journal of Dairy Science, 103(4), 3856–3866. https://doi.org/10.3168/jds.2019-17145
Caffarini, J. G., Bresolin, T., & Dórea, J. R. R. (2022). Predicting ribeye area and circularity in live calves through 3D image analyses of body surface. Journal of Animal Science, 100(9), skac242. https://doi.org/10.1093/jas/skac242

Chapa, J. M., Maschat, K., Iwersen, M., Baumgartner, J., & Drillich, M. (2020). Accelerometer systems as tools for health and welfare assessment in cattle and pigs – A review. Behavioural Processes, 181, 104262. https://doi.org/10.1016/j.beproc.2020.104262

Cominotte, A., Fernandes, A. F. A., Dórea, J. R. R., Rosa, G. J. M., Ladeira, M. M., van Cleef, E. H. C. B., & Machado Neto, O. R. (2020). Automated computer vision system to predict body weight and average daily gain in beef cattle during growing and finishing phases. Livestock Science, 232, 103904. https://doi.org/10.1016/j.livsci.2019.103904

Devlin, J. (2018). BERT: Pre-training of deep bidirectional transformers for language understanding. arXiv preprint arXiv:1810.04805.

Dórea, J. R. R., and Armentano, L. E. (2017). Effects of common dietary fatty acids on milk yield and concentrations of fat and fatty acids in dairy cattle. Animal Production Science, 57(11), 2224–2236. https://doi.org/10.1071/AN17335

Dosovitskiy, A., Beyer, L., Kolesnikov, A., Weissenborn, D., Zhai, X., Unterthiner, T., Dehghani, M., Minderer, M., Heigold, G., Gelly, S., Uszkoreit, J., & Houlsby, N. (2021). An image is worth 16x16 words: Transformers for image recognition at scale. arXiv preprint arXiv:2010.11929.

Evans, D. G. (1978). The interpretation and analysis of subjective body condition scores. Animal Science, 26(2), 119–125. https://doi.org/10.1017/S0003356100039520

Ferreira, R. E. P., Bresolin, T., Rosa, G. J. M., & Dórea, J. R. R. (2022). Using dorsal surface for individual identification of dairy calves through 3D deep learning algorithms. Computers and Electronics in Agriculture, 201, 107272. https://doi.org/10.1016/j.compag.2022.107272

Ferreira, R. E. P., Lee, Y. J., & Dórea, J. R. R. (2023). Using pseudo-labeling to improve performance of deep neural networks for animal identification. Scientific Reports, 13(1), 13875.

Ferreira, R. E. P., de Luis Balaguer, M. A., Bresolin, T., Chandra, R., Rosa, G. J. M., White, H. M., & Dórea, J. R. R. (2024). Multi-modal machine learning for the early detection of metabolic disorder in dairy cows using a cloud computing framework. Computers and Electronics in Agriculture, 227, 109563. https://doi.org/10.1016/j.compag.2024.109563

Higaki, S., Menezes, G. L., Ferreira, R. E. P., Negreiro, A., Cabrera, V. E., & Dórea, J. R. R. (2024). Objective dairy cow mobility analysis and scoring system using computer vision-based keypoint detection technique from top-view 2D videos. Journal of Dairy Science. https://doi.org/10.3168/jds.2024-25545

Hossain, M. E., Kabir, M. A., Zheng, L., Swain, D. L., McGrath, S., & Medway, J. (2022). A systematic review of machine learning techniques for cattle identification: Datasets, methods and future directions. Artificial Intelligence in Agriculture, 6, 138–155. https://doi.org/10.1016/j.aiia.2022.09.002

Hu, H., Dai, B., Shen, W., Wei, X., Sun, J., Li, R., & Zhang, Y. (2020). Cow identification based on fusion of deep parts features. Biosystems Engineering, 192, 245–256. https://doi.org/10.1016/j.biosystemseng.2020.02.001

Ilestrand, M. (2017). Automatic eartag recognition on dairy cows in real barn environment. Linköping University, Computer Vision.

Johnston, A. M., & Edwards, D. S. (1996). Welfare implications of identification of cattle by ear tags. Veterinary Record, 138(25), 612–614. https://doi.org/10.1136/vr.138.25.612

Haisan, J., Sun, Y., Guan, L. L., Beauchemin, K. A., Iwaasa, A., Duval, S., Barreda, D. R., & Oba, M. (2014). The effects of feeding 3-nitrooxypropanol on methane emissions and productivity of Holstein cows in mid lactation. Journal of Dairy Science, 97(5), 3110–3119. https://doi.org/10.3168/jds.2013-7834

Lewis, P., Perez, E., Piktus, A., Petroni, F., Karpukhin, V., Goyal, N., Küttler, H., Lewis, M., Yih, W., Rocktäschel, T., Riedel, S., & Kiela, D. (2020). Retrieval-augmented generation for knowledge-intensive NLP tasks. In Advances in Neural Information Processing Systems (Vol. 33, pp. 9459–9474). https://proceedings.neurips.cc/paper/2020/hash/6b493230205f780e1bc26945df7481e5-Abstract.html

Manzanilla-Pech, C. I. V., Stephansen, R. B., & Lassen, J. (2023). Genetic parameters for feed intake and body weight in dairy cattle using high-throughput 3-

dimensional cameras in Danish commercial farms. Journal of Dairy Science, 106(12), 9006–9015. https://doi.org/10.3168/jds.2023-23405

Phyo, C. N., Zin, T. T., Hama, H., & Kobayashi, I. (2018). A hybrid rolling skew histogram-neural network approach to dairy cow identification system. 2018 International Conference on Image and Vision Computing New Zealand (IVCNZ), 1–5. https://doi.org/10.1109/IVCNZ.2018.8634739

Pupo, M. R., Ferraretto, L. F., & Nicholson, C. F. (2025). Effects of feeding 3-nitrooxypropanol for methane emissions reduction on income over feed costs in the United States. Journal of Dairy Science. https://doi.org/10.3168/jds.2024-25502

Qiao, Y., Su, D., Kong, H., Sukkarieh, S., Lomax, S., & Clark, C. (2019). Individual cattle identification using a deep learning based framework. IFAC-PapersOnLine, 52(30), 318–323. https://doi.org/10.1016/j.ifacol.2019.12.558

Qiao, Y., Su, D., Kong, H., Sukkarieh, S., Lomax, S., & Clark, C. (2020). BiLSTM-based individual cattle identification for automated precision livestock farming. 2020 IEEE 16th International Conference on Automation Science and Engineering (CASE), 967–972. https://doi.org/10.1109/CASE48305.2020.9217026

Qiao, Y., Kong, H., Clark, C., Lomax, S., Su, D., Eiffert, S., & Sukkarieh, S. (2021). Intelligent perception for cattle monitoring: A review for cattle identification, body condition score evaluation, and weight estimation. Computers and Electronics in Agriculture, 185, 106143. https://doi.org/10.1016/j.compag.2021.106143

Radford, A., Kim, J. W., Hallacy, C., Ramesh, A., Goh, G., Agarwal, S., Sastry, G., Askell, A., Mishkin, P., Clark, J., Krueger, G., & Sutskever, I. (2021). Learning transferable visual models from natural language supervision. In Proceedings of the 38th International Conference on Machine Learning (Vol. 139, pp. 8748–8763). https://proceedings.mlr.press/v139/radford21a.html

Rumelhart, D. E., Hinton, G. E., & Williams, R. J. (1986). Learning internal representations by error propagation, parallel distributed processing, explorations in the microstructure of cognition, ed. D. E. Rumelhart & J. McClelland (Vol. 1). Biometrika, 71, 599–607.

Sharma, A., Randewich, L., Andrew, W., Hannuna, S., Campbell, N., Mullan, S., Dowsey, A. W., Smith, M., Hansen, M., & Burghardt, T. (2025). Universal bovine identification via depth data and deep metric learning. Computers and Electronics in Agriculture, 229, 109657. https://doi.org/10.1016/j.compag.2024.109657

Shen, W., Hu, H., Dai, B., Wei, X., Sun, J., Jiang, L., & Sun, Y. (2020). Individual identification of dairy cows based on convolutional neural networks. Multimedia Tools and Applications, 79(21), 14711–14724. https://doi.org/10.1007/s11042-019-7344-7

Smink, M., Liu, H., Döpfer, D., & Lee, Y. J. (2024). Computer vision on the edge: Individual cattle identification in real-time with ReadMyCow system. Proceedings of the IEEE/CVF Winter Conference on Applications of Computer Vision (WACV), 7056–7065.

Sun, Y., Wang, X., & Tang, X. (2013). Deep convolutional network cascade for facial point detection. Proceedings of the IEEE Conference on Computer Vision and Pattern Recognition, 3476–3483.

Tsai, D.-M., & Huang, C.-Y. (2014). A motion and image analysis method for automatic detection of estrus and mating behavior in cattle. Computers and Electronics in Agriculture, 104, 25–31. https://doi.org/10.1016/j.compag.2014.03.00

Van Gastelen, S., Dijkstra, J., Binnendijk, G., Duval, S. M., Heck, J. M. L., Kindermann, M., Zandstra, T., & Bannink, A. (2020). 3-Nitrooxypropanol decreases methane emissions and increases hydrogen emissions of early lactation dairy cows, with associated changes in nutrient digestibility and energy metabolism. Journal of Dairy Science, 103(9), 8074–8093. https://doi.org/10.3168/jds.2019-17936

van der Voort, M., Jensen, D., Kamphuis, C., Athanasiadis, I. N., De Vries, A., & Hogeveen, H. (2021). Invited review: Toward a common language in data-driven mastitis detection research. Journal of Dairy Science, 104(10), 10449–10461. https://doi.org/10.3168/jds.2021-20311

Vaswani, A., Shazeer, N., Parmar, N., Uszkoreit, J., Jones, L., Gomez, A. N., Kaiser, Ł. ukasz, & Polosukhin, I. (2017). Attention is all you need. In Advances in Neural Information Processing Systems (Vol. 30). https://proceedings.neurips.cc/paper/2017/hash/3f5ee243547dee91fbd053c1c4a845aa Abstract.html

Velez, J. F., Sanchez, A., Sanchez, J., & Esteban, J. L. (2013). Beef identification in industrial slaughterhouses using machine vision techniques. Spanish Journal of Agricultural Research, 11(4), 945–957.

Vyas, D., McGinn, S. M., Duval, S. M., Kindermann, M. K., & Beauchemin, K. A. (2018). Optimal dose of 3-nitrooxypropanol for decreasing enteric methane emissions from beef cattle fed high-forage and high-grain diets. Animal Production Science, 58(6), 1049–1055. https://doi.org/10.1071/AN15705

Wang, B., Li, X., An, X., Duan, W., Wang, Y., Wang, D., & Qi, J. (2024). Open-set recognition of individual cows based on spatial feature transformation and metric learning. Animals, 14(8). https://doi.org/10.3390/ani14081175

Wu, Y., Guo, H., Li, Z., Ma, Q., Zhao, Y., & Pezzuolo, A. (2021). Body condition score for dairy cows method based on vision transformer. 2021 IEEE International

Workshop on Metrology for Agriculture and Forestry (MetroAgriFor), 37–41. https://doi.org/10.1109/MetroAgriFor52389.2021.9628437

Zhu, J., Lacroix, R., & Wade, K. M. (2023). Automated extraction of domain knowledge in the dairy industry. Computers and Electronics in Agriculture, 214, 108330. https://doi.org/10.1016/j.compag.2023.108330

Zin, T. T., Misawa, S., Pwint, M. Z., Thant, S., Seint, P. T., Sumi, K., & Yoshida, K. (2020). Cow identification system using ear tag recognition. 2020 IEEE 2nd Global Conference on Life Sciences and Technologies (LifeTech), 65–66. https://doi.org/10.1109/LifeTech48969.2020.1570625232

Zin, T. T., Phyo, C. N., Tin, P., Hama, H., & Kobayashi, I. (2018). Image technology based cow identification system using deep learning. Proceedings of the International Multiconference of Engineers and Computer Scientists, 1, 236–247.

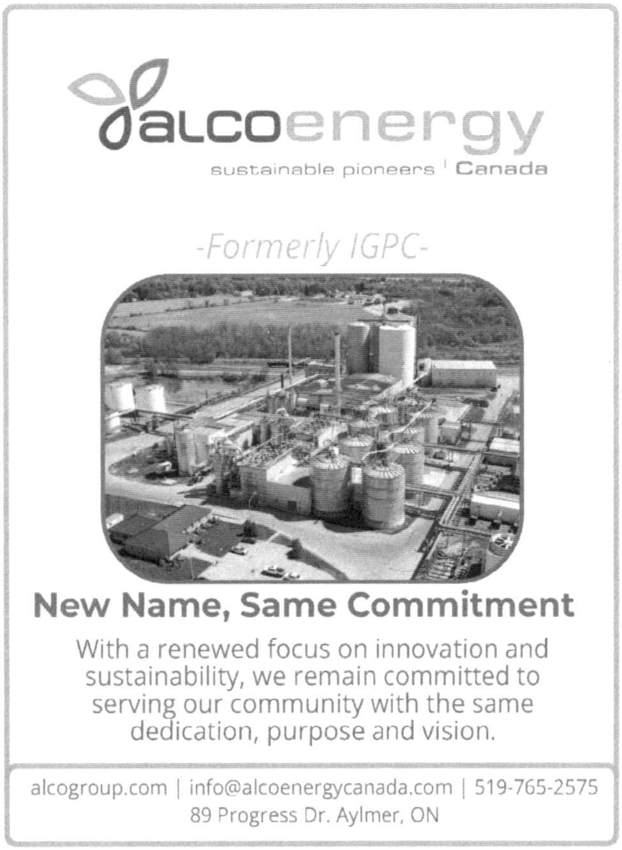
alcoenergy
sustainable pioneers | Canada
-Formerly IGPC-
New Name, Same Commitment
With a renewed focus on innovation and sustainability, we remain committed to serving our community with the same dedication, purpose and vision.
alcogroup.com | info@alcoenergycanada.com | 519-765-2575
89 Progress Dr. Aylmer, ON

PARTNERING IN
MICROBIAL SOLUTIONS FOR
A CHANGING WORLD
Lallemand Animal Nutrition provides a full range of specific, natural microbial products, services and solutions that improve animal well-being and performance while respecting animals, people and the environment.
CONNECT WITH US!
LALLEMAND ANIMAL NUTRITION
www.lallemandanimalnutrition.com
LALLEMAND

Research proven nutrition
to combat rising ingredient costs
Enzymes
Amino Acids
Organic Acids
Natural Betaine
Vitamin E
Direct-Fed Microbials
Medium-Chain Fatty Acids
Organic Trace Minerals
Bypass Soybean Meal
DCAD
HALCHEMIX
www.halchemix.ca
800-540-4756

Canadian Roots. Global Impact.
Bringing Nutrition Science to the Farmgate.
Cbs Bio Platforms
www.cbsbioplatforms.com

ANCC

2025 ANIMAL NUTRITION CONFERENCE OF CANADA

CNAC

2025 COLLOQUE DE NUTRITION ANIMALE DU CANADA

Graduate Student Posters
Affiches des étudiants

Sponsored by/Parrainé par: trouw nutrition

a Nutreco company

Graduate Student Posters
Les affiches des étudiants de cycle supérieur

Visit the Graduate Student Poster Exhibition to learn from the next generation of livestock nutritionists. Posters will be on display in the Fallsview Studio ABC throughout the ANCC. You can also view each student's 3-minute video abstract by visiting the conference website or using the QR code.

Veuillez visiter l'exposition d'affiches pour étudiants de cycle supérieur pour apprendre de la prochaine génération de nutritionnistes du bétail. Les affiches seront exposées dans le Fallsview Studio ABC pendant toute la durée du CNAC. Vous pouvez également visionner le résumé vidéo de 3-minutes (en anglais) de chaque étudiant sur le site web du colloque ou en utilisant le code QR.

Monogastric posters / *Les affiches sur les monogastriques*

M1 – Prince Agyemang, McGill University
Towards net-zero broiler production: An assessment of the potential of commercial protease to reduce environmental impact of broiler production in Canada
Vers une production de poulets à chair sans impact sur l'environnement : une évaluation du potentiel de la protéase commerciale pour réduire l'impact environnemental de la production de poulets à chair au Canada

M2 - Reza Bakhshalinejad, University of Guelph
Effects of feeding hydroxychloride or organic sources of zinc, copper, and manganese to broiler breeder hens on bone characteristics of offsprings when fed hydroxychloride or inorganic sources of the same minerals
Effets de l'alimentation des poules reproductrices de poulets à chair avec de l'hydroxychlorure ou des sources organiques de zinc, de cuivre et de manganèse sur les caractéristiques osseuses de leur progéniture lorsqu'elles sont nourries avec de l'hydroxychlorure ou des sources inorganiques des mêmes minéraux

M3 – Anderson Borba, University of Saskatchewan
A lysine dose-response study using dietary levels established based on individual pig requirements
Une étude dose-réponse de la lysine utilisant des niveaux alimentaires établis en fonction des besoins individuels des porcs

M4 – Laurence Cheng, University of Guelph
A systematic review and meta-analysis of efficacy of exogenous fibre enzymes in modulating growth performances and nutrient digestibility in weanling pigs
Une revue systématique et une méta-analyse de l'efficacité des enzymes de fibres exogènes dans la modulation des performances de croissance et de la digestibilité des nutriments chez les porcelets sevrés

M5 – Atul Jadhav, McGill University
Feeding optimized chelated minerals doses (Zn, Fe, & Mn) enhances embryonic growth, nutrient absorption, and regulates immune responses by balancing inflammation and antioxidant activity in chicken embryos.
L'apport de doses optimisées de minéraux chélatés (Zn, Fe et Mn) améliore la croissance embryonnaire et l'absorption des nutriments, et régule les réponses immunitaires en équilibrant l'inflammation et l'activité antioxydante chez les embryons de poulet.

M6 – Kemi Kanami, University of Manitoba
Assessing the pre- and probiotic effects of bacterial cellulose on laying hen's gut microbiome and egg production
Évaluation des effets prébiotiques et probiotiques de la cellulose bactérienne sur le microbiome intestinal des poules pondeuses et la production d'œufs.

M7 – Spencer Leuschner, University of Guelph
Pre-exposure of Zebrafish (Danio rerio) larvae and ZF-4 fibroblast cells to vitamin D (VD3) increases toxicity to bacterial lipopolysaccharide (LPS)
La préexposition des larves de poisson zèbre (Danio rerio) et des cellules de fibroblastes ZF-4 à la vitamine D (VD3) augmente la toxicité du lipopolysaccharide bactérien (LPS)

M8 – Lonita Manoa, University of Guelph
Egg production and quality in mid-lay hens fed varying supplementation levels of American cranberry (Vaccinium macrocarpon) pomaces.
Production et qualité des œufs chez les poules pondeuses à mi- ponte nourries avec des niveaux variables de suppléments de grignons de canneberge américaine (Vaccinium macrocarpon).

M9 – Sylvester Mutinda, University of Guelph
Significance of low crude protein diets fortified with amino acids in broiler chickens feeding program–a meta-analysis
Importance des régimes pauvres en protéines brutes enrichis en acides aminés dans le programme d'alimentation des poulets de chair - une méta-analyse

M10 – Poonam Neupane, University of Manitoba
Utilizing exogenous enzyme technology to produce novel bioactive fiber components and enhance the value of feed ingredients for poultry
Utilisation de la technologie des enzymes exogènes pour produire de nouveaux composants de fibres bioactives et améliorer la valeur des ingrédients alimentaires pour la volaille

M11 – Felix Njeri, University of Guelph
Thermomechanical and enzyme-facilitated processing of soybean meal enhanced amino acid digestibility in broiler chickens
Le traitement thermomécanique et enzymatique du tourteau de soja améliore la digestibilité des acides aminés chez les poulets de chair

M12 – Melissa Parent, University of Guelph
Biochar as a dietary treatment for the prevention of boar taint
Le biochar comme traitement diététique pour la prévention de l'odeur de verrat

M13 – Italo Pereira, University of Saskatchewan
Effect of microencapsulated multivitamin and mineral premixes on growth performance, plasma vitamin and mineral levels in broiler chickens challenged with coccidiosis
Effet des prémélanges multivitaminés et minéraux microencapsulés sur la croissance et les taux plasmatiques de vitamines et de minéraux chez les poulets de chair atteints de coccidios

M14 – Guyue Renke, University of Guelph
Assessment of the protective effects of Reishi, Laminarin, and Barley β-Glucans for Zebrafish (Danio rerio) Larvae immune challenged with bacterial lipopolysaccharide (LPS)
Des effets protecteurs du reishi, de la laminarine et des β-glucanes d'orge sur les larves de poisson zèbre (Danio rerio) immunodéprimées par le lipopolysaccharide bactérien (LPS)

M15 – Alexander Wickhorst, University of Guelph
Enzymatic hydrolyses of yeast enhanced its nitrogen solubilization and kinetics in a simulated pig digestion model
Les hydrolyses enzymatiques de levure ont amélioré sa solubilisation de l'azote et sa cinétique dans un modèle de digestion porcine simulée

M15 – Mingli Xu, University of Guelph
Meta-analyses of copper digestive utilization with phytase in weanling pigs
Méta-analyses de l'utilisation digestive du cuivre avec la phytase dans l'évaluation, chez les porcelets sevrés

Ruminant posters / *Les affiches sur les ruminants*

R1 – Emma Fehr, University of Saskatchewan
The effect of feed sulfur and water sulfate on ruminal hydrogen sulfide concentrations and gastrointestinal tract permeability in beef heifers
Effet de l'alimentation en soufre et en sulfate d'eau sur les concentrations d'hydrogène sulfuré dans le rumen et la perméabilité du tractus gastro-intestinal chez les génisses de boucherie

R2 – Jacob Foster, Dalhousie University
Assessing the impact of increasing LinPRO-R dosage on the rumen microbiome dynamics and methane production in vitro
Évaluation de l'impact de l'augmentation de la dose de LinPRO-R sur la dynamique du microbiome du rumen et la production de méthane in vitro

R3 – Weiguang Hao, University of Alberta
Effects of stocking rate on activity behavior and growth performance of beef heifers grazing tame pasture
Effets du taux de chargement sur l'activité, le comportement et la croissance des génisses de boucherie en pâturage sur des prairies artificielles

R4 – Rogelio Ramos-Tatiane, University of Lethbridge
Effect of a seaweed blend on growth performance, feeding behaviour, and ruminal fermentation in heifers fed backgrounding and finishing feedlot diets
Effet d'un mélange d'algues sur la croissance, le comportement alimentaire et la fermentation ruminale chez les génisses nourries avec des régimes de semi-finition et de finition en parc d'engraissement

R5– Tatum Schooley, University of Guelph
Reduce, reuse, recycle: Using bentonite to de-water milk processing byproducts for dairy cow diets
Réduire, réutiliser, recycler : Utilisation de la bentonite pour déshydrater les sous-produits de la transformation du lait destinés à l'alimentation des vaches laitières

SPONSORED BY / COMMANDITÉ PAR

trouw nutrition
a Nutreco company

Towards net-zero broiler production: An assessment of the potential of a commercial protease to reduce the environmental impact of broiler production in Canada

Vers une production de poulets de chair nette zéro : Évaluation du potentiel d'une protéase exogène commerciale pour réduire l'impact environnemental de la production de poulets de chair au Canada

Prince Agyemang[1], Ebenezer M. Kwofie[1], Ludovic Lahaye[2], Melissa Otis[2], Émilie Fontaine[3], Glenmer Tactacan[3], Marcia de Souza Viera[3] and Greg Thoma[3]

[1]Department of Bioresource Engineering, McGill University, 21111 Lakeshore, St-Anne-de-Bellevue, QC

[2]Jefo Nutrition Inc, 5020 Jefo Avenue, CP 325, St Hyacinthe, QC

[3]AgNext, Colorado State University, 350 West Pitkin Street, Fort Collins, CO 80523

prince.agyemang@mail.mcgill.ca

Abstract

The objective of the current study was to leverage life cycle assessment and meta-analytical data on trials conducted worldwide to quantify the impacts of using an exogenous protease in broiler feed in Eastern and Western Canadian production value chains. The assessment was defined using five impact categories (climate change, fossil resource extraction, agricultural land use, eutrophication, and acidification potential) following the Livestock Environmental Assessment Performance guidelines and ISO 14040/44 protocols. The results showed a reduction between 1.6–3.8% per kg cumulative feed across the entire broiler life cycle (1–42 days) in all impact categories when the exogenous protease was used to reformulate standard diets. At the farm stage, the estimated climate change impact scores for broilers fed standard diets ranged between 1970–2223 g CO2-eq/kg liveweight broiler chicken. Production scenarios where broiler chickens were fed reformulated diets showed reductions in climate change impact between 6.0–6.4 % per kg liveweight. Other impact categories reduced between 7.4–8.2%, mainly due to decreased land use changes related to soybean, improvement in performance (~3% higher liveweight), lower nitrogen in excreta (-2.7%) and lower phosphorus run-off (-4%), when compared to the standard production system. Suppose this exogenous protease becomes a primary ingredient supplemented at a rate of 125 g/t of complete feed supplied to all broiler chickens in Canada, it could offset an estimated 159,000–170,000 tons of CO2-eq emissions. Supplementing broiler diets with exogenous protease presents a viable strategy to improve productivity and reduce cost while moving towards sustainable broiler chicken production.

SPONSORED BY / COMMANDITÉ PAR

trouw nutrition

a Nutreco company

Effects of feeding hydroxychloride or organic sources of zinc, copper, and manganese to broiler breeder hens on bone characteristics of offsprings when fed hydroxychloride or inorganic sources of the same minerals

Effets de l'alimentation des poules reproductrices de poulets à chair avec de l'hydroxychlorure ou des sources organiques de zinc, de cuivre et de manganèse sur les caractéristiques osseuses de leur progéniture lorsqu'elles sont nourries avec de l'hydroxychlorure ou des sources inorganiques des mêmes minéraux

Reza Bakhshalinejad[1], Stephanie Torrey[2], Elijah G. Kiarie[1]*
[1]Department of Animal Bioscience, University of Guelph, Guelph, ON, Canada
[2]Trouw Nutrition North America, Puslinch, ON, Canada
**rbakhsha@uoguelph.ca*

Abstract

This study evaluated the effects of feeding hydroxychloride (HC) or organic (OR) sources of zinc (Zn), copper (Cu), and manganese (Mn) to broiler breeder hens (BB) on bone characteristics in offspring at 21 and 42 days of age (DOA). Ross 708 BB were assigned to two diets: (1) HO, a blend of 80% HC and 20% OR, and (2) OR, 100% OR sources. Breeders were housed in 24 pens (17♀ and 2♂ per pen) and 12 replicate pens per diet. Hatching eggs were collected at 46, 54, and 62 weeks of age, incubated, and hatched chicks sexed, weighed, and placed in 24 sex-segregated pens (40 birds/pen) while maintaining breeder diet identification. Offspring from BB fed either HO or OR diets were further assigned to receive either HC or inorganic (InO) TM sources, resulting in four offspring dietary treatments ($n = 6$; three pens per sex). Tibia and femur dry matter and ash content, along with bone breaking strength, were assessed at 21 and 42 DOA. Results indicated that maternal diet and age, chick diet and sex did not significantly influence tibia and femur dry matter, ash content, or breaking strength at 21 DOA ($P \geq 0.05$). However, maternal age significantly affected tibia and femoral breaking strength at 42 DOA ($P \leq 0.05$), which 62-week-old BB exhibiting greater strength. Additionally, chick diet influenced tibia and femur ash content, with HC supplementation yielding higher ash content ($P \leq 0.05$). These findings underscore the influence of maternal age and TM sources on broiler bone quality, with potential implications for optimizing mineral nutrition strategies in poultry production.

SPONSORED BY / COMMANDITÉ PAR

A lysine dose-response study using dietary levels established based on individual pig requirements

Une étude dose-réponse de la lysine utilisant des niveaux alimentaires établis en fonction des besoins individuels des porcs

A. Borba[1,2]; E. B. Bosso[1]; D. A. Marçal[1]; A. M. Veira[1]; E. O. Frank[1]; L. G. Barbosa[1]; S. J. Sardinha[1]; S. V. Ferreira[3]; D. A. Columbus[2]; L. Hauschild[1]

[1]São Paulo State University (Unesp), Faculty of Agricultural and Veterinary Sciences, Jaboticabal, São Paulo, Brazil

[2]Department of Animal and Poultry Science, University of Saskatchewan, Saskatoon, SK

[3]DanBred Brazil, Passos de Minas, Minas Gerais, Brazil anderson.borba@unesp.br

Abstract

The estimation of nutritional requirements for swine is directly related to genetic potential and variability between animals. However, there is a lack of studies that have used data from the actual population to establish the range of levels to be evaluated in the estimation of the optimal level. Therefore, the objective of this study was to estimate the optimal level of SID lysine for entire male pigs from 50 to 75 kg using levels established based on individual requirements estimate in real time. Ninety entire male pigs (DanBred Hybrid × Duroc DanBred), with an initial weight of 50.78 ± 1.01 kg were distributed based on initial weight into six treatments, with fifteen replications. A factorial equation using predicted feed intake and weight gain was used to estimate the individual daily lysine requirement at the beginning of the study. Based on the level that corresponded to the 50th percentile of the population (0.94%), six dietary SID lysine levels were set (0.52, 0.66, 0.80, 0.94, 1.08 and 1.22%). Pigs were housed collectively in a climate-controlled barn equipped with 6 automatic and intelligent precision feeders that provided the diet assigned to each animal. Pigs were weighed individually, and body composition was measured. The ideal lysine levels for average daily gain, feed-to-gain ratio, and protein deposition were estimated using quadratic, linear broken line, and quadratic broken line models. Based on the best fitted model, the ideal dietary SID lysine levels for average daily gain, feed-to-gain ratio, protein deposition were 0.95, 1.09, and 0.98%, respectively.

SPONSORED BY / COMMANDITÉ PAR

trouw nutrition

a Nutreco company

A systematic review and meta-analysis of efficacy of exogenous fibre enzymes in modulating growth performances and nutrient digestibility in weanling pigs

Une revue systématique et une méta-analyse de l'efficacité des enzymes de fibres exogènes dans la modulation des performances de croissance et de la digestibilité des nutriments chez les porcelets sevrés

Laurence Cheng[1], Jiali Chen[2], Min Wang[3], Ming Z. Fan[4]
[1]Ph.D. Student, Department of Animal Biosciences, University of Guelph, Guelph, ON
[2]Post-doctoral Scholar, Department of Animal Biosciences, University of Guelph, Guelph, ON
[3]Technician, Department of Animal Biosciences, University of Guelph, Guelph, ON
[4]Professor, Department of Animal Biosciences, University of Guelph, Guelph, ON
lcheng@uoguelph.ca

Abstract

Dietary fiber is an anti-nutritive factor that reduces nutrient digestibility, profitability and contributes to greenhouse gas emissions in swine production. Exogenous fiber enzymes have been used in the swine industry for over two decades, but their efficacy in modulating growth performance and nutrient digestibility in weanling pigs remains inconsistent. We hypothesized no significant differences in the *in vivo* efficacy endpoints in responses to exogenous fibre enzyme supplementation. This study conducted a meta-analysis of 35 studies (2003–2024) to examine the effects of exogenous fiber enzyme supplementation on growth performance and nutrient digestibility in weanling pigs fed commercial diets containing oil meals, cereal grains, by-products and co-products. Results showed no differences ($P>0.05$) between non-enzyme and enzyme-supplemented groups in initial body weight (7.91±0.35 vs. 7.68±0.28 kg) and dietary contents (as-fed) of net energy (2463±63 vs. 2462±63 kcal/kg), crude fat (7.98±2.72 vs. 8.12±2.73%), neutral-detergent fiber (NDF) (12.16±0.77 vs. 13.14±0.85%) and lignin (0.82±0.23 vs. 0.84±0.28%). Supplementation of exogenous fiber enzymes of cellulases and hemi-cellulases in single and/or in combination had no effects ($P>0.05$) on growth performance indicators of average daily gain, feed intake, and gain-to-feed ratio. Apparent total tract digestibility of dry matter (82.8±0.8 vs. 83.3±0.9%), NDF (53.5±2.3 vs. 55.6±2.5%) and crude fat (69.2±8.8 vs. 70.3±9.3%) remained unchanged ($P>0.05$). Our meta-analysis results suggest that currently available commercial exogenous fiber enzymes have limited efficacy for improving growth performances and nutrient digestibility in weanling pigs. More efficacious exogenous fiber enzymes or their modifications need to be further developed to improve profitability and sustainability in swine production. `

Feeding optimized chelated minerals doses (Zn, Fe, & Mn) enhances embryonic growth, nutrient absorption, and regulates immune responses by balancing inflammation and antioxidant activity in chicken embryos

L'apport de doses optimisées de minéraux chélatés (Zn, Fe et Mn) améliore la croissance embryonnaire et l'absorption des nutriments, et régule les réponses immunitaires en équilibrant l'inflammation et l'activité antioxydante chez les embryons de poulet

Atul Jadhav, Ayobami Aboderin*, Lim Mei Lynn*, Bushansingh Baurhoo*‡, Raj Duggavathi*, Alexander Yitbarek*§*

**Department of Animal Science, McGill University, Ste-Anne-de-Bellevue, QC, Canada; ‡Belisle Solution and Nutrition Inc., Saint-Mathias-sur-Richelieu, QC, Canada; §Department of Animal and Food Sciences, University of Delaware, Newark, DE, USA*

Abstract

This study aimed to determine the optimal in ovo doses of chelated zinc (Zn), iron (Fe), and manganese (Mn) that positively affect embryonic health and development. 260 fertilized broiler eggs were randomly assigned to 13 treatment groups (n = 20 each). The control group received normal saline, while the other 12 groups were injected with four different doses of each mineral: Zn (20, 40, 60, 80 µg/egg), Fe (100, 300, 500, 700 µg/egg), and Mn (10, 20, 30, 40 µg/egg). Minerals were injected into the amnion on embryonic day (ED) 15, and samples were collected on ED 20. Gene expression was quantified by qRT-PCR using the 2-ΔΔCt method, and statistical analysis was performed using one-way ANOVA followed by Tukey's post hoc test ($p < 0.05$).

Our results showed that zinc treatments at 40, 60, and 80 µg/egg significantly upregulated mTOR expression, indicating its role in autophagy and protein synthesis, thereby promoting embryonic development. The 40 µg Zn/egg dose showed the highest expression of SGLT1 and PepT1, enhancing glucose and protein absorption. Iron treatment at 500 µg Fe/egg resulted in significant upregulation of mTOR, indicating cellular growth involvement. Among manganese treatments, 20 µg Mn/egg significantly increased SGLT1 and PepT1 expression, suggesting better nutrient absorption. Additionally, 20 µg Mn/egg dose significantly downregulated IL-1β and IL-2 while upregulating IL-10, suggesting improved immune homeostasis. In conclusion, the optimal in ovo doses identified for enhancing nutrient absorption, immune, and oxidative stress regulation were 40 µg Zn/egg, 500 µg Fe/egg, and 20 µg Mn/egg. Further studies will focus on the combined effect of these minerals in challenging conditions.

Assessing the pre- and probiotic effects of bacterial cellulose on laying hen's gut microbiome and egg production

Évaluation des effets prébiotiques et probiotiques de la cellulose bactérienne sur le microbiome intestinal des poules pondeuses et la production d'œufs.

Kemi S. Kanani[1], Yanxing Niu[1], Qingshan Meng [2], Jinguang Hu [2], and Anna Rogiewicz[1]
[1]Department of Animal Science, University of Manitoba, Winnipeg, MB
[2]Department of Chemical and Petroleum Engineering, University of Calgary, Calgary, AB
anna.rogiewicz@umanitoba.ca

Abstract

The growing interest in sustainable materials prompted us to explore whether by-products or waste from the agri-food industry could be transformed into bacterial cellulose (BC) - a bioactive biofilm known for its outstanding characteristics. This interdisciplinary study investigated the potential for producing BC from hemp fiber and tea biomass using a symbiotic culture of bacteria and yeast (SCOBY), the same culture used in kombucha fermentation, a beverage recognized for its health benefits. We hypothesized that BC, as a feed additive, could support the health of the gastrointestinal tract in laying hens.

In addition to characterizing its physicochemical properties, we evaluated BC derived from tea fermentation in a laying hen study to assess its effects on gut health, physiological function, egg production, and egg quality. Twenty-four-week-old Dekalb hens were fed either a Control or two experimental diets containing 0.1% or 0.5% dried BC for three weeks (Period 1), followed by the Control diet for an additional three weeks (Period 2). Egg production and egg quality parameters in Period 1 were not affected by BC, while in Period 2, feed intake tended to decrease in hens previously fed diets containing BC (p=0.078). The results suggest that the relative abundance of beneficial bacteria in the cecal microbiota tended to increase in birds fed BC diets, indicating a potential positive effect. However, the need for drying and milling may present challenges for its application as a feed additive.

SPONSORED BY / COMMANDITÉ PAR

trouw nutrition

a Nutreco company

Pre-exposure of zebrafish (Danio rerio) larvae and ZF-4 fibroblast cells to vitamin D (VD3) increases toxicity to bacterial lipopolysaccharide (LPS)

La préexposition des larves de poisson zèbre (Danio rerio) et des cellules de fibroblastes ZF-4 à la vitamine D (VD3) augmente la toxicité du lipopolysaccharide bactérien (LPS)

S. Leuschner[1], Umesh K. Shandilya[1], D. Huyben[1], B. Bridle[2], N.A. Karrow[1]
[1] Department of Animal Biosciences, University of Guelph, Guelph, ON, N1G2W1
[2] Department of Pathobiology, University of Guelph, Guelph, ON, N1G2W1
sleuschn@uoguelph.ca

Abstract

Vitamin D (VD) is recognized for its immunomodulatory properties and is critical to fish health, however, higher VD concentrations may be toxic. In this study, we hypothesized that pre-exposure to high sublethal concentrations of VD3 would decrease viability of zebrafish (ZF) larvae and ZF-4 fibroblast cells to bacterial lipopolysaccharide (LPS) immune challenge. Larval ZF were pre-exposed to VD3 (0, 5, 10, 20 ng/µL) at 72 h post fertilization (hpf) then LPS challenged (0, 40, 50, 60, 70, 80, 90, 100 µg/mL) at 96 hpf and survival was assessed at 120 hpf. Results revealed that LPS and interactions with VD3 dose-dependently reduced survival of larval Zf ($p<0.05$). Viability of ZF-4 fibroblast cells was also assessed after VD3 pre-exposure for 24 h followed by 24 h LPS challenge. Similarly, LPS and interactions with VD3 dose-dependently reduced ZF-4 cell viability ($p<0.05$). qPCR was performed on RNA extracted from larvae at 72 hpf, and 96 hpf after 24 h exposure to VD3 (0, 5, 10 ng/µL) to examine the expression of enzymes involved in vitamin D metabolism (*CYP2R1, CYP27A1, CYP27B1, CYP24A1*). VD3 significantly affected the expression of *CYP2R1, CYP27B1 and CYP24A1* ($p<0.05$). These findings demonstrate that exposure to high sublethal concentrations of VD3 may be detrimental to Zf that have been subjected to an innate immune challenge. Also, for the first time, VD metabolism enzymes were found to be expressed in Zf as early as 72 hpf and were found to be regulated by VD3.

SPONSORED BY / COMMANDITÉ PAR

Egg production and quality in mid-lay hens fed varying supplementation levels of American cranberry (vaccinium macrocarpon) pomaces.

Production et qualité des œufs chez les poules pondeuses à mi-ponte nourries avec des niveaux variables de suppléments de grignons de canneberge américaine (Vaccinium macrocarpon)

Lonita Manoa[1], Veronica Cheng[1], Jessica Gasarabwe[1], R. Patterson[2], Kelly Ross[3], Moussa S. Diarra[4], and Elijah G. Kiarie1

[1]Department of Animal Biosciences, University of Guelph, Guelph, ON, Canada

2CBS Bio platforms, Calgary, AB, Canada

[3]Summerland Research and Development Centre, Agriculture and Agrifood Canada (AAFC), Summerland, BC, Canada

[4]Guelph Research and Development Centre, AAFC, Guelph, ON

research@uoguelph.ca

Abstract

The nutritive and functional values of polyphenol rich fruit pomaces in poultry diets are getting increased interest. We investigated the effects of feeding cranberry pomace (CP) on egg production and quality. From 45 to 56 weeks of age (WOA), 160 Lohmann LSL lite hens (4 hens/cage) had *libitum* access to water and allocated five diets (n=8): a basal control diet (CON, 0.0% CP) and CON with 0.5, 1.0, 2.5 or 5.0% CP at the expense of corn. Egg production was recorded daily, feed intake (FI) weekly and egg weight (EW) bi-weekly to calculate hen day egg production (HDEP), egg mass (EM) and feed conversion (FCR). Egg shell thickness (EST), shell breaking strength (ESBS), yolk color (YC), and albumin height (AH) were analyzed on eggs collected in weeks 4, 8, and 10 respectively. Results showed that the HDEP, EW, EM, feed, metabolizable energy, crude protein and calcium intakes displayed a quadratic ($P < 0.01$) response to increasing CP concentration. Specifically, HDEP and EM values were highest in 2.5% CP-fed birds compared to hens fed 0 or 5.0% CP. Increasing CP resulted in a linear decrease ($P < 0.001$) in EST. However, ESBS response interacted with bird age and there was no ($P > 0.05$) dose response of CP on YC and AH. In conclusion, feeding up to 2.5% enhanced HDEP, whereas a higher inclusion level (5.0%) resulted in poorer HDEP. Moreover, CP affected eggshell quality. Further research is needed to investigate the long-term cost benefits of feeding CP in laying hens.

Keywords: Cranberry pomace; Hens; Egg; Production and quality

Significance of low crude protein diets fortified with amino acids in broiler chickens feeding program–a meta-analysis

Importance des régimes pauvres en protéines brutes enrichis en acides aminés dans le programme d'alimentation des poulets de chair - une méta-analyse

Mutinda Sylvester., Ellis J., and Kiarie, E. G.
Department of Animal Biosciences, University of Guelph, Guelph, ON, Canada

Abstract

Adopting a "less is more" approach in poultry nutrition has become increasingly sustainable over the years. The implementation of low crude protein (LCP) formulation strategies is widely associated with improved nitrogen utilization efficiency, reduced nitrogen excretion and ammonia emissions, enhanced litter quality, and lower feed costs. However, reducing dietary crude protein (CP) by more than 3-4% often results in compromised growth performance. Supplementation with crystalline amino acids (CAA) can help alleviate these negative effects. A meta-analysis was conducted to determine the optimal extent of CP reduction while maintaining sustainable growth and production in broiler chickens. The study analyzed 120 treatment means from 12 studies in a mixed effect model. Findings indicated that reducing dietary CP by more than 2%, regardless of CAA supplementation, led to inconsistencies such as increased abdominal fat, higher feed conversion ratio, decreased in body weight, and reduced feed intake. These challenges were likely linked to suboptimal or imbalanced essential amino acid profiles in the diet, among other factors. LCP diets generally resulted in lower ammonia emissions, reduced plasma uric acid levels, and proved to be a more effective strategy for minimizing ammonia output in poultry barns compared to conventional methods. Specifically, a 1% reduction in CP led to a 9% decrease in litter nitrogen content. With growing consumer demand for high-quality poultry products, further research is needed to optimize LCP formulation strategies. This must be in done alignment with regulatory restrictions on antibiotic growth promoters (AGP) while ensuring optimal gut health and sustainable poultry production.

* Corresponding author: smutinda@uoguelph.ca

Key words: amino acids, Nitrogen, proteins, meta-analysis

Utilizing exogenous enzyme technology to produce novel bioactive fiber components and enhance the value of feed ingredients for poultry

Utilisation de la technologie des enzymes exogènes pour produire de nouveaux composants de fibres bioactives et améliorer la valeur des ingrédients alimentaires pour la volaille

Poonam Neupane[1], Yanxing Niu[1], Rob Patterson[2], Bogdan A. Slominski[1], and Anna Rogiewicz1
Department of Animal Science, University of Manitoba, Winnipeg, MB
2CBS Bio-Platforms Inc., Calgary, AB
anna.rogiewicz@umanitoba.ca

Abstract

Alternative feed ingredients not only provide essential nutrients and energy for monogastric animals but also hold potential as valuable functional components capable of enhancing poultry health and performance. This research applies multienzyme technology to enzymatically modify feed ingredients and fractionate fibre components from these feedstuffs, generating bioactive feed ingredients that can positively impact monogastric animal health and production.

The detailed composition of total non-starch polysaccharides (NSP), the key components of dietary fibre, was analyzed in locally available alternative feed ingredients. The results revealed variations among different feedstuffs, not only in total NSP content but also in their constituent sugar composition. In vitro incubation studies were conducted to assess the extent of depolymerization of NSP in these feedstuffs using multicarbohydrase preparations. Variation in NSP breakdown into smaller, more bioactive molecules was observed among test feed ingredients, resulting from differences in fibre structure.

The effects of bioactive fibre-derived components and diets containing alternative feedstuffs supplemented with enzyme preparations will be evaluated to assess laying performance, nutrient retention, egg quality, and gut health and function in laying hens.

SPONSORED BY / COMMANDITÉ PAR

trouw nutrition
a Nutreco company

Thermomechanical and enzyme-facilitated processing of soybean meal enhanced amino acid digestibility in broiler chickens

Le traitement thermomécanique et enzymatique du tourteau de soja améliore la digestibilité des acides aminés chez les poulets de chair

Felix Njeri, and Elijah. G. Kiarie*,*
University of Guelph
Department of Animal Biosciences, University of Guelph

Abstract

Processed soybean meal (PSBM) with reduced anti-nutritional factors is widely used in piglet diets and may enhance gut resilience in newly hatched chicks. However, accurate feed formulation requires determining its digestible amino acid (AA) and metabolizable energy values. This study evaluated the apparent (AID) and standardized ileal digestibility (SID) of AAs and apparent metabolizable energy (AME) in thermomechanical and enzyme-facilitated PSBM (AlphaSoy) compared to conventional soybean meal (SBM) in broilers. A total of 270 one-day-old Ross 708 male chicks were fed a commercial starter diet until day 13 (n=6). On day 14, birds were assigned to one of three semi-purified diets: a nitrogen-free diet (NFD) to estimate endogenous AA losses, or test diets containing 20% crude protein from either PSBM or SBM as the sole AA source. Diets included 0.5% titanium dioxide as an indigestible marker. Excreta were collected from days 17–20, and ileal digesta on day 21. On a dry matter basis, PSBM had 4.5% higher gross energy (4,481 vs. 4,290 kcal/kg) and 1.4% more crude protein (49.6% vs. 48.9%) than SBM. The AID and SID of most AAs were 2-4% higher ($P < 0.05$) in PSBM, with the largest difference observed for His ($P = 0.002$). PSBM-fed birds had greater nitrogen-corrected AME (2,543 vs. 2,205 kcal/kg, $P < 0.004$). These findings indicate that thermomechanical and enzyme-facilitated processing improves nutrient digestibility and energy utilization, enhancing PSBM's value in broiler diets and its potential application in the poultry industry.

Keywords: Processed soybean meal, broiler nutrition, amino acid digestibility, metabolizable energy, poultry feed formulation, anti-nutritional factors, enzyme-treated soybean meal.

Biochar as a dietary treatment for the prevention of boar taint

Le biochar comme traitement diététique pour la prévention de l'odeur de verrat

Melissa Parent[1], Christine Bone[1], E. James Squires[1]
Department of Animal Biosciences, University of Guelph, Guelph, ON
mparen02@uoguelph.ca

Abstract

Boar taint is an unpleasant odour in meat from intact male pigs and is prevented through castration, but this dramatically reduces production efficiency and lean yield. We have previously used activated charcoal (AC) in the diet to control boar taint as an alternative to castration, but AC is not an approved feed ingredient. Biochar, an approved feed ingredient, has binding characteristics comparable to AC, and we hypothesized that adding biochar to finishing diets would control boar taint similarly to AC. The objective was to add AC or biochar to finishing diets fed to slaughter weight boars and assess the incidence of boar taint, growth performance, and relationship between treatment response and maturity status, in two genetic lines. We performed two animal trials using [(Yorkshire x Landrace) x Duroc] boars: first, 50 boars from AGC were divided into 5% AC (n=17), 5% biochar (n=17) or control (n=16) groups; the second trial used 52 PIC boars fed a 5% biochar diet. Diets were fed for 4 weeks (treatment period) followed by a 2-week recovery period. Hormone assays were used to assess boar taint and maturity status. Biochar and AC prevented boar taint in 83% of boars and all late maturing animals responded to treatment. Biochar did not affect growth (average daily gain and feed conversion ratio), in both genetic lines ($p>0.05$) and is therefore a potentially broadly applicable treatment for boar taint. Future research will optimize biochar dosage and treatment duration and develop prediction models to identify animals that respond to treatment.

SPONSORED BY / COMMANDITÉ PAR

trouw nutrition
a Nutreco company

Effect of microencapsulated multivitamin and mineral premixes on growth performance, plasma vitamin and mineral levels in broiler chickens challenged with coccidiosis

Effet des prémélanges multivitaminés et minéraux microencapsulés sur la croissance et les taux plasmatiques de vitamines et de minéraux chez les poulets de chair atteints de coccidios

Italo S. R. Pereira[1], Ludovic Lahaye2, Deborah Adewole[1]
[1] Department of Animal and Poultry Science, University of Saskatchewan, Saskatoon, SK, Canada, S7N 5A8.
2Jefo Nutrition Inc., Saint-Hyacinthe, QC, Canada, J2R 2E7.

Abstract

Coccidiosis impairs poultry performance by reducing nutrient absorption, increasing oxidative stress, and damaging intestinal health. Vitamins and minerals possess antioxidant properties that may enhance immunity. Microencapsulated vitamin-mineral premixes (PVTM) offer protection and slow release during feed processing and digestion. This study evaluated effects of free vs. 2 formulations of PVTM (differing notably in vitamins-minerals proportions) on coccidiosis challenged broilers. A 4×2 factorial experiment was conducted using 560 mixed-sex broilers (10 replicates/treatment; 7 birds/cage). Broilers were fed four corn-soybean meal diets: T1 (free vitamins-minerals, VTM1), T2 (microencapsulated VTM1 providing 75% of VTM1), T3 (PVTM formulation 2, dosage 1), and T4 (PVTM formulation 2, dosage 2, reduced). On d14, birds received either saline or a 12× dose of coccidiosis vaccine. Measurements included weekly body weight and feed intake, organ weight, excreta and blood assessment on d21, and blood and bone samples collected on d28. The three diets containing PVTM significantly improved ($P<0.05$) growth performance. Plasma Mo and Se levels were significantly reduced ($P<0.05$) by coccidiosis, while higher plasma Mo was observed for T4. T4 reduced ($P<0.05$) plasma vit A and marginally reduced ($P<0.1$) bursa weight. A significant interaction ($P<0.05$) was observed for plasma Co in which birds fed T4 had lower level compared to challenged birds fed T1. Tibia breaking strength and excreta blood were unaffected ($P>0.05$). These results suggest PVTM even in lower dosages altered vitamin mineral metabolism and enhanced growth performance under disease stress. Further analyses on nutrient digestibility, gut morphology, immunoglobulin levels, and antioxidant status are underway.

SPONSORED BY / COMMANDITÉ PAR

trouw nutrition
a Nutreco company

Assessment of the protective effects of reishi, laminarin, and barley β-glucans for zebrafish (Danio rerio) larvae immune challenged with bacterial lipopolysaccharide (LPS)

Des effets protecteurs du reishi, de la laminarine et des β-glucanes d'orge sur les larves de poisson zèbre (Danio rerio) immunodéprimées par le lipopolysaccharide bactérien (LPS)

G. Renke[1], U.K. Shandilya1, D. Huyben, N.A. Karrow[1]
[1]Department of Animal Biosciences, University of Guelph, Guelph, ON, N1G2W1.
grenke@uoguelph.ca

Abstract

β-Glucans are natural polysaccharides derived from fungi, algae, and cereals, known for their immunomodulatory properties in vertebrates. This study evaluates the potential of β-glucans from Reishi (Ganoderma lucidum), Laminarin (Laminaria spp.), and Barley (Hordeum vulgare) in mitigating inflammatory stress in zebrafish (Danio rerio) larvae following lipopolysaccharide (LPS) exposure. We hypothesized that pre-exposure to sublethal concentrations of β-glucans (50–200 μg/mL) would provide protection against LPS-induced inflammatory stress.In the in vivo experiment, zebrafish larvae at 72 hours post-fertilization (hpf) were immersed in β-glucan solutions (Reishi: 50–200 μg/mL; Laminarin: 50–200 μg/mL; Barley: 50–200 μg/mL), followed by a 24-hour LPS challenge (100 μg/mL) at 96 hpf. Larval survival rates were assessed at 120 hpf. In addition, zebrafish fibrobalst cell line (ZF-4)were exposed to β-glucans (50–200 μg/mL) *in vitro* for 24 hours to assess its their impact on cell viability. In vivo results demonstrated that Laminarin and Barley β-glucans significantly enhanced larval survival ($p<0.05$) under LPS-induced stress, whereas Reishi β-glucan did not confer notable protection. However, preliminary findings indicated that all β-glucans increased Zf-4 cell viability, with the most significant enhancement observed in the 50 μg/mL in Reishi-treated group. This suggests that while Reishi β-glucan did not improve survival in vivo, it may support cellular resilience.These findings highlight the potential of specific β-glucans in supporting immune resilience in aquatic species. Future research will focus on β-glucan-LPS binding assays and transcriptomic analysis to elucidate underlying molecular mechanisms, providing insights for sustainable aquaculture applications.

SPONSORED BY / COMMANDITÉ PAR

trouw nutrition
a Nutreco company

Enzymatic hydrolyses of yeast enhanced its nitrogen solubilization and kinetics in a simulated pig digestion model

Les hydrolyses enzymatiques de levure ont amélioré sa solubilisation de l'azote et sa cinétique dans un modèle de digestion porcine simulée

Alexander Wickhorst, Felix Muchiri, and Elijah G. Kiarie*
**MSc. Student*
Department of Animal Bioscience, University of Guelph, Guelph, ON
awickhor@uoguelph.ca

Abstract

Yeast based feed ingredients are used for nutritional functional attributes in farm and companion animals nutrition. Recent innovation in preparation of yeast feed products involves enzymatic hydrolysis of yeast cell walls to enhance functionality. However, little is known on the effect of hydrolysis on the yeast protein quality. We determined *in vitro* gastric and intestinal nitrogen (N) solubility and digestion kinetics of hydrolyzed yeast (HY) compared to unhydrolyzed yeast (UY). Samples were digested with pepsin at pH 3.5 and 39°C for 0, 0.5, 1, and 1.5 hours (gastric phase) followed by pancreatin and bile extract digestion at pH 6.8 for 0, 0.5, 2, 4, and 6 hours. Data analysis considered yeast type and gastric pH and their interactions as fixed factors. Crude protein (CP) digestion kinetics were modeled using the Gompertz equation and classified into three categories based on intestinal digestion time: CP_fast (0–0.5 h), CP_slow (0.5–4 h), and CP_resistant (>4 h). The concentration of CP was similar between HY and UY (43.2 ±0.30 *vs.* 42.1 ±0.20 % DM), however, compared to UY, HY had higher concentration of soluble CP and lower concentration of fiber. Gastric N was 222% higher for HY than UY ($P < 0.05$). At intestinal phase, HY had 24.7% higher N solubility resulting in 28.9% reduction in resistant CP compared to UY. In conclusion, yeast hydrolysis enhanced nitrogen solubility and digestion kinetics. *In vivo* validation is recommended to confirm these findings.

Keywords: in vitro digestibility, hydrolyzed yeast, gastric pH,

SPONSORED BY / COMMANDITÉ PAR

Meta-analyses of copper digestive utilization with phytase in weanling pigs

Méta-analyses de l'utilisation digestive du cuivre avec la phytase dans l'évaluation, chez les porcelets sevrés

Mingli Xu[1], Jiali Chen[2], Laurence Cheng[3], Min Wang[4], Ming Z. Fan[5]
[1]M.Sc. Student, Department of Animal Biosciences, University of Guelph, Guelph, ON
[2]Post-doctoral Scholar, Department of Animal Biosciences, University of Guelph, Guelph, ON
[3]Ph.D. Student, Department of Animal Biosciences, University of Guelph, Guelph, ON
[4]Technician, Department of Animal Biosciences, University of Guelph, Guelph, ON
[5]Professor, Department of Animal Biosciences, University of Guelph, Guelph, ON
mxu14@uoguelph.ca

Abstract

Exogenous phytases have been used in swine nutrition for over four decades to improve phosphorus utilization, but their impact on dietary Cu digestibility in weanling pigs remains inconsistent. A literature reviewing-based meta-analysis was conducted to determine the effect of dietary exogenous phytase supplementation (500–1000 FTU/kg) on the digestive utilization of dietary Cu in the weanling pigs. This Meta-analysis study included work published between 1992 and 2023 on weanling pigs fed dietary Cu levels within Cu requirements without or including exogenous phytases. Data normality was checked by the Shapiro-Wilk test. The Kruskal-Wallis contrast showed no difference in the apparent fecal Cu digestibility (16.8±5.2 vs. 22.0±6.4%; P=0.53; n=23 vs. 10) between the non-phytase and the phytase groups. There was an exponential relationship [Y=exp(0.175±0.059x); R2=0.36; P<0.05; n=23] between the contents of the apparent fecal digestible Cu and the total dietary Cu in the non-phytase group. There were linear relationships (y=0.339±0.119x; r2=0.28, P<0.05; n=23 for the non-phytase group vs. y=0.275±0.110x; r2=0.44; P < 0.05; n=10 for the phytase group) between the contents (mg/kg DMI) of the apparent fecal digestible Cu and the total dietary Cu, thus allowing estimation of the true fecal Cu digestibility values in the two groups. The Kruskal-Wallis contrast showed difference in the fecal endogenous Cu output (2.331±0.759 vs. 0.656±0.967 mg/kg DMI; n=23 vs. 10; P=0.0052) between the non-phytase and the phytase group. Thus, our meta-analyses indicate that exogenous phytases did not enhance apparent or true fecal Cu digestibility but significantly reduced fecal endogenous Cu loss in weanling pigs.

The effect of feed sulfur and water sulfate on ruminal hydrogen sulfide concentrations and gastrointestinal tract permeability in beef heifers

Effet de l'alimentation en soufre et en sulfate d'eau sur les concentrations d'hydrogène sulfuré dans le rumen et la perméabilité du tractus gastro-intestinal chez les génisses de boucherie

E.C. Fehr1, K. S. Hare1, J.A. Johnson[1], E.R. Snyder[2], and G.B. Penner[1]
[1]Department of Animal and Poultry Science, University of Saskatchewan, Saskatoon, SK, Canada S7N 5A8
[2]Department of Large Animal Clinical Sciences, Western College of Veterinary Medicine, University of Saskatchewan, Saskatoon, SK, Canada, S7N 5A8

Abstract

Consumption of high-sulfate water and high-sulfur feed by cattle results in production of ruminal hydrogen sulfide gas (H2S), increasing the risk of developing sulfur-induced polioencephalomalacia (PEM). However, it is unclear whether water sulfate or feed sulfur contribute more to elevated ruminal H2S concentrations. The hypothesis of this study was that high sulfur feed would have a greater influence on H2S concentrations in the rumen and permeability of the gastrointestinal tract than high sulfate water. Eight ruminally cannulated heifers were used in a replicated 4×4 Latin square design with a 2×2 factorial treatment arrangement. Main factors included low (LW; 2,400 mg/L) or high (HW; 3,450 mg/L) water sulfate and diets with low (LF; 0.28% S) or high (HF; 0.40% S on a DM basis) sulfur. Water intake, dry matter intake (DMI), ruminal H2S gas concentrations, and total tract and post-ruminal gastrointestinal permeability were evaluated. Water intake and DMI were not affected ($P \leq 0.23$). Mean and area under the curve H2S concentrations were greater for HF ($P<0.01$) and HW ($P \leq 0.02$) relative to LF and LW. The area under the curve for the H2S concentration tended to be greater for HFHW ($P \leq 0.08$) than HFLW and LFHW, with LFLW being the least. Total-tract ($P \leq 0.12$) and post-ruminal ($P \leq 0.18$) permeability were not affected. While neither water sulfate or feed sulfur reduced feed or water intake or altered gastrointestinal permeability, high sulfur in feed may pose a greater risk for S-induced PEM than high water sulfate due to increased concentration of H2S gas in the rumen.

SPONSORED BY / COMMANDITÉ PAR

trouw nutrition
a Nutreco company

Assessing the impact of increasing LinPRO-R dosage on the rumen microbiome dynamics and methane production in vitro

Évaluation de l'impact de l'augmentation de la dose de LinPRO-R sur la dynamique du microbiome du rumen et la production de méthane in vitro

Jacob Foster[1], Stephanie Collins[3]
[1]Student, Department of Animal Science and Aquaculture, Dalhousie University Agricultural Campus, Truro, NS
[2]Research Scientist, Agriculture and Agri-Food Canada, Sherbrooke Research and Development Centre, Sherbrooke, QC
[3]Associate Professor, Department of Animal Science and Aquaculture, Dalhousie University Agricultural Campus, Truro, NS
jc933700@dal.ca

Abstract

Methane from North American Dairy cows has a greenhouse gas intensity of 1.3kg CO2e/kg milk drawing increasing attention with concerns about enivronmental impact and production efficiency. Since rumen microbial dynamics are driven by dietary factors, feeding strategies have the greatest potential to reduce methane, increasing production efficiency by altering community diversity or shifting fermentation to divert nutrients away from methanogens. We aimed to determine if LinPRO-R, an extruded product combining fat from linseed in a protected starch-protein complex, would reduce methanogenesis through the addition of dietary fats, and by shifting starch fermentation from the rumen. LinPRO-R was tested at 0%, 2%, 4.5%, and 9% inclusion levels in a TMR using rumen simulation technology, to assess impact on microbiome diversity and methane production. Using a replicated completely randomized design, 16 fermenters were monitored for pH and redox daily over two 14-day periods to ensure system stability. On three consecutive days, fermenter liquid and feed bag solids were sampled for microbial analysis using targeted 16S rRNA sequencing. Total gas was measured and subsamples were analysed using gas chromatography to determine methane produced. LinPRO-R did not alter fermenter pH ($p = 0.560$), redox potential ($p = 0.553$), total gas production ($p = 0.444$), nor methane proportion ($p = 0.521$). This study demonstrates that LinPRO-R can be added to dairy diets at inclusion levels up to 9% without disrupting rumen stability, or increasing methane emissions in vitro. This work contributes to the exploration of feeding strategies aimed at mitigation of emissions dairy production.

SPONSORED BY / COMMANDITÉ PAR

Effects of stocking rate on activity behavior and growth performance of beef heifers grazing tame pasture

Effets du taux de chargement sur l'activité, le comportement et la croissance des génisses de boucherie en pâturage sur des prairies artificielles

Weiguang Hao[*1], A. Macias Franco[ƚ1], A. E. Moreira da Silva[ƚ1], V. Ramirez Sepulveda[†1], S. Lasso Ramirez[*1], M. C. Londono Mendez[*1], E. Santos[§1], G.O. Ribeiro[2], S. Terry[3], K. Wood[4], and, G. M. da Silva[§1]*

*[1]*PhD student, [ƚ]Postdoctoral, [†]MSc student and [§]Assistant Professor, Department of Agricultural, Food & Nutritional Science, of University of Alberta, Edmonton, AB*

[2]Assistant Professor, Department of Animal and Poultry Science, University of Saskatchewan, Saskatoon, SK

[3]Research Scientist, Lethbridge Research and Development Centre, Agriculture and Agri-Food Canada, Lethbridge, AB

[4]Associate Professor, Department of Animal Biosciences, University of Guelph

weiguan2@ualberta.ca, PhD student*

Abstract

This study evaluated the effects of stocking rates on performance and behavior of grazing beef heifers. A total of 42 heifers (24 Angus and 18 Hays Converter) were stratified by initial body weight and breed and randomly assigned to low (0.45 AUM/acre) and high stocking rates (0.93 AUM/acre) targeting 30% and 70% forage utilization, respectively. Heifers were managed with temporary double-lined electric fences and maintained in experimental tame pastures for 20 d adaptation, followed by two, 40 d evaluation periods. Body weight was measured at the beginning (d 0), middle (d 60), and end of the experiment (d 100). Activity behavior was constantly recorded using pedometers from d 0 to d 60 of the experiment. Forage botanical composition and biomass were measured bi-weekly. Stocking rates had no effect (P = 0.15) on average daily gain. There were no differences (P = 0.26) in final body weight, but low stocking rate heifers were heavier (P – 0.02) at 60 d. Animals under high stocking rate, had greater (P < 0.05) number of steps, motion index, transitions and standing time, while lying time was lower (P < 0.05) compared to low stocking rate. These findings suggest that despite differences in activity levels, animals effectively adapted to both stocking rates, mitigating potential performance differences. The increased physical activity observed in the high stocking rate group may reflect compensatory grazing behavior to meet nutrient requirements. These are preliminary data and continued evaluation will be performed in 2025 and 2026.

SPONSORED BY / COMMANDITÉ PAR

trouw nutrition

a Nutreco company

Effect of a seaweed blend on growth performance, feeding behaviour, and ruminal fermentation in heifers fed backgrounding and finishing feedlot diets

Effet d'un mélange d'algues sur la croissance, le comportement alimentaire et la fermentation ruminale chez les génisses nourries avec des régimes de semi-finition et de finition en parc d'engraissement

Tatiane R. Ramos[1], Kim Stanford[2], Stephanie A. Terry[3]
[1]Master's Student, Department of Biological Science, University of Lethbridge, Lethbridge, AB, Canada
[2]Associate Professor, Department of Biological Science, University of Lethbridge, Lethbridge, AB, Canada
[3]Research Scientist, Agriculture and Agri-Food Canada, Lethbridge Research and Development Centre, Lethbridge, AB, Canada
tatiane.rogelioramos@agr.gc.ca

Abstract

Seaweed as a feed additive for ruminants has gained attention due to its bioactive compounds, which may influence ruminal microbiota, reduce enteric methane (CH4) production, and affect animal performance. This study examined the effects of supplementing 50 g/heifer/day of a seaweed blend (red, brown, and green species) in a total mixed ration (TMR) compared to a control diet (without seaweed) on growth performance, feeding behavior, and ruminal ammonia and volatile fatty acid (VFA) production. A total of 112 Angus × heifers (Initial BW 324.5 ± 32.0 kg), (100 intact, 12 ruminally cannulated), were used in a backgrounding and finishing feedlot study. Each feeding phase was divided into four, 4-week periods. At the end of each period, heifers were weighed on two consecutive days and ruminal samples were collected from the cannulated heifers. Feeding behavior was assessed using the GrowSafe system. Heifers were evenly distributed across four pens, with each pen assigned to one of the two treatments (n = 56 heifers/treatment). Treatment effects were evaluated using the PROC MIXED procedure of SAS, with statistical significance declared at $P < 0.05$. The seaweed blend had no effect on growth performance (P > 0.05) or feeding behavior (P > 0.05) across either feeding phase; however, during the finishing phase, ruminal ammonia concentrations (mM) were greater ($P < 0.05$) in heifers receiving the seaweed blend compared to the control. Additionally, total VFA production (mM) and propionate percentage were reduced ($P < 0.05$) in the seaweed blend supplemented heifers, while butyrate, isobutyrate, and isovalerate proportions were increased ($P < 0.05$). In conclusion, supplementing 50 g/heifer/day of a seaweed blend did not affect heifer growth performance or feeding behavior, but it modified ruminal fermentation parameters during the finishing phase.

SPONSORED BY / COMMANDITÉ PAR

trouw nutrition
a Nutreco company

Reduce, reuse, recycle: Using bentonite to de-water milk processing byproducts for dairy cow diets

Réduire, réutiliser, recycler : Utilisation de la bentonite pour déshydrater les sous-produits de la transformation du lait destinés à l'alimentation des vaches laitières

Tatum E. Schooley, A. Heebner, B.E. Abbassi, P.F. Wu, M.A. Steele, J.P. Cant, T.C. Wright, K.M. Wood

Abstract

Milk processing results in a large volume of byproducts, including solids non-fat (SNF) and cheese whey. Half of these are processed into food or feed, but disposal of the surplus presents environmental concerns. These byproducts make valuable feed ingredients, however the low DM% complicates handling. New techniques in de-watering milk byproducts with bentonite and nanofiltration could produce feed ingredients for ruminant diets. Objectives included determining the maximum inclusion of liquid ingredients, estimating nutritive value relative to HMC, and assessing pathogenic microbial growth from whey byproducts. Cheese whey was mixed with 30g/L of sodium bentonite clay and allowed to settle into separated lactose (SEP-LAC) and whey protein plus bentonite (WPB) over 4-6h. TMR was mixed with 12-16% synthetic lactose (SYN-LAC), used to simulate SEP-LAC DM% after nanofiltration. Samples were stored in buckets drilled to resemble a particle separator to assess liquid seepage and collection over 24h. SEP-LAC, SYN-LAC and WPB had similar available energy to HMC (TDN 88.52%, 88.28%, 87.04% and 83.051%). Microbial analysis found no *coliforms*, *E. coli*, *Listeria*, or *Salmonella* in SEP-LAC or WPB samples. The seepage test determined SYN-LAC can be included up to 15% in a TMR without significant liquid seepage ($P < 0.05$). Revalorizing milk byproducts could reduce food waste and facilitate sustainable disposal of whey and SNF surplus. Additionally, replacing HMC with lactose sugar could increase ruminal pH and prevent SARA caused by high starch diets in early lactation cows and feedlot steers. Further research is needed to assess inclusion of this byproduct in cattle diets.

SPONSORED BY / COMMANDITÉ PAR

ANCC CNAC

MARK YOUR CALENDARS!

The Animal Nutrition Conference of Canada 2026 will take place **May 5 to 7** at Westin Hotels & Resorts, **Edmonton, Alberta**

MARQUEZ VOS CALENDRIERS!

Le Colloque de nutrition animale du Canada 2026 aura lieu du **5 au 7 mai** au Westin Hotels & Resorts, **Edmonton (Alberta)**

THANK YOU TO OUR SPONSORS MERCI À TOUS LES COMMANDITAIRES

PRE-CONFERENCE PRÉ-COLLOQUE

dsm-firmenich

HUVEPHARMA

KEMIN

NEOGEN

PLATINUM PLATINE

KEMIN
Compelled by Curiosity™

GOLD OR

ADM

CANOLAMAZING
Discover the amazing value of canola meal

dsm-firmenich

Elanco

O&T Farms
Naturally Better

MEDIA PARTNERS PARTENAIRES MÉDIAS

burleigh dodds SCIENCE PUBLISHING

Engormix

sermo.farm
CULTIVATED CONTENT

SILVER ARGENT

Alltech

Jefo

trouw nutrition
a Nutreco company

WLT Distributors Inc.

BRONZE BRONZE

AGRIBUTION

Barentz.
Pestell

Cbs Bio Platforms

Diamond V

HALCHEMIX
research proven nutrition

LALLEMAND
LALLEMAND ANIMAL NUTRITION

MORE THAN JUST FEED

O&T Farms
Naturally Better

PROBIOTECH

Sollio Agriculture

SUPPORTING PARRAINAGE

ADISSEO

ALCO energy

ANS
Agri-Nutrient Solutions

Blachford

Canadian Journal of Animal Science

COUNTRY JUNCTION FEEDS

Diamond V

FLORADALE
FEED MILL LIMITED

FNF

HAMLET PROTEIN

Masterfeeds
an Alltech company

msp[RS] RESISTANT STARCH

NSC MINERALS

ROBINSON

Romer Labs

SGS

Star Quality Samplers Inc